Jin Kuk Kim · Sabu Thomas
Prosenjit Saha
Editors

Multicomponent Polymeric Materials

Springer

Editors
Jin Kuk Kim
School of Materials Science and Engineering
Gyeongsang National University
Jinju, Gyeongnam
Republic of Korea (South Korea)

Prosenjit Saha
Indian Institute of Engineering Science and Technology (IIEST), Shibpur
Howrah
India

Sabu Thomas
School of Chemical Sciences
Mahatma Gandhi University
Kottayam, Kerala
India

ISSN 0933-033X ISSN 2196-2812 (electronic)
Springer Series in Materials Science
ISBN 978-94-017-7323-2 ISBN 978-94-017-7324-9 (eBook)
DOI 10.1007/978-94-017-7324-9

Library of Congress Control Number: 2016940347

Printed on acid-free paper

This Springer imprint is published by Springer Nature
The registered company is Springer Science+Business Media B.V. Dordrecht

Springer Series in Materials Science

Volume 223

The Springer Series in Materials Science covers the complete spectrum of materials physics, including fundamental principles, physical properties, materials theory and design. Recognizing the increasing importance of materials science in future device technologies, the book titles in this series reflect the state-of-the-art in understanding and controlling the structure and properties of all important classes of materials.

More information about this series at http://www.springer.com/series/856

Contents

Part I Multicomponent Polymer Material Processing

1 **Multi Component Materials** 3
Prosenjit Saha, Sukanya Chowdhury, Min Cheol Kim and Jin Kuk Kim

2 **Design for Multicomponent Materials** 13
Jaideep Adhikari, Sukanya Chowdhury, Prosenjit Saha and Jin Kuk Kim

3 **Design of Mold for Multicomponent Material** 37
Soo-Jin Park and Ki-Seok Kim

4 **Injection Molding for Multicomponent Materials** 79
Soo-Jin Park and Seul-Yi Lee

5 **Extrusion of Multicomponent Product** 109
Aloke Kumar Ghosh

6 **Compression for Multiphase Products** 133
Ganesh Chandra Nayak and G. Hatui

7 **Paints and Coating of Multicomponent Product** 157
P. Sathish Kumar, R. Rajasekar, Samir Kumar Pal, Ganesh Chandra Nayak and S.M.R. Syed Ismail

Part II Applications of Multicomponent Product

8 **Multilayer Polymer Films** 229
A.R. Ajitha, M.K. Aswathi, Hanna J. Maria, Joanna Izdebska and Sabu Thomas

9 **Hybrid Systems for Multi-layer Fuel and Air Hoses in Automobiles** 259
Narendra Reddy

10 Multi Layer Pipes 279
R. Rajasekar, K.S.K. Sasi Kumar and P. Sathish Kumar

11 Multilayer (Fuel) Storage Tank 301
R. Rajasekar, K.V. Mahesh Kumar,
K. Krishnamurthy and P. Sathish Kumar

12 Multilayer Bottles 325
Rane Ajay Vasudeo, A.R. Ajitha, Shivendra S. Patil, Krishnan Kanny
and Sabu Thomas

13 Multiphase Materials for Tire Applications 349
R. Rajasekar, C. Sivasenapathy, C. Moganapriya and A. Sivakumar

14 Interfacial Compatibilization of Multilayered Products 369
V.K. Abitha, S. Lakshmi Priya, Sainath Jadhav and Sabu Thomas

15 Multilayer Nanowires and Miscellaneous Multilayer Products 383
A.R. Ajitha, P.S. Sari, Hanna J. Maria and Sabu Thomas

Contributors

V.K. Abitha School of Chemical Sciences, Mahatma Gandhi University, Kottayam, Kerala, India

Jaideep Adhikari Indian Institute of Engineering Science and Technology, Shibpur, Dr. M.N. Dastur School of Materials Science and Engineering, Howrah, WB, India

A.R. Ajitha International and Inter University Centre for Nanoscience and Nanotechnology, Mahatma Gandhi University, Kottayam, Kerala, India

M.K. Aswathi International and Inter University Centre for Nanoscience and Nanotechnology, Mahatma Gandhi University, Kottayam, Kerala, India

Sukanya Chowdhury Environmental Science Department, Asutosh College, University of Calcutta, Kolkata, India

Aloke Kumar Ghosh Department of Inorganic and Physical Chemistry, Indian Institute of Science, Bengaluru, Karnataka, India

G. Hatui Department of Applied Chemistry, Indian School of Mines, Dhanbad, India

Joanna Izdebska Department of Printing Technology, Warsaw University of Technology, Warsaw, Poland

Sainath Jadhav Department of Polymer and Surface Engineering, Institute of Chemical Technology, Matunga, Mumbai, India

Krishnan Kanny Composite Research Group, Department of Mechanical Engineering, Durban University of Technology, Durban, South Africa

Jin Kuk Kim Department of Materials Engineering and Convergence Technology, Gyeongsang National University, Jinju, Gyeongnam, South Korea

Ki-Seok Kim Department of Chemistry, Inha University, Nam-gu, Incheon, South Korea

Min Cheol Kim Department of Materials Engineering and Convergence Technology, Gyeongsang National University, Jinju, Gyeongnam, South Korea

K. Krishnamurthy School of Building and Mechanical Sciences, Kongu Engineering College, Erode, Tamilnadu, India

Seul-Yi Lee Department of Chemistry, Inha University, Nam-gu, Incheon, South Korea

K.V. Mahesh Kumar School of Building and Mechanical Sciences, Kongu Engineering College, Erode, Tamilnadu, India

Hanna J. Maria School of Chemical Sciences, International and Inter University Centre for Nanoscience and Nanotechnology, Mahatma Gandhi University, Kottayam, Kerala, India

C. Moganapriya Department of Mechanical Engineering, Kongu Engineering College, Erode, Tamil Nadu, India

Ganesh Chandra Nayak Department of Applied Chemistry, Indian School of Mines, Dhanbad, Jharkhand, India

Samir Kumar Pal Department of Mining Engineering, Indian Institute of Technology—Kharagpur, Kharagpur, West Bengal, India

Soo-Jin Park Department of Chemistry, Inha University, Nam-gu, Incheon, South Korea

Shivendra S. Patil Institute of Chemical Technology, Nathalal Parekh Marg, Matunga, Mumbai, India

S. Lakshmi Priya International and Interuniversity Centre for Nanoscience and Nanotechnology, Kottayam, Kerala, India

R. Rajasekar Department of Mechanical Engineering, School of Building and Mechanical Sciences, Kongu Engineering College, Erode, Tamil Nadu, India

Narendra Reddy Center for Emerging Technologies, Jain University, Ramanagara District, Bengaluru, India

Prosenjit Saha Department of Materials Engineering and Convergence Technology, Gyeongsang National University, Gyeongnam, Jinju, South Korea; Dr. M.N. Dastur School of Materials Science and Engineering, Indian Institute of Engineering Science and Technology, Shibpur, Howrah, West Bengal, India

P.S. Sari International and Inter University Centre for Nanoscience and Nanotechnology, Mahatma Gandhi University, Kottayam, Kerala, India; LIMATB Laboratory, Université de Bretagne Sud, Lorient, France

K.S.K. Sasi Kumar Department of Mechanical Engineering, Kongu Engineering College, Erode, Tamil Nadu, India

P. Sathish Kumar Department of Mining Engineering, Indian Institute of Technology Kharagpur, Kharagpur, West Bengal, India

A. Sivakumar Department of Mechanical Engineering, Kongu Engineering College, Erode, Tamil Nadu, India

C. Sivasenapathy Department of Mechanical Engineering, Kongu Engineering College, Erode, Tamil Nadu, India

S.M.R. Syed Ismail Rubber Technology Centre, Indian Institute of Technology—Kharagpur, Kharagpur, West Bengal, India

Sabu Thomas International and Inter University Centre for Nanoscience and Nanotechnology, Mahatma Gandhi University, Kottayam, Kerala, India; School of Chemical Sciences, Mahatma Gandhi University, Kottayam, Kerala, India

Rane Ajay Vasudeo Composite Research Group, Department of Mechanical Engineering, Durban University of Technology, Durban, South Africa

Part I
Multicomponent Polymer Material Processing

Chapter 1
Multi Component Materials

Prosenjit Saha, Sukanya Chowdhury, Min Cheol Kim and Jin Kuk Kim

1.1 Introduction and Definition

Polymeric materials can be differentiated in many ways from other conventional materials such as metals, and glasses. The research based on development of multicomponent polymeric materials was introduced to meet the increasing demand for improved multifunctional engineering polymeric materials with reduced processing cost, and broad utilization ability obtained from mixing several homopolymers. Polymers can be mixed with many materials through various mechanisms to fabricate diversified engineering materials. Polymer may be cross-linked, plasticized, reinforced, or blended to develop an optimized product to fulfill the industrial demands. The concept of mixing two or more different homopolymers has been recognized since the historic ages. Composite brick reinforced with straw, pottery made with reinforced goat-hair [1] are few of the examples of some oldest polymeric multicomponent or hybrid materials [2]. In general multicomponent materials can be classified in three different categories: reinforced composites, laminated composites, and hybrid composites [3]. A generalized working definition

P. Saha (✉) · M.C. Kim · J.K. Kim (✉)
Department of Materials Engineering and Convergence Technology,
Gyeongsang National University, Gyeongnam Jinju 660-701, South Korea
e-mail: senjitiitkgp@gmail.com

J.K. Kim
e-mail: rubber@gnu.ac.kr

P. Saha
M.N. Dastur School of Materials Science and Engineering,
Indian Institute of Engineering Science and Technology-Shibpur,
Howrah 711 103, West Bengal, India

S. Chowdhury
Environmental Science Department, Asutosh College,
University of Calcutta, Kolkata 700026, India

J.K. Kim et al. (eds.), *Multicomponent Polymeric Materials*,
Springer Series in Materials Science 223, DOI 10.1007/978-94-017-7324-9_1

for multicomponent polymeric materials would be, "a combination of two or more materials in a pre-determined configuration and scale, optimally serving a specific engineering need" [4]. As per the classification provided by Ashby, the polymeric multicomponent materials can be divided in four classes as shown in Fig. 1.1 [5]. It has been evident that polymeric multicomponent materials can be called by many names. The term polymer blends has been widely used to describe basic structure-property relation of multi-materials. Another name that has been used frequently for multicomponent polymer materials is "polymer alloys" that exhibit a distinct phase separation behavior like most of the polymer materials. However, the term "multicomponent polymer materials" used in this present research will present a new and broader aspect covering all the collective names and methods to describe the details of materials fabricated from two or more polymers.

1.2 Why Multicomponent Materials?

The properties of basic engineering materials has been mapped and it showed the ranges of mechanical, thermal, chemical, electrical, and optical properties (Fig. 1.2). The maps also indicate some empty spaces or areas of material's structure-property responses. Some of these empty areas are fundamentally inaccessible due to existing limitation related to the size of atoms and nature of interacting forces that bind the atom together. However, surprisingly the other empty areas also could not be filled with existing materials even though, in principle, that might possible. There are two possible approaches; one is to make new materials in a traditional way to fulfill the empty areas of the property map, although that could be an expensive and uncertain process. And the alternative method is to combine two or more existing materials to allow superposition of their properties by developing multicomponent materials. This approach will allow the scientists to explore the ways to fill the empty spaces with preplanned design and comprehensive understanding on structure-property relationship.

There are substantial possibilities that multicomponent materials prepared from two or more polymeric materials will exhibit and reflect the properties of their component polymers. The primary aim to fabricate multicomponent materials has been to reduce processing for improved performance of engineered products. The designers are interested to select this technique to incorporate maximum functions into a material to acquire higher performance characteristics. The possible justifications to select multicomponent materials by materials scientist are as follows:

1. Multicomponent materials can induce an improvement of performances of a part of a material as per requirement, such as increase of strength, stiffness, or service life. Reduction of mass, overall dimensions can also be imparted by this technique.
2. Often it can be required by the designers to induce more and more functionalities to a material. Car windscreen is an ideal example for multifunctionalities.

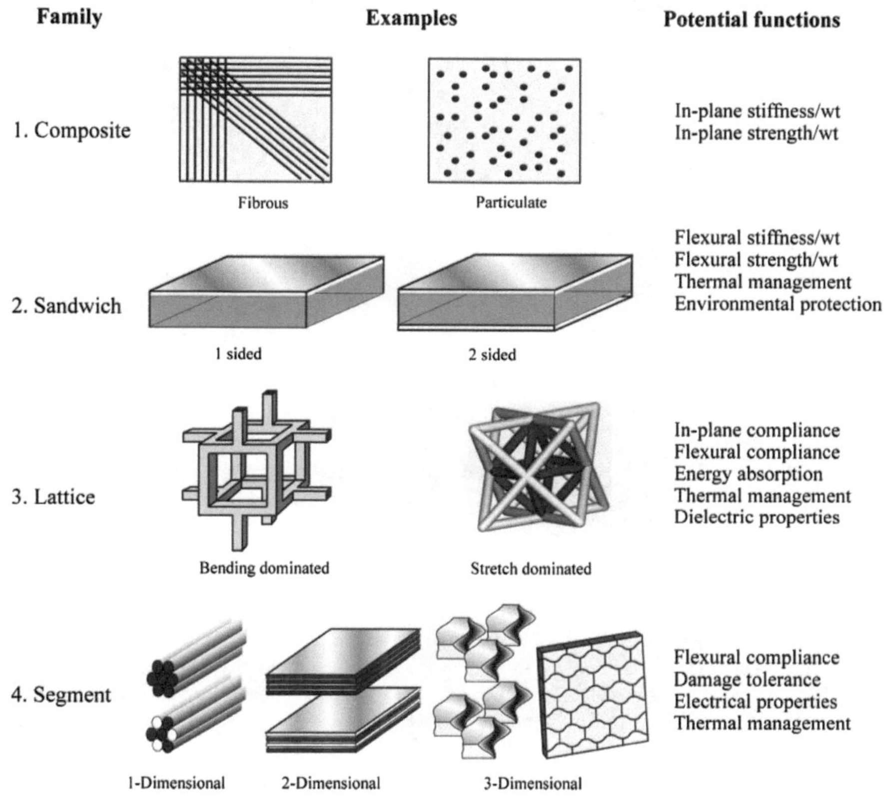

Fig. 1.1 Classification of multicomponent polymer materials [5]

Primary function of a car windsheet is to protect the driver from exposures of outside environment, however, in practical use it can also contribute to increase the overall stiffness to car chassis, to protect passenger inside from sun rays, and to act as self-destruction material. No monolithic polymeric materials would be able to provide all such characteristics in a single shot, unless multicomponent material is used.

3. Reduction of manufacturing and operational cost is another important advantage with multicomponent materials. This factor is directly related to reduced energy consumption for practical applications.
4. Complex geometries can be achieved when using composite materials. This would facilitate the manufacturing process to reduce the number of parts production to produce a single unit. A complex design of aircraft can be made by multicomponent material technique without assembling a large number of parts by different materials.
5. Use of multicomponent materials can avoid the overdimensioning of a part.

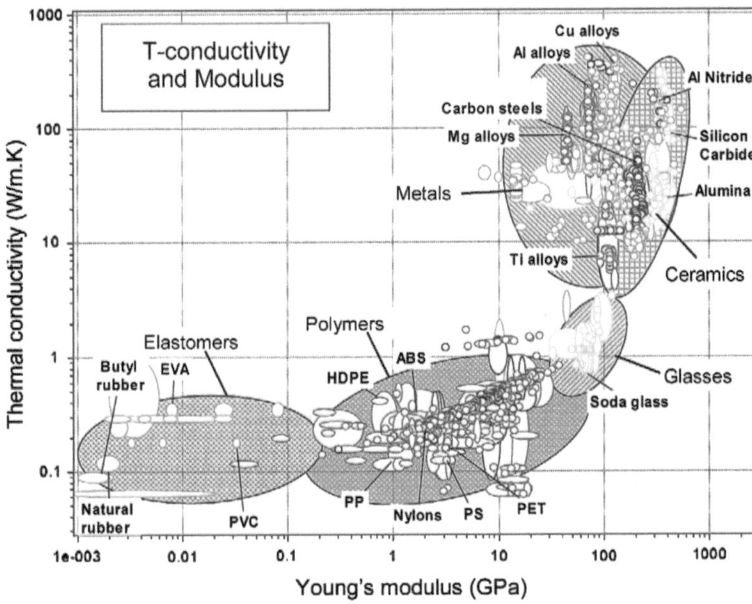

Fig. 1.2 Map for interrelation of thermal conductivity and Young's modulus for 2300 materials from different classes. A large area of this map is empty, for example, there are no materials with high thermal conductivity with low modulus. The basic challenge to fabricate polymer multicomponent materials is to fill the empty spaces [2]

These reasons describe why a search for a multicomponent material is essential when a single monolithic material is found unsuccessful. Henceforth, it establishes the background for extensive research for multicomponent materials in recent times.

1.3 Recent Technologies

The processing of polymeric materials such as plastics, elastomer, and composites in existing industrial practice can be characterized by vast categories of different techniques and methods. These methods are summarized in the following sections.

Techniques that target continuous manufacturing of a product basically include extrusion, extrusion covering, film blowing, and calendaring. On the other side, techniques that aim to shape a deformable polymer preformed against a mold surface include sheet thermoforming, and blow molding. Techniques aim to completely fill-up a mold cavity include casting, compression molding, transfer molding, injection molding, and reaction injection molding. Techniques aiming to gradual build-up of a polymer layer against a mold surface include coating and rotational molding.

1.3.1 Extrusion

Extrusion of polymeric materials mainly involves a continuous shaping of a polymer fluid by a suitable die, and subsequent solidification of the fluid into a solid product. This process exhibits a uniform cross-section of the final product. In general practice, thermoplastic polymers are fed in powder or pellet form upon heated to a fluid state and pumped through a die by a screw extruder. The polymer is then solidified by cooling after exiting from die. Uncured rubbers and some of the thermosets on the other hand can be extruded and solidified by heat curing method.

1.3.2 Extrusion Covering

Extrusion covering refers to a process that entirely surrounds a continuous substrate with a thin layer of polymeric material. The most common use of this process can be observed to provide an insulating layer or a protective jacket on power or electronic communication wires and cables. Plasticized polyvinylchloride (PPVC) and polyethylene (PE) are mostly used as basic materials for thermoplastic outer cover in extrusion covering process. Other polymeric materials such as polyamides, fluoropolymers, and cellulosics can also be used for electrical insulation or for materials with transparency property requirements. In case of rubber, generally an unvulcanized rubber cover is applied in extrusion covering through continuous curing by steam, molten salt, and heating with high frequency.

1.3.3 Film Blowing

In this process a tube of molten thermoplastic is extruded followed by continuous inflating of the tube up to several times of its initial diameter, to form a thin tube shaped product. The final product can be used as a slit or flat film. Polyethylene of various qualities such as LDPE, LLDPE, HMWHDPE, etc. are commonly used as blowing resin. Other resins that can be used alone or in combination with polyethylene for barrier applications are polyamide, ionomers, polyvinylidenechloride, polyvinylalcohol, and compolymers such as EVOH, and EVA. Films produced from blowing can be used in diversified applications. The areas include agricultural, industrial, packaging, and covering materials for greenhouses, silage, solar ponds, flat cars, etc.

1.3.4 Calendering

The calendaring process is considered to be useful for rapid production of flat films or sheets with controlled thickness from plastics or rubbers. During calendaring a fixed amount of plasticized material is continuously fed between two counter rotating cylinders to entrain on the surface of the cylinders and subsequently squeezed in the wedge. The thickness of the sheet corresponds approximately to the shortest distance between two rolls (gap). Most commonly used calendered materials are vinyl thermoplastics and vulcanizable rubbers (PVC and copolymers). Polyethylene and ABS are also used for special requirement for calendering. Films and sheets both can be formed by calendering. The thickness for film material normally lies bellow 0.05 mm, whereas the sheet thickness can be increased as about 6 mm.

1.3.5 Sheet Thermoforming

Sheet thermoforming includes the heating of a flat thermoplastic sheet to a softened state above their glass transition temperature for non-crystallized polymer or near the melting temperature for crystallized polymers. The final step of sheet thermoforming involves the deformation (forming) of the softened sheet to a required shape followed by solidification through cooling with acceptable mechanical properties. Thermoplastics in noncrystallized form are relatively easier to thermoform due to their higher elastic properties over a wide range of temperature above glass transition temperature. Such materials include high impact polystyrene, and copolymers such as ABS. Acrylic resin and cellulosics are also used for thermoforming in transparency applications. Sometimes crystallized polymers such as HDPE, PP, and polyamide may also be used for thermoforming however they need special process control. Machine covers, skin or blister packs, housings, light diffusers, furniture facing, machine liners are the common products prepared from thermoforming.

1.3.6 Blow Molding

Blow molding process includes the inflation of a softened thermoplastic hollow perform against the cooled surface of a closed mold. The thermoplastic solidifies into hollow product. Generally, two types of blow molding are found to apply for multicomponent material manufacturing process. Those are (a) extrusion blow molding, where preform is produced by extrusion and (b) injection blow molding, where, preform is produced by injection mold.

Materials used for blow molding need relatively larger molecular weight and a good stretchability over a wide range of temperature window. High molecular weight polyethylene is the most used material for extrusion blow molding. Thermoplastic polyester is primarily used in injection blow molding.

1.3.7 Casting

Globally, the "casting method" refers to the polymerization of a liquid resin directly into the shape of a product and where the resin and the product are subjected to minimal stresses (gravity flow, atmospheric pressure, slow chemical reaction). Commonly, casting method involves the pouring of a liquid resin system into mold (open or two-piece mold). In some cases, the chemical reaction may take place during a casting process that converts a low-molecular weight monomer into a high molecular weight thermoplastic. Such examples are acrylics (PMMA) and polyamides. For some different cases, polymerization may take place simultaneously in the casting process, leading to thermosets (such as polyurethane resins, and epoxy resins).

1.3.8 Compression Molding

The compression molding process involves the exposure of thermosetting resin to sufficient heat to soften or plasticize it. The fluid plastic is then held at the molding temperature under pressure (2000–4000 psi) for a sufficient time length for the subsequent polymerization or curing of the materials to make it hard and rigid. The hard materials are then removed from the mold cavity. In compression molding the material can be processed into a desired shape using press containing two-part closed mold under required pressure and temperature. Different thermosetting resin such as phenolics, polyester, epoxides, ureas, melamines, and silicones are generally used for compression molding. A large number of compression molded products are found applications in electrical and electronic sector. Other applications include heat resistance and structural resistance.

1.3.9 Transfer Molding

Transfer molding principle is mostly similar to the compression molding as in both cases the same two classes of materials, temperature-activated thermosets, and vulcanizable rubbers are used for product formation. In transfer molding, the mold halves are brought together under pressure to charge the molding materials into a pot. The material within the pot is then driven out of the pot by runners and gates into a mold cavity using plungers. However, transfer molding differs from compression molding as follows. In compression molding, the plastic is heated to a point of plasticity in the pot before it reaches or forced into closed mold. On the other hand in transfer molding, the plastic is transferred into heated closed mold through a narrow gate for subsequent heating and curing. Materials used for transfer molding must have substantially high pre-curing fluidity to facilitate the flow from the loading area to cavities. Unlike compression molding, transfer molding permits the fabrication of relatively complex parts, or parts featuring fragile inserts.

1.3.10 Injection Molding

Basic injection molding process involves the following stages:

- Plasticizing the materials by heating and melting.
- Injecting the plasticized materials under pressure for a fixed time interval in a closed mold where the plastic starts solidification.
- Maintaining the pressure on the injected materials to resist the backflow of the melt and to compensate volume decrease of the melt during solidification.
- Cooling the mold until complete solidification.
- Open up the mold parts ejection of materials and cleaning of mold for next use.

Injection molding consists of two basic parts, (a) an injection unit for melting and transferring of melt into mold, and (b) a clamping unit for closing and opening the mold.

Main advantage of injection molding is, it permits to fabricate the most complex shapes with accurate dimensional control. Generally speaking, all thermoplastics are found suitable for injection molding; however, good fluidity is the necessary property for a thermoplastic to be used in injection molding.

1.3.11 Reaction Injection Molding

This is relatively a new molding process that involves a rapid mixing, in precise proportions, of two or more highly reactive low-molecular weight liquids and the subsequent injection of the mixture in a closed mold. Polymerization has been found to take place in the mold rapidly to yield a solid product in a total cycle time of the order of few minutes. This special molding process was developed for polyurethane materials and still mainly used for it. This method is less expensive compared to injection molding process, and this process is suited basically for manufacturing relatively thin part such as exterior automobile body panels, fenders, furnitures, and electronic cabins.

1.3.12 Coating

Coating can be defined as the formation of an extra layer of polymer on the multicomponent materials. The process involves "planar coating" applied for flat film and sheet substrates, and "contour coating" applied for general three dimensional substrates. Coating is found useful for additional material performances such as appearance, hydrophobicity, electrical insulation, and stability. The adhesion between substrate and coated layer is one of the most important aspects for coating that determines the coating methods and process control. Increase the adhesion can be obtained by chemical treatment, exposure to gas flame, corona discharge, and by use of anchoring material for good bonding. Different methods are being used to apply for coatings to substrates of all size, and types depending upon the requirement and applications. The method generally includes one major thermoplastic or thermoset resin, a mixture of solvents, one or more pigments, and several additives.

1.3.13 Rotational Molding

Rotational molding also known as rotomolding or rotational casting, is a process for making hollow, seamless items of all sizes and shapes. The molded multicomponent products range from domestic tank to industrial containers. The process involves charging of suitable thermoplastic material into an open mold followed by double rotation of the closed mold around two orthogonal axes to create tumbling action. Furthermore, the heating of exterior of the mold is carried out while rotating in a chamber. Heat transfer through the mold surface initiates melting of inner thermoplastic material and it's sticking to the mold surface. The heating continues until the whole material inside the mold consolidates, and homogenizes through intermolecular diffusion to produce a smooth inner surface. The final stage is to cool the mold surface that causes solidification of the polymeric shell to allow unloading of material from mold.

1.4 Future Trends for Multicomponent Material Fabrication

A great future is waiting for multicomponent polymeric materials with wide scope of applications in engineering as well as industrial fields especially in composites, blends, and compatibilized systems. The basic advantage of multicomponent system lies in the offering of versatile and new range of materials manufacturing and processing according to the need from end-users. Possible extraordinary applications of multicomponent polymeric materials include novel mechanical, optical, and chemical performances. Furthermore, recent advancement of nanomaterials research opens up wide possibilities of manufacturing multicomponent materials using nanoparticles for commercial applications.

References

1. Lee SM (1995) Dictionary of Compos. Mater Technol Technomic Publishing, Lancaster
2. Ashby MF, Bréchet YJM (2003) Acta Mater 51:5801
3. Barbero EJ (1999) Introduction to composite materials design. Taylor & Francis, London
4. Kromm FX, Quenisset JM, Harry R, Lorriot T (2003) Mater. Design 24:235
5. Sirisalee P, Ashby MF, Parks GT, Clarkson PJ (2006) Adv Eng Mater 8:48

Chapter 2
Design for Multicomponent Materials

Jaideep Adhikari, Sukanya Chowdhury, Prosenjit Saha and Jin Kuk Kim

2.1 Introduction

Designing is the primary phase of the life cycle of a product or material. Therefore, rationalization between process design and subsequent mechanisms along with the performance of the material has been tried out by several researchers for many years [1]. Defining the architecture of the multi components is the first step towards designing which includes refinement and optimization of the process in brief [2]. Material selection is also one of the most important stages of product design keeping in view about its end applications. Improvement in property of materials inevitably increases its cost. So, designers nowadays tend to correlate between cost and technical specifications of a material by optimizing several key parameters. To make a product commercially viable thus integration between cost and property as a single function is required [3].

Designing is a function of constraints as certain dimensions remain fixed according to the end uses. Performance of material is evaluated according to their mechanical, electrical, thermal, chemical, and optical properties. Mapping between these properties through different ranges reveals that a single material is not suffi-

J. Adhikari · P. Saha (✉)
Indian Institute of Engineering Science and Technology—Shibpur,
Dr. M.N. Dastur School of Materials Science and Engineering,
Howrah 711 103, WB, India
e-mail: senjitiitkgp@gmail.com; prosenjit@matsc.iiests.ac.in

S. Chowdhury
Environmental Science Department, Asutosh College,
University of Calcutta, Kolkata 700026, India

P. Saha · J.K. Kim (✉)
Department of Materials Engineering and Convergence Technology,
Gyeongsang National University, Gyeongnam, Jinju 660-701, South Korea
e-mail: rubber@gnu.ac.kr

J.K. Kim et al. (eds.), *Multicomponent Polymeric Materials*,
Springer Series in Materials Science 223, DOI 10.1007/978-94-017-7324-9_2

cient for fulfillment of every aspects of its application. Thus, multicomponent materials come into the scenario which copes up a system inadequacy about a specific property by incorporation of different material into the system which is superior in that particular property [4]. So combining several different properties into a single material is the main characteristics of multicomponent material. The subject is of great relevance as real life structures or problems involve multiple components and thus they are interrelated since change in one component has its effect on the others also. Thus a general methodology is required which can tackle several shortcomings considering the multi component system as a whole [5].

2.1.1 Fundamentals of Materials Processing and Design

The shift from conventional materials towards high end polymeric multicomponent materials is based on the availability of subsequent economic advantages. The primary parameters that ensure the substantial changes in technological trend for designing multicomponent materials mainly include processing and fabrication, corrosion resistance, low specific weight, diverse options of colours, low thermal and electrical conductivity, high dielectric strength and good toughness with degradation resistance to alkali, acid and moisture etc. [6] Multicomponent polymeric material processing basically deals with advanced polymer blends, composites, nano composites and interpenetrating polymer networks. A large variety of these multicomponent comes by combining different polymers or by adding different types of fillers ranging from nanometer, micrometer to larger particulates. The filler may be fiber or combination of both particulate and fiber as reinforcement leading to something like multi component system. Fillers improve tensile and compressive strength, abrasion resistance, dimensional stability etc. of the materials. Apart from filler some foreign substances or additives viz. plasticizers, colorants, stabilizers, fire retardant, antistatic agents, cross linking agents; blowing agents (for cellular or expanded products) are also intentionally introduced to the system for having a modified property. Different potential processing techniques like compression molding, injection molding, thermoforming, pultrusion etc. are successfully employed to shape the polymers. Modification of these processes is successfully carried out by first modelling them appropriately. Commercial software packages with advanced modelling successfully demonstrate different macroscopic parameters viz. stress, flow rate, viscosity etc. But often correlation between morphological properties with multicomponent has been neglected in the design stages [7].

Traditional approach for extending material property space suggests modification of new polymer chemistry with new composition. Imparting such new modification approach is quite costlier, time consuming, and uncertain, therefore using multi-component material with superposition of properties of different material has been tactfully used for targeted properties [4]. For starting a design problem a designer starts with basic steps which are illustrated in Fig. 2.1. Design requirements involve mainly four stages [8]

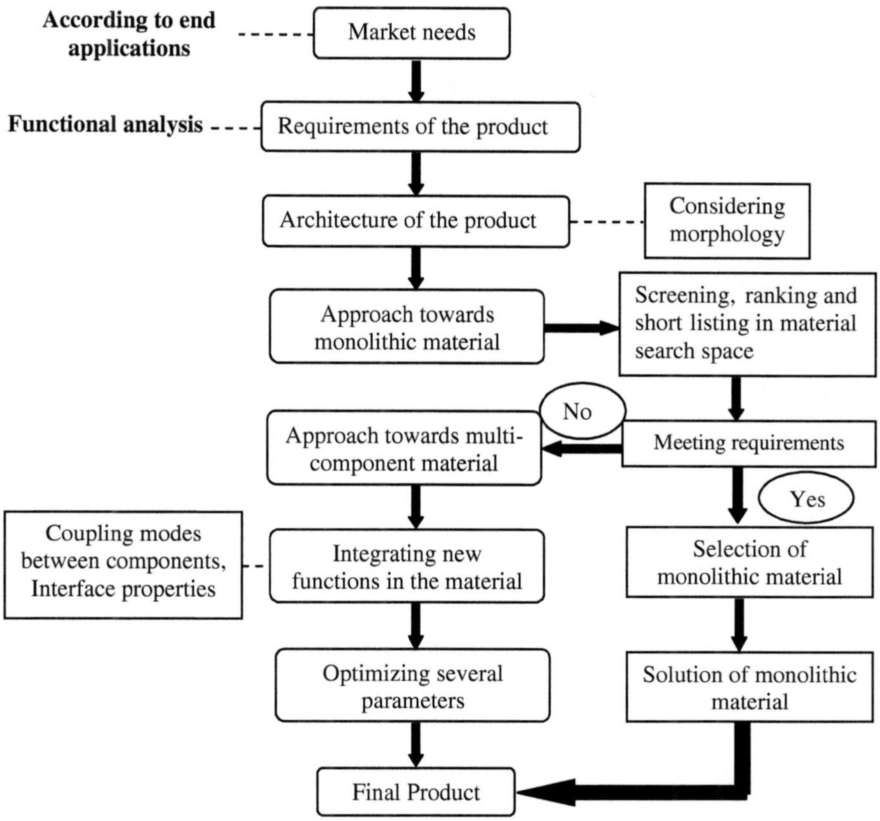

Fig. 2.1 Basic steps for designing

- Establishment of correlation between design requirements and specification of materials.
- Elimination of unsuitable candidates (materials).
- Gradation of greatest potential material.
- Selecting ideal material considering strengths, weaknesses, and history of use and future potentials.

Designer needs to evaluate and select the monolithic materials as an answer of a set of material requirements, however upon failure of such strategy a multi material must be designed by integrating new functions available to multicomponent materials. Thus, optimizing the several parameters of a multi material must fulfil the requirements of end-users.

Necessity of multicomponent materials:

Multicomponent materials are combination of materials that cannot be disassembled and the materials must be in a reasonable fraction of volume and smaller in dimension in comparison to the system in which these are engaged [1]. The basic reason to go for a multicomponent material has already been discussed earlier.

Selection strategy of materials plays a vital role in designing the process. An optimization tool performs by analyzing a set of given inputs with their constraint equation followed by converting them to a set of outputs. Some previously known outputs of the materials and processes are used to perform as a transform functions and fit them to a best possible desired design requirements. A designer considers some macro objectives for a competitive product such as (a) minimizing mass; (b) minimizing volume (sometimes); (c) minimizing manufacturing costs and increasing productivity; (d) minimizing environmental impact etc. On the basis of these guidelines designer theoretically performs three tasks:

- Incorporating user needs into material features (technical and non-technical requirements).
- Formulation of performance matrices on basis of material key features inventory.
- A search procedure-structured material selection method
  - (a) Explore a solution space,
  - (b) Identify materials that meet the constraints,
  - (c) Rank them by their ability to meet the requirements.

An introduction of term strategy opens up the scope for material selection. Thus material selection is not just only confined to a method but a well-defined plan and action on a long term basis keeping in mind on the success of the product in the market. Consequently it also effects on the success of production houses against competitive forces in terms of market evaluation. Strategy is like a transfer function that converts a set of inputs into a set of outputs. Selection strategy in respect of material selection is briefly discussed in Fig. 2.2.

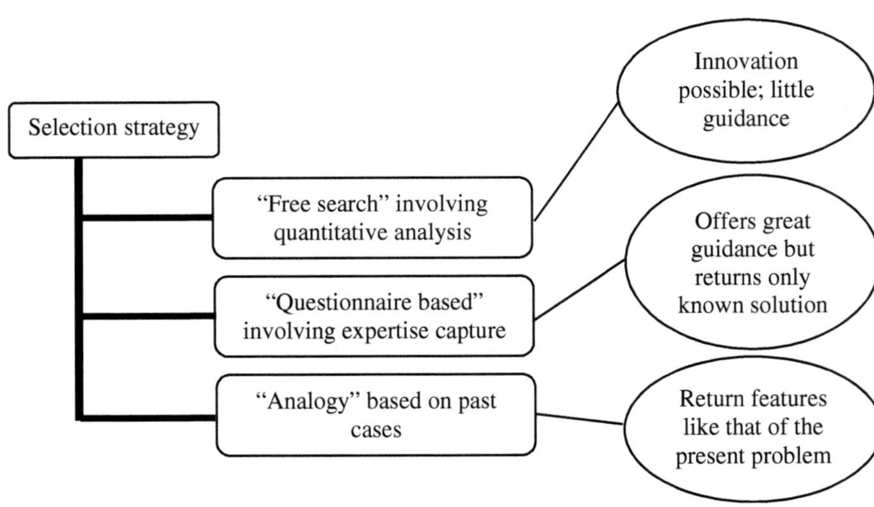

Fig. 2.2 Selection strategy [8]

Quantitative analysis is fast, efficient, flexible etc. but requires very precise degree of input. Questionnaires are more or less structured set of decisions usually works like a set of family tree. But questions should be properly targeted in each step; otherwise it will lead to a vague conclusion. Previously solved problems are used as library of transfer function of third case but difficulty rise in correlating these solutions with the present design problem.

After strategy it is necessary to consider some selection rules for choosing an appropriate multi component system, in view of its end application. Selection of a multi component material and process can be explained through an inverted pyramid process as shown in Fig. 2.3.

Final selection of material and process is performed on the basis of ranking and screening out of unpotential candidate. Researchers proposed different screening methods for selection of materials. The screening methods are mainly-

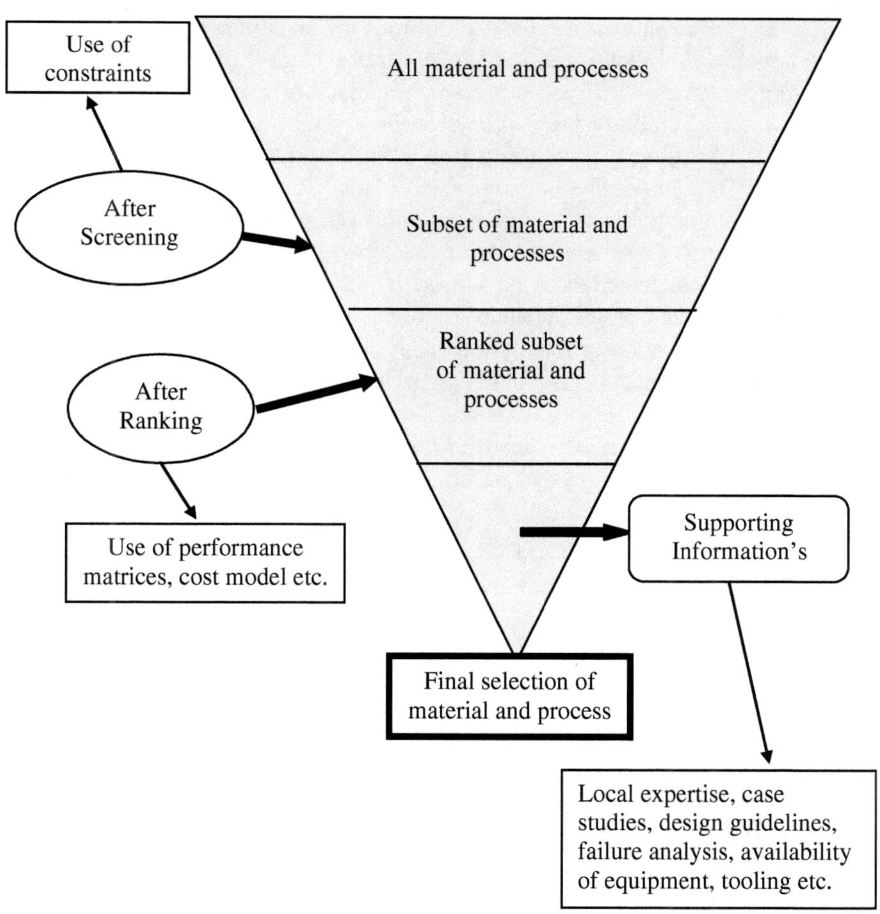

Fig. 2.3 Steps for selecting a material and process [9]

- Cost per unit property method (considers most critical property ignoring others).
- Chart method (useful for initial screening with respect to mechanical performance).
- Questionnaire method (ranking cannot be possible but useful for achieving optimal design solution).
- Materials in production selection tools (low translation percentage between user interaction and material property).
- Artificial intelligence method (appropriate tools for searching) [10].

Performance matrices are used for ranking of materials. Considering environmental effects nowadays it is essential to develop a sustainable and recyclable product. Reduction of waste stream, lesser use of earth resources are also considered as constraints for material selection rules. So, for development of a sustainable product apart from economic aspects social and environmental aspects also come under the selection guidelines [11]. An economical aspect mainly includes purchase, process, transport, disposal cost (Fig. 2.4).

Selection of material is a step forward towards achieving the goal. The very next stage is optimization. Optimization of several parameters is done by incorporating several tools to achieve a well-processed product life cycle. It is important to conceptualize and embody these processes in a single design. More than often incorporation of different system in a single material improves its properties in a particular way but deteriorate in another way. Thus the need of optimizing several parameters is felt and the short comings of multi-component material are searched. The process is briefly discussed through a flow chart in Fig. 2.5. Search for solution of multi component materials is performed in view of components, architecture, adapted interfaces and pre-established solutions.

Different configuration of multi component material is carried out under several optimization tool packages some of them are tools for dividing up material requirements, topological and screening tools, architecture selection and optimization tools etc. Multi criteria decision making (MCDM), Technique of Ranking Preferences by Similarity to Ideal Solution (TOPSIS) method are widely used in

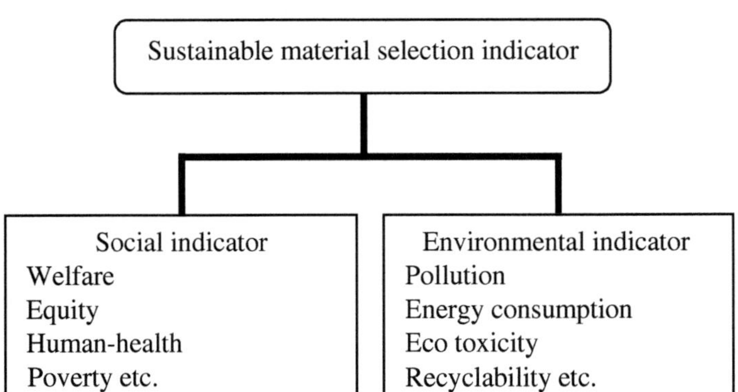

Fig. 2.4 Sustainable material selection indicator [11]

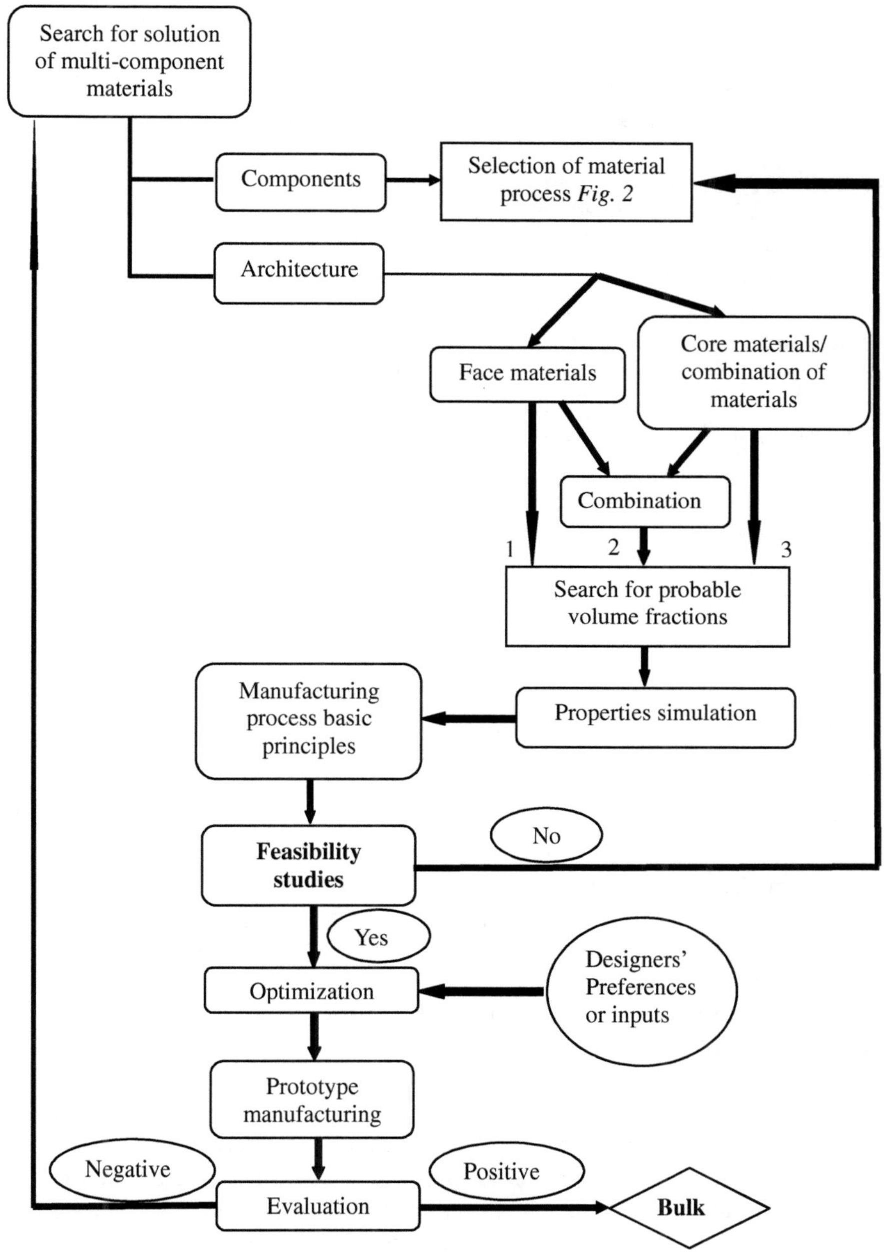

Fig. 2.5 Solution of multi component material

serving superior material requirements. Engineering problem with high degree of complexities has been sorted out nowadays with hybrid methods such as TOPSIS along with Analytical Hierarchy Process (AHP) or sometimes in combination of

Genetic Algorithm (GA) with Artificial Neural Network (ANN) for optimization of a process. ELimination Et Choix Traduisant la REalité (ELimination and Choice Expressing REality) or ELECTRE method is also a multi criteria decision analysis tool often used for material selection of bipolar plates for polymer electrolyte membrane fuel cell (PEMFC) [12, 13]. One recent approach in process control for optimizing polymer composites properties is performed by fuzzy logic controller system. Reduction of temperature variation is obtained by fuzzy logic controller [14] across melt flow of polymer extrusion. A synopsis of process monitoring by Raman spectroscopy, and adaptive Neurofuzzy approach helps to successfully monitor molecular changes in real time. This soft computing technique has been successfully applied to solve the problems of polymers materials during manufacture of thermoset based composites typically for epoxy resins [15]. Researchers have also tried fuzzy logic for prediction of total specific pore volumes of derived polymers [16]. ANN based soft sensor are utilized by a group of researchers [17] for control on viscosity of real time industrial polymerization process of polyethylene terephthalate (PET). Some commercial software packages like ABAQUS (a finite elemental analysis suite), MOLDFLOW (used for injection moulding) are often used for assistantship in micro injection moulding. Optimization of moulding tools is performed with this kind of commercial tools [18]. Temperature distribution can be simulated by ABAQUS package tool in micro injection moulding system. Above all success of a process also depends on the clarity of design, any shortcomings or design errors in selection rules or any other steps during processing of materials leads towards a totally different and undesired product.

2.1.1.1 Influence of Rheology on Design

Rheology is the science of deformation and flows under external stress. It is critically performed for understanding and interpreting mechanical behavioural pattern under a given temperature. The deformation generally caused by the external forces is of three types-tensile force, compressive force and shear force. Under processing and during performance a polymeric multi-component experiences these forces. Flow behaviour of polymeric materials is also governed by the rheological properties while processing under the molten state. Rheology is an important parameter for designing as it correlates with the structure of a multi-component. Rheological properties of a material are sensitive to structure in certain aspects and are easier to use than the conventional analytic method viz. nuclear magnetic resonance [19]. Rheological properties are dependent on several conditions mainly (a) strain rate, (b) polymeric creep (under a static load, the deformation of polymeric material increases with time, (c) Stress relaxation (During a constant deformation process the stress required to maintain the deformation decreases with time, (d) mechanical strain energy of a system can which is convertible to heat energy is accessed by the measure of internal friction. Any polymer system typically shows an in-phase as well as an out-of-phase components when subjected to sinusoidal stress. The phase lag (angle) between the stress and strain is a measure of the internal friction. At normal temperature creep, stress relaxation phenomena are insignificant to metals so they are usually neglected during design consideration at room temperature. They are taken into account when they are close to their recrystallization temperature (usually 0.4–0.6 T_m). But for multi component polymeric material particularly for structural applications they should be taken into account at the design stages [20].

For a heterogeneous system like multi component polymeric material structure and property can be correlated with rheological measurements. According to force exerted during tests rheological properties can be divided into two types: Static rheological measurements performed under steady shear flow motion i.e. under certain stress or strain and Dynamic rheological measurement performed under oscillatory shear flow i.e. under periodic stress or strain. During static rheological testing a multi component polymeric materials continuous effect of stress is hardly obtained as this testing leads to change even breakage of macromolecular segment. Birefringence, dichroism are optical properties of a material usually performed to predict the orientation of the aggregates. So static rheology is integrated with optics to detect shear stability and orientation of the aggregates and segments in a flow field as flow birefringence and flow dichroism [21].

Dynamic rheological testing can be effectively carried out for the measure structure/morphology of multi-component as these testing are done under small strain amplitude. Dynamic rheology testing is believed to be a preferential method because of materials exposed to the testing processes is not destroyed under small-strain amplitude. But care has to be taken such that the working condition remains under linear viscoelastic region.

For determining the linear viscoelastic region (LVR) amplitude sweep is performed over a stress or strain range. Stability can also be predicted from a plot of complex shear modulus versus shear stress under amplitude sweep. After prediction of LVR frequency sweep is performed to determine the nature of material. Storage modulus G' is evaluated from amplitude sweep and loss modulus G'' from frequency sweep. The complex modulus that is overall material resistance to deformation $G^* = G' + iG''$. The ratio of loss modulus to storage modulus is predicted as tan delta parameter which is equal to G''/G' and gives a measure of the viscous portion to the elastic portion of a polymeric melt. If

Tan (delta) > 1 ($G'' > G'$) (liquid or 'sol')
Tan (delta) = 1 ($G' = G''$) (viscoelastic or 'gel point')
Tan (delta) < 1 ($G'' < G'$) (solid or 'gel')

Tan (delta) is especially helpful in measuring sol-gel transition point. Viscosity, strain rate, Deborah number is some very common rheological parameters and should be considered while designing multi-component.

Smart fluids such as magneto rheological suspension consist of micrometer-sized particles (which can be magnetized), dispersed in suitable carrier liquids. In the absence of an external magnetic field this material behaves like Newtonian fluid but apparent suspension viscosities can be increased by several orders of magnitude under external magnetic field. Magneto-moments aligned in the direction of external field and causes strong magnetic interactions which in terms causes aggregation of complex networks. Thus these structures can withstand a certain amount of shear stress and can be characterized by a high yield stress values. Magneto rheological fluids finds their usage in automotive clutches, brakes for exercise equipment, polishing fluids, seat dampers, prosthetic knee damper, actuator systems, shock absorbers etc. [22, 23]. Thermo-oxidation temperature region of polymeric material (generally for polyolefin) can be predicated by dynamic rheological behaviour of multi component systems such as high density polyethylene (HDPE)/carbon black (CB) composite system [24].

- Electrically conductive polymer matrix composites have attracted researchers for its unique electrical and mechanical properties. Filler concentration and the state of the filler dispersion are the two most important influencing parameters for these composites. For conductive composites a continuous conductive chain is required which is dependent on some critical concentration of fillers. This phenomenon is taken care in terms of percolation theory. Dynamic rheological measurements believed to be a good method for dealing with filler dispersion. Homo-polymers in low-frequency regions exhibits the dynamic storage modulus (G′) and dynamic loss modulus (G″) are proportional to the frequency squared (ω^2) and to ω. $G' \infty\ \omega^2$ and $G'' \infty\ \omega$. For filled polymer melts dynamic viscoelastic functions exhibit a special response. Thus estimation of the agglomeration for conductive polymer composites through dynamic rheological measurements is carried out by correlating electrical percolation with viscoelastic percolation [25].
- Viscosity, surface tension and conductivity are three main solution parameters of nano fiber preparation by electro spinning technique. Generally higher viscosity smoothers formation of fibers. Rheological study of polymer solutions in the bulk and at the interface finds correlations between those properties. For chitosan or alginate filled with poly(ethylene oxide) (PEO) polyblend it is found that interfacial rheological parameters have much more influence than the bulk. Conductive system in the bulk displaying predominant plasticity over elasticity is found to be suitable for electro spinnability [26].
- Composites manufactured from compression moulding technique undergo an anisotropic viscous behaviour of matrix material. Non-Newtonian flow behaviour is often modelled for successful distribution of polymeric melt in the mould while fabricating composites like sheet moulding composites (SMC) [27]. Compression moulded polymer matrix are often porous. Mould composite friction effects are usually neglected in rheological studies thus new modelling technique proposed by the researchers taking compressibility and mould-composite friction into account. Outer layers with higher polymer density of fiber reinforced composites act as lubricating layers that are squeezed and sheared near the mould surfaces. However, the core of the composite filled with fiber deforms like extensional plug flow [28].

- In shear induced crystallization of polymers the effect of viscosity is negligible but with increasing crystallization, viscosity also increases. So crystallization time is modelled with rheological properties. It is performed mainly for polyolefins [29].
- Phase separation of poly blends, different chain segments of block co-polymer (micro-phases) have also relationship with their rheological behaviours.
- Rheological measurements while liquid to solid transformation takes place are an important parameter as it involves gelation. The prediction of gel point gives access to working time (which can be modified by delaying gelation) for fabrication of multi-component.

Rheological measurements are made in conjunction with the product quality and efficiency (ex. normally if thermosetting resin is stored for a longer period of time its viscosity increases). Viscosity is the principal parameters for rheological measurements. A designer can predict product dimension from viscosity such as liquid, semi-solid. Designers can predict pumpability and pourability of polymeric material, performance in a dipping or coating operation, or the ease with which it may be handled, processed, or used. Detecting changes in colour, density, stability, solids content, and molecular weight can also be interpreted by rheological measurements. Thus rheological experiments are useful for (a) characterization, (b) determination of processability, (c) failure analysis (sometimes for extrusion, injection moulded samples), (d) simulation (gives data as input for modelling and simulation).

2.1.1.2 Influence of Shear Rate on Design

A liquid is a material in which strain is a function of both stress and time. Shear rate is defined as the ratio of shear stress to strain rate. Thus shear rate or specifically shear strain rate is an influential parameter in engineering design. For a Newtonian fluid shear rate versus shear stress plot shows a straight line passing through the origin. Dilute suspension of particles in this type of fluid also tend to behave like Newtonian fluids but higher volume fraction of particles shifts the flow behaviour to non-Newtonian. Particle size and size distribution are influenced by shear thickening behaviour of material which in terms depends on shear rate. At high shear rates shear thinning material resemble Newtonian fluid and at low shear rates shear thickening material resemble Newtonian flow. Pseudoplastics are shear thinning material in which long chain molecules of the system is believed to achieve a stable configuration through chain entanglement and molar cohesion. Thus high rate of shear decreases the apparent viscosity (shear stress/shear rate) which can be taken into account for spraying, brushing of paints. In PP/NBR blends; decrease in viscosity is evidenced by increase in shear rate. Dialtant is shear thinning materials where crystallization or structure formation occurs at high rate of shearing and thus the apparent viscosity of the system increases [30]. Polymer processing techniques such as extrusion, injection moulding requires high shear rates but finished part quality is often controlled by low shear rates. Complex fluids including polymer melts and solutions, block copolymers, polyelectrolytes, surfactants, suspensions, emulsions etc. are non-Newtonian in nature. Modelling of these complex materials is characterized by testing the response of a material to large shear rate called as LAOS (Large Oscillatory Shear Flows). This technique records shear stress for large strain rate which is then modelled by using fourier concepts. In most processing operations the deformations are large and rapid thus it exhibits nonlinear material properties. Thus to control the system response under this dynamic situation LAOS based modelling is very useful [31]. Shear rate is important as it can predict the miscibility of polymeric blend. A study reveals that microemulsion consists of a ternary blend of poly(ethyl ethylene) (PEE), poly(dimethyl siloxane) (PDMS) and a PEE-PDMS diblock copolymer exhibits Newtonian flow at low shear rates, at intermediate shear rates anisotropy in the morphology (leads towards shear thinning) is observed. Further increment in shear rate causes flow induced phase separation of blend and if shear rate is extremely high it resembles of immiscible blend [32]. Multicomponent polymer system scatters light under the influence of shear fields and the scattering mainly depends on shear rate. Rheo-optic technique uses electromagnetic radiation (optics) in studying the deformation and flow (rheology) of polymers with particular emphasis on polymeric solids [33]. Generally increment in strain rate results in decrement of ductility. But modulus and yield or tensile strength of a material increases with an increase of shear rate.

2.1.1.3 Flow Performance and Design

Flow behavior is an indirect measure of product consistency and quality as in industrial scaling operation batch to batch variation should be minimal. By tuning the processing conditions mainly temperature, strain rate and molecular composition wide ranges of molecular morphologies are possible, such as spherulitic, shish–kebab, or row-nucleated structures [29]. Prediction of flow performance while designing gives the scope that molecular morphology can be altered in desired ways. Flow behavior is an indirect measure of product consistency and quality and is responsive to properties such as molecular weight and molecular weight distribution. When the temperature of melt increases free volume around the polymer chain also increase which in terms facilitates in easier flow. Complex structure of polymeric chain exhibits stronger effect of temperature on flow behaviour. Two types of flows are commonly studied for non-Newtonian fluids: simple shear and simple elongational (extensional) flow. Simple shear is uniform flow where each fluid element on same stream line undergoes same deformation and the distance between them remains unchanged. In extensional (or elongational) flow during processing, material undergoes stretching along the streamlines due to extensional deformation and the distance between particles on the same streamline changes. For industrially purposes Melt Flow Index (MFI) is carried out by the grams of polymer extrude per 10 min from a die of prescribed dimensions according to an ASTM standard under the action of a specified load. Melt flow index has direct correlation with molecular weight, density, viscosity, shear rate. Orientation of multi-component can be tailored to an external flow field viz. steady shear, extrusion, oscillatory. In modelling aspects tensor coupled with course mesh size finite element calculation helps in prediction of planar flow and orientation. Flow analysis helps in obtaining in plane stiffness which helps in designing of complex shapes aided by computer based model. 3D micromechanical models for laminates like mould flow packages involves flow behaviour and prediction of mechanical properties [34]. Designers often use plasticizers, a high boiling liquid which works by lowering glass transition temperatures and improving the flow characteristics of the system.

2.1.1.4 Elasticity and Design

Elasticity is very useful in determining the mechanical properties of polymeric system especially in structural load bearing application. Polymers show excessive elastic deformation due to its inadequate stiffness. So, for controlling the failure of system designers play with elastic modulus of the material. Designers rely on factor of safety (ratio of ultimate stress to the actual stress) calculation for implementation of FRCs in structural application. For rubber processing elastomers are generally thermoplastic in nature which exhibits low modulus, elasticity at least twice of their original length at room temperature (when stretched) and has the ability to return to their approximate original length when stress is released. TPEs (thermoplastic elastomers) have the properties and performance of rubber, but they are processed like plastic. A thermoplastic elastomer has two segments viz. hard phase and a soft phase. Plastic properties of a thermoplastic elastomer are governed by the hard phase of the materials and soft phase of the material determines the elastic property. Behaviour of the hard phase is governed by processing temperatures, tensile strength, tear strength, peel strength, chemical and fluid resistance, adhesion to inks, adhesives, and over-molding substrates. Elastomeric property of the material is considered on the basis of lower service temperature limits, hardness, flexibility, compression set and tensile set.

Thus modulus of elasticity plays a vital role in designing this multi-component. Apart from thermoplastic elastomers, elasticity also influences designing of elastomeric alloys. Commercial elastomeric alloys are of two types melt-processable rubbers (MPRs) and thermoplastic vulcanizates (TPVs).

Resistance offered by polymer melt against stretching is termed as melt strength. It depends on the molecular chain entanglements of the polymer and its resistance to untangling under strain. Melt strength of the polymer is dependent on molecular weight, molecular weight distribution, and molecular branching. Increment of any property increases the melt strength. This relation is valid at low shear rates. In design aspects melt strength also affects in drawdown of extrusion coating. In co-extrusion process balance of melt strength can predict interfacial properties. Sometimes melt strength is related to the extensional viscosity of the polymer. A good elastic polymer material should have good damping properties and can be used in a variety of applications to absorb shock, dissipate heat, isolate vibration, or damp noise. It should have a high damping coefficient, indicating that the material will bounce back or return energy to the system. It needs to be work in a wide range of temperatures and environments, including exposure to chemical.

2.1.1.5 Molecular Weight and Design

Molecular weight and its distribution is an important parameter in determining of material bulk properties. Generally high molecular weight permits higher degree of chain entanglement which in term associated with material high melting or softening temperature and high tensile strength. Failure or degradation of material is depicted as loss of material load bearing capacities. For example atactic polystyrene mechanical properties increases with increase in molecular weight below the glass transition temperature (T_g). Researchers correlated impact strength and toughness with molecular weight in glassy state of the material, impact strength increases with molecular weight. Fracture toughness is strongly dependent on molecular weight distribution [35].

Relationship between molecular structure and mechanical properties is very complicated. Evolution of excellent catalyst has minimized the effect of tacticity in high molecular weight polymers (isotacticity is maintained in commercial polymers). Flory proposed an empirical equation for predicting mechanical property by using molecular weight as an influential parameter.

$$P = A + B/M \qquad (2.1)$$

where P stands for mechanical properties, M for molecular weight, and A, B are constants. A, B values are varied according to the property taken. If B is positive, mechanical properties decrease with increase of molecular weight and vice versa. Sometimes a more complicated empirical relationship is used where M is replaced by $M^2 + C$ [36]. Glass transition is another important parameter which is also dependent on molecular weight. An empirical expression relating the inverse relations between T_g (glass transition temperature) and $\overline{M_n}$ (number average molecular weight) is given by Blanchard et al. [37]

$$T_g = T_g^{\infty} - K/\overline{M_n} \qquad (2.2)$$

where T_g^{∞} is the T_g of an infinite polymer and k is an empirical constant. According to Flory-Fox glass transition temperature is the temperature where free space volume for molecular chain movement reaches its minimum value. In a polymer system chain end segments are restricted only at one end whereas internal segments constrained at both ends thus their mobility is more imparted than the end segments. As the number of chain ends increases (which means a decrease in $\overline{M_n}$) free volume increases thus T_g decreases. But in crosslinking intermolecular connection is formed and this network structure results in increment of glass transition temperature. In broader aspects mechanical properties and resistance to chemical attack can be altered by molecular weight.

2.1.1.6 Chemical Changes and Design

Performance, durability and long term behavior of fiber reinforced composites for structural applications are tested under several environmental conditions. Simulated aggressive environments [38] are generally considered and the performance over a period of 6 months or more is monitored for a structural material. Generally performance of material is evaluated under neutral pH (fresh water); alkaline pH (saturated calcium hydroxide $Ca(OH)_2$ solution (pH = 12.5)); acidic pH (hydrochloric acid, HCl, solution (pH = 2.5)). Apart from that material performances under simulated seawater, moist alkaline soil with microorganisms, UV radiation are sometimes also monitored. Simulated ocean water with a pH of 7.25 is used for simulating the marine environment. The major ingredients of this solution are NaCl (24.53 g/L), $MgCl_2$ (5.20 g/L), Na_2SO_4 (4.09 g/L, $CaCl_2$ (1.16 g/L), KCl (0.695 g/L), and $NaHCO_3$ (0.201 g/L) in water according to ASTM D1141. Chemical changes are considered as a performance indicator while designing for long term applications. Apart from that degradation studies (swelling and dissolution) are also taken into account while designing for end product disposal and recyclability (if possible) of the product. Degradation is basically a separation of chain segments-termed as scission. Scission effect or bond rupture can take place due to radiation effects, chemical reaction effects and thermal effects.

2.1.1.7 Physical State and Design

Temperature and pressure is the two main pillars controlling physical state of a material. Miscibility of most polymer-polymer system decreases with increase in temperature. Thermodynamic properties of polymer blend as a function of pressure and temperature are explained with the help of random phase approximation, self-consistent-field theory, and Flory—Huggins theory [39].

2.1.1.8 Other Parameters

Apart from the above mentioned parameters miscibility and phase separation and segregation also plays an important role while designing multicomponent system. Interfacial tension reduction, coalescence suppression, increase in compatibility are some approach towards obtaining miscible polymer blends.

Elastic modulus is a pressure dependent property of material. Increase in pressure decreases the free volume available in the structure and thus increases the modulus of elasticity. Ferroelectric polymers (from the polyvinylidene fluoride

(PVDF) family) exhibits piezoelectricity under mechanical stimulus. Mechanical stimulus causes change in the dipole densities and creates moderate piezoelectric coefficients in comparison to ceramic piezoelectric [40]. Piezoelectric materials are of three types.

- Bulk piezoelectric (molecular structure and orientation).
- Polymer piezoelectric composites (PVDF).
- Voided charged polymers [41].

In designing of shape memory polymer (SMP) physical state is an important parameter. To exhibit shape memory effect a material (generally elastomer) under thermal or other stimuli maintains the deformed state in a temperature region according to the relevance of the application [42].

2.1.2 Material Selection Approaches (Example) [43]

An example can be taken to explain the material selection approach for plastic material in medical device design:

Basics of polymers and their additives, properties, process-abilities are used to design the morphology of a multicomponent. Thoroughly defined applications are required to select appropriate candidate materials and careful consideration must be taken while combining different configuration. Final selection should be obtained by testing after narrowing of choices. Selection approach starts with-

1. Market trends and challenges, followed by
2. Design requirements in context of the application

Objectives of the material will vary from product to product and thus the features governing factor may not be the same for all material within the same classes also. As medical product requirement varies according to end application. Several tests are performed on the material according to the physical contact with human body which has been tabulated in table [1]. If floor panels of refrigerated trucks are considered than objectives varies accordingly, minimizing mass; minimizing cost; adequate thermal insulation to keep goods refrigerated and minimizing heat loss are the key objectives for this application. Likewise for medical products general objectives are to search on questionnaire basis (a selection strategy) on expertise advice considering several constraints.

Environmental constraints-

- Biocompatibility
- Contact ability (with body tissues or drugs-along with duration)
- Instrument/device usage-Single use/multi use
- Sterilization required or not (with frequency of sterilization)
- Reaction with other chemicals/solvent/vapours (ex. hospital cleaning)

- Attachment mode of device with body
- Exposure to humidity and temperature with time period (maximum and minimum)
- Dimensional stability under different condition
- Tolerances
- UV resistance (a requirement or not)
- Visibility under a fluoroscope or X-ray (a requirement or not)
- Flame retardability (a requirement or not)
- Colour of the material (If it is an influential parameter)
- Outdoor application (if any)

Dimensional and mechanical constraints:

- Parts dimensions (diameter, length, width, thickness)
- Load bearing capacity-time period-continuous or intermittent load
- Maximum stress on the part, kind of stress (tensile, flexural, etc.)
- Toughness or impact resistance-during application
- Wear properties
- Electrical insulator (if required or not)
- Available manufacturing processes
- Target cost of the component
- Projected life of the part or design
- other mechanical properties
- Static dissipation/conductivity/dielectric properties (if required)

3. Screening

(a) Biocompatibility of the polymeric material is the first requirements for a material to be considered for medical device/biomedical uses.

According to application several types of polymers can be chosen-

Device/Instrument directly not connected with human body, e.g. syringes, blood storage bags, glucose drip bags (PVC, PA, PE, PS, Epoxy resins)
Device/Instrument in short period of contact with human body, e.g. catheters, feeding tubes, drainage tubes, surgical instruments (Silicone rubber, Natural rubber, PVC, Polyurethane, PE, PP, Polyester, PEEK, Polyphenylsulfone, Nylon, Teflon)
Device/Instrument in medium period of contact with human body, e.g. cultures, ligatures (Nylon, PP, Polyester)
Device/Instrument in long period of contact with human body, e.g. implants, drug delivery devices (PE, UHMWPE, PET, Silicone rubber,

Polyurethane, PMMA, Polysulphones, Hydrogels Polyphosphazenes, Thermoplastic elastomers, Polydimethylsiloxane)

(b) Drug flow path (device performance under flow of drugs)
(c) Chemical inertness, sterilization
(d) Resistance to hospital cleaners (chemicals) against routine cleaning of devices
(e) Mechanical properties
(f) Electrical and Thermal Properties (dielectric strength and thermal resistance)
(g) Dimensional stability, durability, aesthetics, radiopacity, conductivity, manufacturing feasibility

Finally correlation between technical performance and economics considering the screening tools:

Several biocompatibility tests are generally performed on polymeric medical devices before applications according to ISO 10993-1. Performance of the product is monitored under some common biological risks viz. cytotoxicity, sensitization, irritation, acute toxicity, sub chronic toxicity genotoxicity, implantation, hemocompatibility, chronical toxicity, carcinogenic, reproduction toxicity, biodegradation (Table 2.1).

4. Ranking of materials

A performance equation is used to determine the ranking of materials.

P = [(F), (G), (P)] where F = Functional requirements, G = Geometrical parameters, P = Material properties.

Miniaturization, portability, improved aesthetics, and environmental responsibilities in medical devices have toughened selecting the right material with optimal product usability, appearance, endurance and performance. Sometimes pre-selection of material is also performed.

2.1.3 *Case Studies*

Tubular filament composite shaft designing [44]:

Specifications: Outside diameter-75 mm, inside diameter-55 mm, length of 1.0 m.
Main constraints: Bending stiffness (Strength, fatigue resistance is not significant since it is a filament composite). Maximum allowable deflection while subjected to a three point bending is specified as bending stiffness.
Problem: Load 900 N is to be applied at the center of the shaft producing a deflection not more than 0.3 mm. Continuous fibers parallel to the tube axis will be used in filament winding process. Possible fibers may be are glass, carbon in standard-, intermediate-, high-modulus grades, Kevlar with maximum allowable fiber volume fraction is to be 60%. Matrix material will be epoxy resin (Fig. 2.6 and Table 2.2).

Table 2.1 Selection of test for medical products as a function of contact method and time with human body [43]

Selection of tests for polymeric medical products ISO 10993-1												
Biological risks										Duration of contact	Contact nature of the product with human body	
Cytotoxicity	Sensitization	Irritation	Acute toxicity	Sub chronic toxicity	Genotoxicity	Impantation	Hemocompatibility	Chronical toxicity	Canceroginity			
√	√	√								Up to 24 h	Skin	Products in contact with the body surface
√	√	√								(>24 h–30 days)		
√	√	√								(>30 days)	Mucous membrane	
√	√	√								Up to 24 h		
√	√	√	*	*	*	*				(>24 h–30 days)	Injured surface	
√	√	√	*	√	√	*		*		(>30 days)		
√	√	√	*							Up to 24 h	Blood stream	Products in contact with interior of body
√	√	√	*	*		*				(>24 h–30 days)		
√	√	√	*	√	√	*		*		(>30 days)	Tissue/Bone/Dentin	
√	√	√	√				√			Up to 24 h		
√	√	√	√	*			√			(>24 h–30 days)	Circulating blood	
√	√	*	√	√	√	*	√	√	√	(>30 days)		
√	√	√	*							Up to 24 h	Tissue/Bone	Implant
√	√	√	*	*	√	√				(>24 h–30 days)		
√	√	√	*	*	√	√		*	√	(>30 days)	Blood	
√	√	√	√				√			Up to 24 h		

√Test to be included to ISO 10993-1

*Additional tests may be applicable

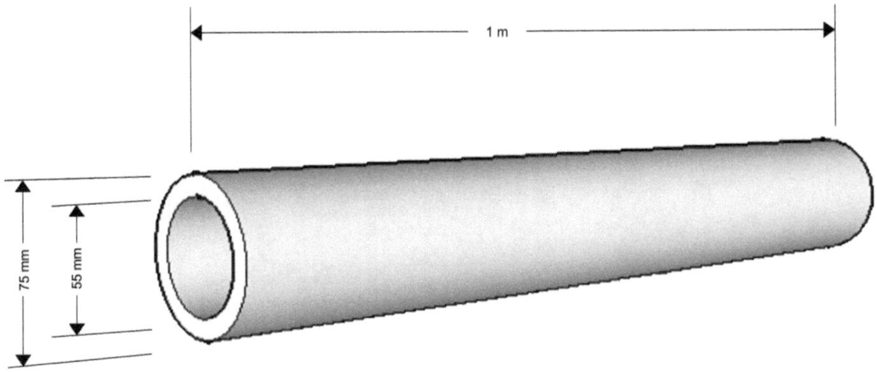

Fig. 2.6 Schematic representation of the drawing objective [44]

Table 2.2 Elastic modulus and density of different materials [44]

Material	Elastic modulus (GPa)	Density (g/cm^3)
Glass fibers	72.5	2.58
Carbon fibers (standard)	230	1.80
Carbon fibers (intermediate)	285	1.80
Carbon fibers (high modulus)	400	1.80
Kevlar	130	1.45
Epoxy	2.4	1.14

Solution: Three point deflection equation is given by $\Delta y = \frac{FL^3}{48EI}$ where F is the applied load, L is the gauge length, E is the modulus of elasticity and I is the moment of inertia (cross-sectional). So for computational necessities longitudinal modulus of elasticity is first needed to be calculated.

For a tubular shaft-

$$I = \frac{\pi}{64}(d_0^4 - d_i^4)$$

$$E = \frac{4FL^3}{3\pi\Delta y(d_0^4 - d_i^4)}$$

Following this problem F = 900 N; L = 1.0 m; Δy = 0.3 mm; d_0 = 75 mm; d_1 = 55 mm.

Longitudinal modulus of elasticity is

$$E = \frac{4(900)(1.0)^3}{3\pi(0.3\times10^{-3})[(75\times10^{-3})^4-(55\times10^{-3})^4]}\,\text{GPa}$$
$$= 56.6\,\text{GPa}$$

In the next step fiber and matrix volume fractions are calculated for all potential candidates by using rule of mixtures.

$$E_c = E_m V_m + E_f V_f = E_m(1 - V_{f)} + E_f V_f$$

E_c is the modulus of elasticity of the composite and $E_{m,}$ E_f are the modulus of elasticity of the matrix material and fiber respectively. V_m and V_f are the volume fractions of matrix material and fiber.

Taking E_c = 56.6 GPa respective volume fractions of different fibers are Tables 2.3, and 2.4.

From this chart it can be concluded that glass fiber cannot be chosen as maximum allowable volume fraction of fiber is 0.6.

So, after eliminating glass fibers from this design problem overall requirement of rest of the fibers and matrix for each composite type is necessary to be estimated.

Table 2.3 Volume fractions of matrix and fibers of the composites

Fiber type	V_m	V_f
Glass fibers	0.2268	0.7732
Carbon fibers (standard)	0.7619	0.2381
Carbon fibers (intermediate)	0.8082	0.1918
Carbon fibers (high modulus)	0.8637	0.1363
Kevlar	0.5752	0.4248

Table 2.4 Required weight of fiber and matrix in respect of a tube (designed)

Fiber type	Fiber volume (cm^3)	Fiber mass (Kg)	Matrix volume (cm^3)	Matrix mass (Kg)
Carbon fibers (standard)	486	0.875	1555	1.773
Carbon fibers (intermediate)	391.5	0.705	1649.5	1.88
Carbon fibers (high modulus)	278.2	0.5	1762.8	2.08
Kevlar	867	1.26	1174	1.338

Total tube volume

$$\begin{aligned} V_t &= \frac{\pi L}{4}(d_0^2 - d_i^2) \\ &= \frac{\pi(100)}{4}\left(7.5^2 - 5.5^2\right)\,\text{cm}^3 \\ &= 2041\,\text{cm}^3 \end{aligned}$$

In this design problem as strength and other design parameters are neglected the criterion for selecting an optimum material is purely based on economical. Assuming process cost of each type of composite remains same, total cost of each type of composite is considered (total cost including fiber and matrix) as selection rule. In general the cost of fibers and epoxy resin follows glass fibers < Epoxy resin < Carbon fibers (Standard) < Carbon fibers (Intermediate) < Kevlar < Carbon fibers (High modulus).

Again in design problem if performance of the shaft against twisting is considered than calculation regarding torsion needs to be estimated [45].

For tubular hollow composite shaft with a constant wall thickness, the torsion, T:

$$\mathrm{T} = \frac{J}{r_0}\tau = \mathrm{T} = \frac{J}{l}G\theta$$

where J is second moment of area given by

$$\mathrm{J} = \frac{\pi}{2}(\mathrm{r}_0^4 - \mathrm{r}_\mathrm{i}^4)$$

τ maximum shear stress at the outer surface of shaft
l gauge length of the tube
r_i inner diameter of the tube
r_o outer diameter of the tube
θ angle of twist in radians
G shear modulus or the modulus of rigidity.

2.2 Summary

Multi-component material designing is a function of constraints. The constraints are governed according to the performance of the material under certain environmental conditions and according to the end applications. Optimization and refinement in process defines the architecture of the material. Material selection is one of the main parameter while optimizing a process it basically includes evaluation of recent market trends and challenges; design requirements in context of the application;

screening and finally ranking of materials. Design of a material is influenced by several parameters like rheology, shear rate, flow performance, elasticity, molecular weight, chemical changes, and physical state of the material.

References

1. Kromm FX, Quenisset JM, Lorriot T, Harry R, Wargnier H (2007) Definition of a multimaterials design method. Mater Des 28:2641–2646
2. Ashby MF (2000) Multi objective optimization in material design and selection. Acta Mater 48:359–369
3. Wargnier H, Kromm FX, Danis M, Brechet Y (2014) Proposal for a multi-material design procedure. Mater Des 56:44–49
4. Ashby MF, Brechet YJM (2003) Designing hybrid materials Acta Materialia 51:5801–5821
5. Chickermane H, Gea HC (1997) Design of multi-component structural systems for optimal layout topology and joint locations. Eng Comput 13:235–243
6. Panar M, Epstein BN (1984) Multicomponent polymeric engineering materials. Science 226:642–646
7. S. Thomas, W. Yang Advances in polymer processing: from macro- to nano- scales 1st edn. Woodhead Publishing Limited
8. Ashby MF, Brechet YJM, Cebon D, Salvo L (2004) Selection strategies for materials and processes. Mater Des 25:51–67
9. Sirisalee P, Ashby MF, Parks GT, Clarkson PJ (2006) Multi-Criteria Material Selection of Monolithic and Multi-Materials in Engineer-ing Design. Adv Eng Mat 8:48–56
10. Jahan A, Ismail MY, Sapuan SM, Mustapha F (2010) Material screening and choosing methods—A review. Materials and Design 31:696–705
11. Zarandi MHF, Mansour S, Hosseinijou SA, Avazbeigi M (2011) A material selection methodology and expert system for sustainable product design. Int J Adv Manuf Technol 57:885–903
12. Shanian A, Savadogo O (2006) A non-compensatory compromised solution for material selection of bipolar plates for polymer electrolyte membrane fuel cell (PEMFC) using ELECTRE IV. Electrochim Acta 51:5307–5315
13. Shanian A, Savadog O (2006) TOPSIS multiple-criteria decision support analysis for material selection of metallic bipolar plates for polymer electrolyte fuel cell. J Power Sources 159:1095–1104
14. Jha MK, Qureshi MF (2015) Development of Interval Type-2 Fuzzy Logic Controller for Polymer Extruder Melt Temperature Control. Int J Innovative Res Sci, Eng Technol 4:592–603
15. Merad La, Merad Lo, Bouchaour M, Benyouce B (2015) Adaptative neurofuzzy approach for *in-situ* polymeric resins as studied by raman spectroscopy. Revue algerienne de physique 2:64–68
16. Nazari A, Khalaj G (2012) Prediction total specific pore volume of geopolymers produced from waste ashes by fuzzy logic. Mater Res 15:242–252
17. Gonzagaa JCB, Meleirob LAC, Kianga C, Filhoc RM (2009) ANN-based soft-sensor for real-time process monitoring and control of an industrial polymerization process. Comput Chem Eng 33:43–49
18. Piotter V, Mueller K, Plewa K, Ruprecht R, Hausselt J (2002) Performance and simulation of thermoplastic micro injection molding. Microsystem Technol 8:387–390
19. Agboola OS, Sadiku ER, Adegbola AT, Biotidara OF (2011) Rheological properties of polymers: structure and morphology of molten polymer blends. Mater Sci Appl 2:30–41
20. Ebewele, R.O. Polymer science and technology. CRC press

21. Min Z, Qiang Z (2008) Correlation between rheological behavior and structure of multi-component polymer systems. Sci China Series B: Chem 51:1–12
22. Klingenber DJ (2001) Magnetorheology: applications and challenges. Am Inst Chem Eng 47:246–249
23. Viota JL, Vicente JD, Durán JDG, Delgado AV (2005) Stabilization of magnetorheological suspensions by polyacrylic acid polymers. J Colloid Interface Sci 284:527–541
24. Wu G, Song Y, Zheng Q, Du M, Zhang P (2003) Dynamic rheological properties for HDPE/CB composite melts. J Appl Polym Sci 88:2160–2167
25. Wu G, Zheng Q (2004) Estimation of the agglomeration structure for conductive particles and fiber-filled high-density polyethylene through dynamic rheological measurements. J Polym Sci: Part B: Polym Phys 42:1199–1205
26. Rošic R, Pelipenko J, Kocbek P, Baumgartner S (2012) The role of rheology of polymer solutions in predicting nanofiber formation by electrospinning. European Polym J 48:1374–1384
27. Dumont P, Orgéas L, Corre SL, Favier D (2003) Anisotropic viscous behavior of sheet molding compounds (SMC) during compression molding. Int J Plast 19:625–646
28. Guiraud O, Dumont PJJ, Orgéas L, Favier D (2012) Rheometry of compression moulded fibre-reinforced polymer composites: rheology, compressibility, and friction forces with mould surfaces. Compos: Part A 43:2107–2119
29. Pogodinaa NV, Lavrenkob VP, Srinivasc S, Winter HH (2001) Rheology and structure of isotactic polypropylene near the gel point: quiescent and shear-induced crystallization. Polymer 42:9031–9043
30. Ghosh P Polymer science and technology-plastics, rubbers, blends and composites 2nd edn. Tata Mcgraw Hill Private Ltd.
31. Hyun K, Wilhelm M, Klein CO (2011) A review of nonlinear oscillatory shear tests: analysis and application of Large Amplitude Oscillatory Shear (LAOS). Progress Polym Sci 36:1697–1753
32. Krishnan K, Chapman B, Bates FS, Lodge TP (2002) Effects of shear flow on a polymeric bicontinuous microemulsion: Equilibrium and steady state behavior. J Rheol 46:529–554
33. Wilkes GL (1974) Rheo-optical methods and their application to polymeric solids. J Macromol Sci, Part C 10:149–261
34. Papathanasiou TD, Guell DC Flow induced alignment in composite materials. Woodhead Publishing Series in Composites Science and Engineering
35. Nicholson LM, Whitley KS, Gates TS Hinkley JA (2000) How molecular structure affects mechanical properties of an advanced polymer. SAMPE 2000: international SAMPE symposium & exhibition, "3A: composites durability", May 21–25, Long Beach, CA
36. Ogawa T (1992) Effects of molecular weight on mechanical properties of polypropylene. J Appl Polym Sci 44:1869–1871
37. Blanchard LP, Hesse J, Malhotra SL (1974) Effect of molecular weight on glass transition by differential scanning calorimetry. Can J Chem 52:3170–3175
38. Saadatmanesh H, Tavakkolizadeh M, Mostofinejad D (2010) Environmental effects on mechanical properties of wet lay-up fiber reinforced polymer. ACI Mater J 267–274
39. Ruegg ML, Reynolds BJ, Lin MY, Lohse DJ, Krishnamoorti R, Balsara NP (2007) Effect of pressure on a multicomponent A/B/A−C polymer blend with attractive and repulsive interactions. Macromolecules 40:355–365
40. Bauer S, Bauer F Piezoelectric polymers and their applications. Piezoelectricity volume 114, Springer Series in Materials Science, pp 157–177
41. Ramadan KS, Sameoto D, Evoy S (2014) A review of piezoelectric polymers as functional materials for electromechanical transducers. Smart Mater Struct 23
42. Lendlein A, Kelch S (2002) Shape-memory polymers. Angew Chem Int Ed 41:2034–2057
43. An Overview of the Plastic Material Selection Process for Medical Devices|White paper February 2013
44. Callister Jr WD, Rethwisch DG Materials science and engineering an introduction 8th edn. Wiley Plus
45. Siddiqui NA, Khan SU, Kim JK (2013) Experimental torsional shear properties of carbon fiber reinforced epoxy composites containing carbon nanotubes. Compos Struct 104:230–238

Chapter 3
Design of Mold for Multicomponent Material

Soo-Jin Park and Ki-Seok Kim

3.1 Introduction

Recently, various materials including polymers, metals and composites have been widely used for manufacturing the parts for household and industrial items. Especially, in these days, multi-materials have been studied extensively as promising new materials to replace traditional single phase materials. Multi-materials are defined as the class of objects consisting of two or more different materials and they are heterogeneous. Multi-material parts can have either continuously various material compositions or discrete sections of different homogenous materials. The former type of multi-material parts is functionally gradient materials. As well known, it is functioned or advanced composite structures in which the composition changes gradually during material processing, resulting in corresponding changes in the properties of the structure compared to single material. The latter type of multi-material parts is the object with distinct interfaces or boundaries separating the various materials. To manufacture the parts using multi-materials, injection molding, extrusion molding, compression molding and transfer molding are useful methods [1–3].

Among various molding process, injection molding is the most widely used polymer-based material processing operation. More than one third of all polymer industry in the USA uses injection molding. In addition, injection molding is an ideal process to fabricate large numbers of geometrically complex parts. An injection molding machine has two main sections: (1) the injection unit and (2) the clamp unit, or press, which houses the mold. As well known, injection molding is a cyclic process and molding process is simple as follows [4, 5].

S.-J. Park (✉) · K.-S. Kim
Department of Chemistry, Inha University, 253, Nam-gu, Incheon 402-751, South Korea
e-mail: sjpark@inha.ac.kr

J.K. Kim et al. (eds.), *Multicomponent Polymeric Materials*,
Springer Series in Materials Science 223, DOI 10.1007/978-94-017-7324-9_3

1. Polymer is fed into a heated screw extruder and then polymer is melted.
2. The molten polymer is injected into a mold.
3. The molten polymer is solidified in the mold by cooling.
4. The shaped polymer parts are ejected from the mold.

Through the above process, many items are injection molded, such as simple drinking cup, mobile phone housings, automobile bumpers, television cabinets, and compact discs, etc. The parts produced by the injection molding process are also becoming commonplace in less obvious applications.

Compared to common injection molding for single material, specially, the most common and economically easy method of performing in-mold assembly for multi-materials is through multi-material molding. This method is usually known as specialized injection molding technique. This means that the various polymers, which consist of the different material sections, are heated to their melting temperatures, and then injected in sequence into a mold. The liquid-like polymers are then solidified into the mold to obtain their desired shapes by cooling system. Furthermore, the largest advantage of multi-material molding is its ability to directly produce fully assembled components by one-step. It indicates that an entire parts consisting of multiple pieces can be produced by a single set of molds, thereby eliminating the need for secondary assembly process and the use of bolts, welds, glue, or other fasteners [6–8].

The properties of a multi-material product strongly depend on its morphological features such as molecular orientations, crystallinity, spherulitic structure, lamellar structure, and residual stress, and miscibility or interface adhesion between different materials and injection molding process. The thermomechanical properties of the product during processing naturally can be controlled by adjusting the morphological details. The processing variables in injection molding are melt temperature, mold temperature, injection pressure, injection speed, mold pressure, hold pressure during cooling, and cooling time. These factors influence the dimensional stability and properties of final products.

To make uniform part by injection molding, there are various factors, such as injection unit, nozzle, clamp unit or pressure, pressure for injection, and mold, etc. Among these factors, the mold is the most important factor in the molding machine to determine the shape of the product, and the mold vents the entrapped gas, cools the product and ejects the product. The mold is simply comprised of mold base, core and cavity that determine the feature of the product, sprue, runner, and gate that deliver the melt to the cavity, cooling system and ejection system, as shown in Fig. 3.1.

In injection molding, final product quality and properties and efficient processing depended on the mold design as well as the control injection molding process. The mold design process is generally the critical path of a new product development. In injection molding process, quality of mold is responsible for the economics of the entire process. To manufacture an injection mold, a high level of skill is required, and the success of an injection molding depends on the mold design and the skill of the moldmaker and the molding technician. It is desirable that some experience of

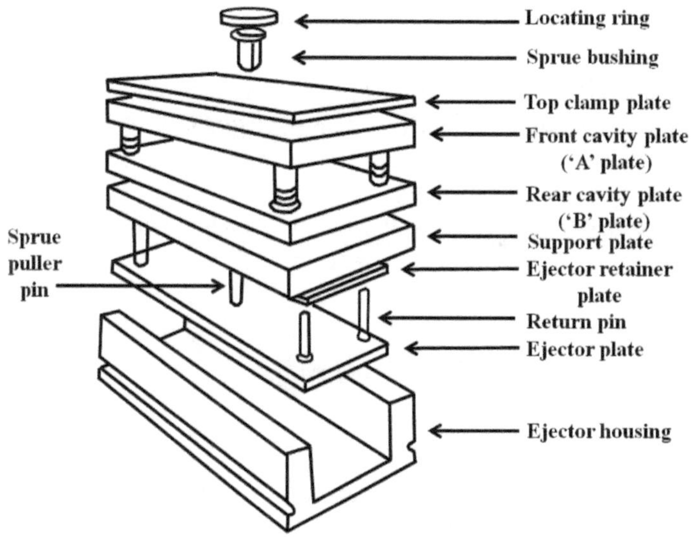

Fig. 3.1 Typical American standard mold base

moldmaking is required to a mold designer. In addition, it is essential for the mold designer to have both experience and knowledge of materials and injection molding process. Therefore, this chapter will describe a knowledge-based parametric design system for mold design, which requires only a minimum set of injection molding parameters to be established before being able to complete the design of the main components of a mold.

Conventionally, mold design has been known as a "mystified" art, requiring years of experience before one can be relatively proficient in it. Due to the initial difficulty in learning it, less and less people are benefiting from the experience and knowledge of the experts in mold design field. Therefore, mold designers are required to possess thorough and broad experience, because detailed decisions require the knowledge of the interaction among various parameters. However, unfortunately, it is presently impossible to cover the growing demand for experienced designers. To change this situation, one method is to use a computer-aided design (CAD) system. Generally, CAD systems have become one of the most important manners to increase productivity. Most CAD software can offer simple and complex geometrical modeling function. Therefore, intelligent CAD systems can assist in the various tasks of the mold design process and are very important to the productivity of the mold-making industry [9–11].

This chapter is focused on the mold design for injection molding. We will describe a brief introduction of the processing methods of polymer-based materials and basic construction for mold. In addition, we will discuss various parameters for mold design and computer-aided design system.

3.2 Injection Mold for Multi-materials

There are many different methods to convert raw plastics form into final products or to fabricate stock plastics materials into finished products. In the industry fields, molding and thermoforming are the method for the mass production processes. Molding includes injection molding, extrusion, compression and transfer molding. In this section, we will focus on the injection molding and basic mold construction for injection molding.

3.2.1 Classification of Molding Process

1. Injection molding

Injection molding is the most important process used to manufacture plastic products. Today, more than one third of all thermoplastic materials are injection molded. Also, injection molding is the best process to fabricate intricate plastics parts required in high volume with high-speed and low-cost molding. The basic principle of injection molding is to inject molten polymer into a closed mold with heating system, cooled mold with cooling system, where it solidifies to give the final product. The molding is recovered by opening the mold to release shaped product. An injection molding machine has basically two main sections: (1) the injection unit and (2) the clamp unit, or press, which houses the mold [12–14].

Almost all thermoplastics-based parts can be obtained by injection molding and some thermosets are also being injection molded with modified equipment. Polyethylene (PE), polypropylene (PP), copolymer of acrylonitrile, butadiene, and styrene (ABS), nylon polyamide (PA), acrylic and polystyrene (PS) [15–20] are amongst the leading thermoplastics used in injection molding. In this process, thermoplastic is fed from the hopper through an opening at the rear of the heated injection barrel. The thermoplastic is forced forward to the front of the heated barrel by the rotation of a reciprocating screw, where the thermoplastic is heated in various stages until it reaches a molten state. The injection screw forces the measured amount of molten resin into the shaped cavity of a closed mold through the nozzle, sprue, runner, gate by a ram action. The molten resin cools and solidifies in the mold cavity by cooling system. After cooling, the mold is opened and the molding is ejected, and finally shaped product can be obtained [21, 22]. The process is described in Fig. 3.2.

In addition, injection molding is an ideal process for manufacturing large numbers of geometrically complex parts. Various items needed from household to industry fields are injection molded: mobile phone housings, appliance housing, automobile bumpers, television cabinets, camera cases, compact discs, lenses, gears, fan blades, spoons, and wastebaskets etc. [23].

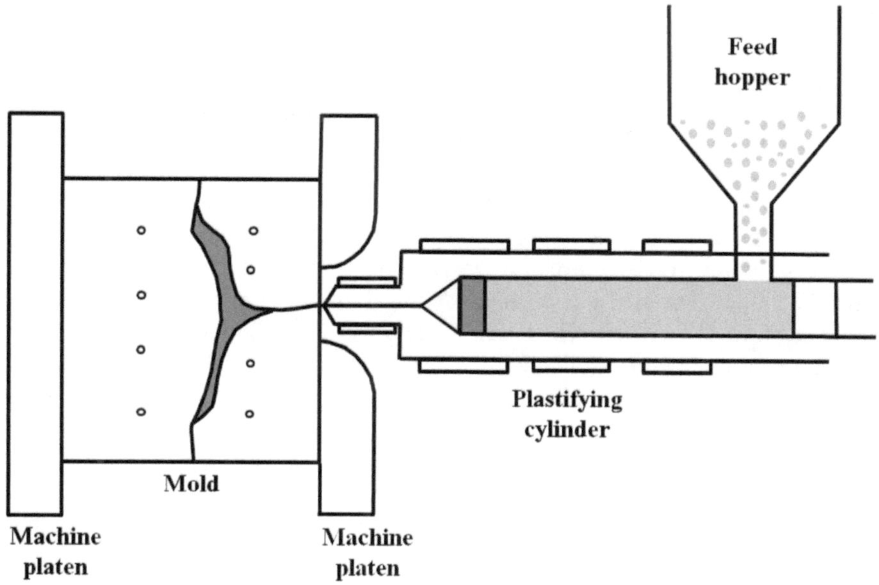

Fig. 3.2 Schematic cross section of the plastifying cylinder and mold

2. Extrusion molding

Extrusion molding is common process in plastic manufacturing with injection molding, nearly 40% plastic product are made from extrusion molding. Extrusion molding has been found widespread application in manufacture of elongated products with various transverse section shapes, because of its related technical and economic advantages by continuous manufacturing. In particular, the extrusion is an important processing technique in many industrial fields, such as ceramics, chemical, food and pharmaceutical. At present, it has also been utilized for fabricating metallic and composite materials.

In principle, the extrusion molding process comprises the forcing of a plastic or molten material through a shaped die by means of pressure. The extrusion process has been used for many years for metals such as aluminum which flow plastically under deforming pressure, and in the earliest form of extrusion process for polymers similar ram-driven machines were used. However, in the modern process, screws are used to progress the polymer in the molten or rubbery state along the barrel of the machine. The single screw machine is the most widely used type. Twin screw types are also used where superior mixing of conveying. The extruder machine consists of an Archimedian screw fitting closely in a cylindrical barrel, with just sufficient clearance to allow its rotation.

It is a process used for fabricating in determined length of thermoplastics with constant cross-section. Polymer pellet or powder are continuously fed into a heated barrel by the action of a rotating screw, where they are heated and melted as they progress through the heated barrel. At the front end of the extruder, the melted polymers are forced through a shaped die that determines the final cross-section of the extrudate and then it is uniformly cooled and carried away on a continuous basis. Finally, length can be cut as desired dimension. Many parts using ABS, PE, PS, and poly(vinyl chloride) (PVC) are extensively fabricated by extrusion molding. Typical product includes piping, drinking straw, window track, wire and cable coating, film and sheet.

The screw of an extruder has one or two flights spiraling along its length. The diameter to the outside of the flight is constant along the length to allow the close fit in the barrel. However, the root is of varying diameter and so the spiraling channel varies in depth. Generally, the channel depth is decreased from feed end to die end, although there are variants for special purposes. The decreasing channel depth cause increasing pressure along the extruder, resulting in driving the melt thorough the die. There are three zones, whose functions are as follows: feed zone, compression zone, and melting zone [24–26].

3. Compression molding

Compression molding is known as the oldest mass production process for polymer products. Compared to other molding process, it is almost exclusively used for thermoset resins, although these are also processed by the injection molding. In compression molding, a pre-weighed and preheated thermoset resins are loaded into a heated mold, the mold is closed and high pressure is applied to the resins. The resins are melted under heat and pressure and flows into all parts of the mold cavity and then an internal chemical reaction is occurred into the mold, leading to crosslinking between resin chains and hardening the resins. The cured thermoset part is released from the mold while still hot and allowed to cool outside the mold [27–29].

The compression molding process is widely used in the automotive, aerospace, sporting goods, and electronics industries to produce large, thin, lightweight and strong parts using fiber-reinforcing into resins. As shown in Fig. 3.3, compression molded parts are formed by squeezing a cold glass fiber reinforced resin charge, known as sheet molding compound (SMC), between two heated cavity surfaces. The fibers as reinforcement are randomly oriented in the plane of the sheet and make up for approximately 25% of the molding compound's volume fraction. Generally, the mold is charged with 1–4 layers of SMC, each layer has about 3 mm thickness [30, 31].

An alternate process is injection-compression molding. As shown in Fig. 3.3, injection-compression is a hybrid molding process which has the features of both injection and compression molding. In this process, a bulk molding compounds (BMC) are injected into the mold and then compressed with pressure. BMC materials typically contain shorter glass fibers than SMC and shows lower structural properties. The main advantages of injection-compression molding are automation process and shorter cycle times [32, 33].

Originally, compression molding is developed to manufacture composite parts for replacing metal applications and mostly used to make larger flat or moderately curved parts such as hoods, fenders, scoops, spoilers, lift gates and the like for automotive end-uses. Although this method is the main processing for thermoset resins, it can

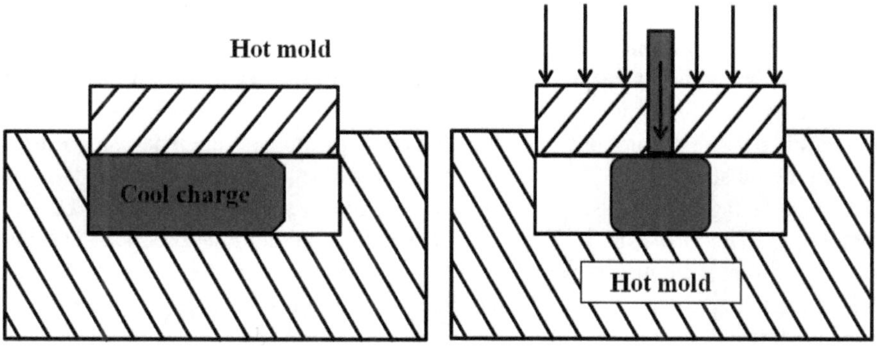

Fig. 3.3 Basic compression molding (*left*) and injection-compression molding process (*right*)

also be employed to make thermoplastic parts using long fiber reinforced thermoplastics, and ultrahigh molecular weight polyethylene (UHMWPE) [34].

4. Transfer molding

This process is a developed technique of compression molding. The molding compound is preheated and loaded into a chamber, which known as the pot. A plunger is then used to force the molten material from the pot through channels, which known as a sprue and runner system into the mold cavities. The mold walls are heated to a temperature above the melting point of the mold material for faster flow of material through the cavities. After cooling, the mold is opened to release the final product. The processes are illustrated in Fig. 3.4.

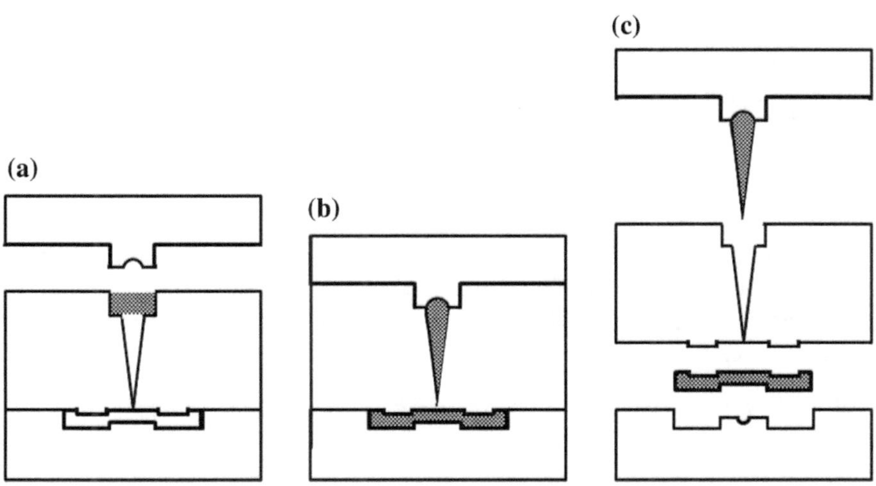

Fig. 3.4 Transfer molding: **a** plug of molding compound inserted; **b** press closes; compound transfers through runners to mold cavities; **c** three-plate assembly opens; moldings recovered from lower daylight, sprue is withdrawn with top member

In general, transfer molding is used to give many small parts more easily and to reduce the risk of damage or movement of thin of delicate mold parts or inserts, because it is claimed to be faster due to better heat transfer thorough the runners.

In cases, transfer molding offers many advantages. As seen previously, no single process is best for all products, and it is the job of the process engineer and designer to find the most suitable process for the particular case. In reaching such a decision, the disadvantages of transfer as opposed to compression molding must be considered: (1) The flow usually gives unwelcome orientation in the product, (2) It also increases wear and maintenance costs, (3) Tooling is rather more complex and hence more expensive, and (4) The runners are scrap, with little chance of a hot runner system [35–37].

3.2.2 Basic Mold Construction of Injection Mold

As mentioned prior section, the injection mold is an assembly of parts containing within it an impression into which plastic material is injected and cooled. It is the impression which gives the molding its form. Therefore, the impression can be defined as that part of the mold which imparts shape to the molding. The injection molding is formed by two mold members: (1) the cavity, which is the female portion of the mold, gives the molding its external form and (2) the core, which is the male portion of the mold, forms the internal shape of the molding [38–40].

1. Cavity and core plates

These are illustrated for a simple hexagonal container, as shown in Fig. 3.5. The basic mold consists of two plates. One plate is sunk the cavity which shapes the outside form of the molding and therefore it is known as the cavity plate. Similarly, the core plate forms the inside shape of the molding. When the mold is closed, the two plates come together forming a space between the cavity and core which is the impression [41].

2. Sprue bush

During the injection process, polymer material is delivered to the nozzle of the injection molding machine as a melt phase and then it is transferred to the impression through a passage. In the case, this passage is a tapered hole within a bush. The polymer material in the passage is named as the sprue, and the bush is called as sprue bush.

3. Runner and gate systems

The polymer materials can be directly injected into the impression through the sprue bush or molds containing several impressions. It can pass from the sprue bush hole through a runner and gate system before entering the impression.

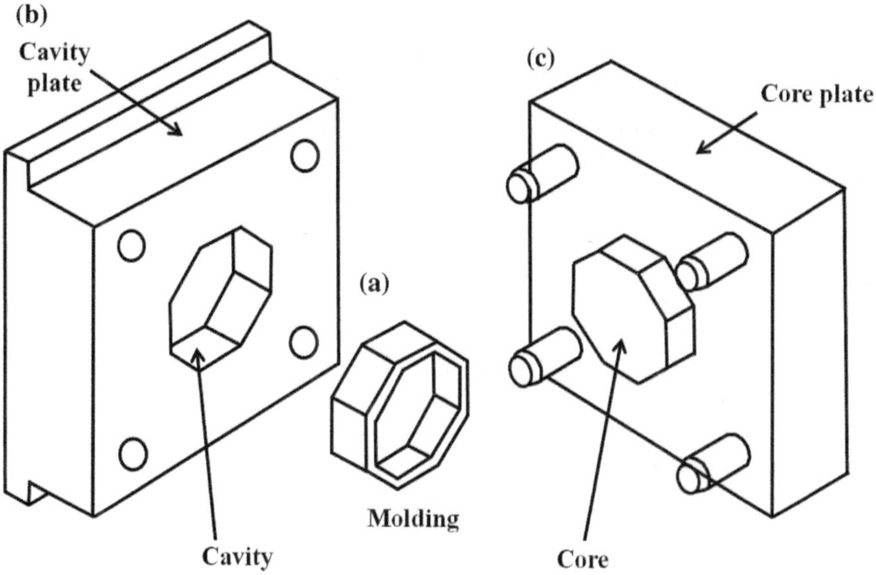

Fig. 3.5 Basic mold consisting of cavity and core plate

4. Register ring

If the polymer materials are to pass without hindrance into the mold, the nozzle and sprue must be correctly aligned. To ensure that the mold must be central to the machine and this can be achieved by including a resister ring.

5. Guide pillars and bushes

To mold and even-walled article, it is necessary to ensure that the cavity and core are kept in alignment. It can be done by incorporating guide pillars on one mold plate and then this mold plate enter corresponding guide bushes in the other mold plate as the mold closes. Guide pillars and guide bushes are mounted on the core side and the cavity side, respectively. The size of the guide pillars should be such that they maintain alignment irrespective of the applied molding force.

6. Fixed half and moving half

The various mold parts fall into two sections or halves. Hence that half attached to the stationary platen of the injection molding machine is called as the fixed half. The other half of the mold attached to the moving platen of the injection molding machine is known as the moving half. It has to be decided in which of the two halves the cavity or core is to be situated. Generally the core is situated in the moving half.

During cooling process, the molding will shrink on to the core and remain with it as the mold opens. This will occur irrespective of whether the core is in the fixed half or the moving half. However, this shrinkage on the core means that some form of ejector system is almost certainly necessary. Motivation for this ejector system is provided if the core is in the moving half. Furthermore, in the case of single-impression basic mold, where a direct sprue feed to the underside of the molding is desired the cavity must be in the fixed half and the core in the moving half [42].

In this section, general mold construction is simply explained and detail discussion for the construction for mold design will further discuss next section.

3.3 Injection Mold Design

In injection molding machine, the mold is one of the most important elements, determining the shape of the final product. Furthermore, the quality including part appearance, strength, toughness, size of the final product and the manufacturing cost are largely determined by the mold [43–45].

It is also a fact that disturbances of the production process are most frequently due to molds which are subject to early wear, cause processing difficulties owing to poor thermal and rheological lay-out, or require a very specific operating range entailing a constancy of parameters that cannot be realized effectively by the machine (Fig. 3.6).

In injection molding process, delay of production is caused by time-consuming mold design and long machining times in the toolmaking shop and the initial costs and hence the financial risk in case of failure increase as a result of the fact that component sizes increase continuously as well as that molds are required to perform more and more different functions [46]. Lately, the demands on the mold increase owing to the growing need for highly quality industrial components. Therefore, there is greater use of special materials which exhibit difficult, specific processing properties.

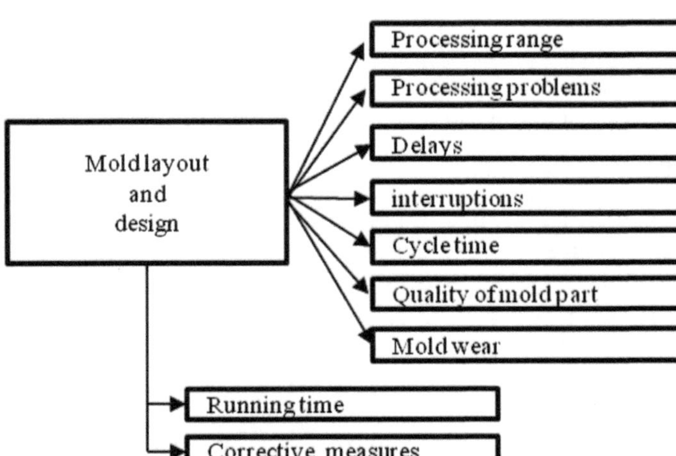

Fig. 3.6 Influences of mold design into following processes

All these facts lead to the conclusion that injection mold can become the most critical factor in the entire production process and must become the chief point of focus for all rationalization efforts. These efforts will mainly have to be directed towards the various aspects of mold design. Therefore we will define the injection mold and discuss the classification of injection mold and various aspects for consideration to making mold in this section.

3.3.1 Injection Mold

It is well known that the mold is fastened to the clamp plates, at its simplest, in two halves. In very broad sense, molds can be classified as cold runner and hot runner molds. Two-plate mold and three-plate mold are most common in cold runner molds. There are other types of cold runner molds like external under-cut mold, internal under-cut mold, side core mold, unscrewing mold, stack mold [47–49].

The Fig. 3.7 shows such a two plate mold. The two plate mold is the most basic and simple type in design. It consists of a front and stationary half basically. The cavity or core can be mounted on either half, depending upon the part design and the location of the location of the knock-out pins. This mold is easily adapted for different designs and all part ejection methods. The Fig. 3.7 shows the sprue leading from the nozzle to the runners which lead on to the gate at the entrance to the cavity or impression.

Two-plate mold has a single parting line, there are two plates in the cavity plate, with the central sprue bushing assembled into the stationary half of the mold, the moving half of the mold contains the cores and ejector mechanism, and in most designs the runner system. The considerations for two-plate mold design will discuss in next section in detail.

Figure 3.8 shows a three-plate mold. It consists of the standard two-plate design with the third movable between the two. This type of mold is used primarily to centre gate or submarine gate parts in multiple cavities. Automatic degating is possible with three-plate molds, but runner scrap is increased. Three-plate molds are needed when the runner system and the cavities are in different planes; two openings are required to remove the moldings and the spure and runners. There is provision to pull the sprue from its channel. When the mold opens the solidified materials feature at the gate and the molding is recovered from one daylight; the sprue is pulled and falls at the other daylight.

The products shown diagrammatically in Figs. 3.7 and 3.8 are similar. The two-plate mold gates from the side and the melt will have to flow round a core to join up on the far side. The three-plate mold gates at two points on the bottom and the melt can flow evenly to form the walls of the molded vessel. The latter will give a molding with better properties, with a lower likelihood of distortion.

The hot manifold mold is similar to the three-plate mold in design. However, the runner system is confined in a block and heated by cartridge heaters. This system will eliminate large quantities of runner scrap and improve automatic cycling. If

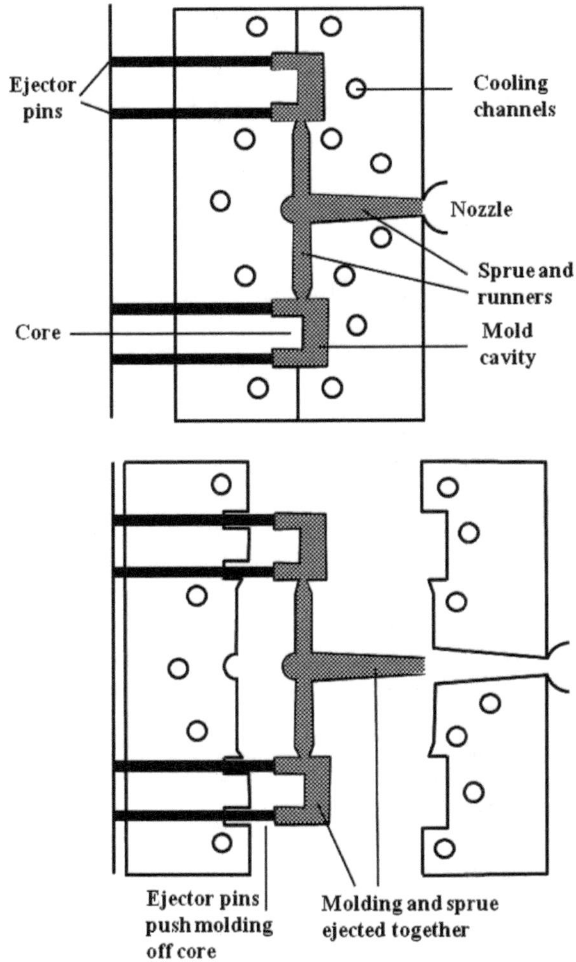

Fig. 3.7 Two-plate mold, closed and opened

heated torpedos are used, they should be made from a good grade stainless steel. Beryllium copper is not recommended.

To obtain optimum results, the hot manifold system should be balanced properly, insulated from other parts of the mold to minimize heat loss or heat transfer to the mold plate, and provide uniform distribution of heat throughout the manifold. In insulated runner mold, this system utilizes the hot manifold concept without the cartridge heaters. The insulated runner system depends entirely on heat generation in plasticizing cylinder and insulation around the runner to maintain proper melt temperature.

When the mold fills the air already in it must be vented. Often this happens spontaneously via the ejector pin clearances, but sometimes marrow vents may be provided, of about 0.025 mm diameter, sufficient to vent the air but not to allow melt to enter them. If the venting is inadequate a number of processing or product

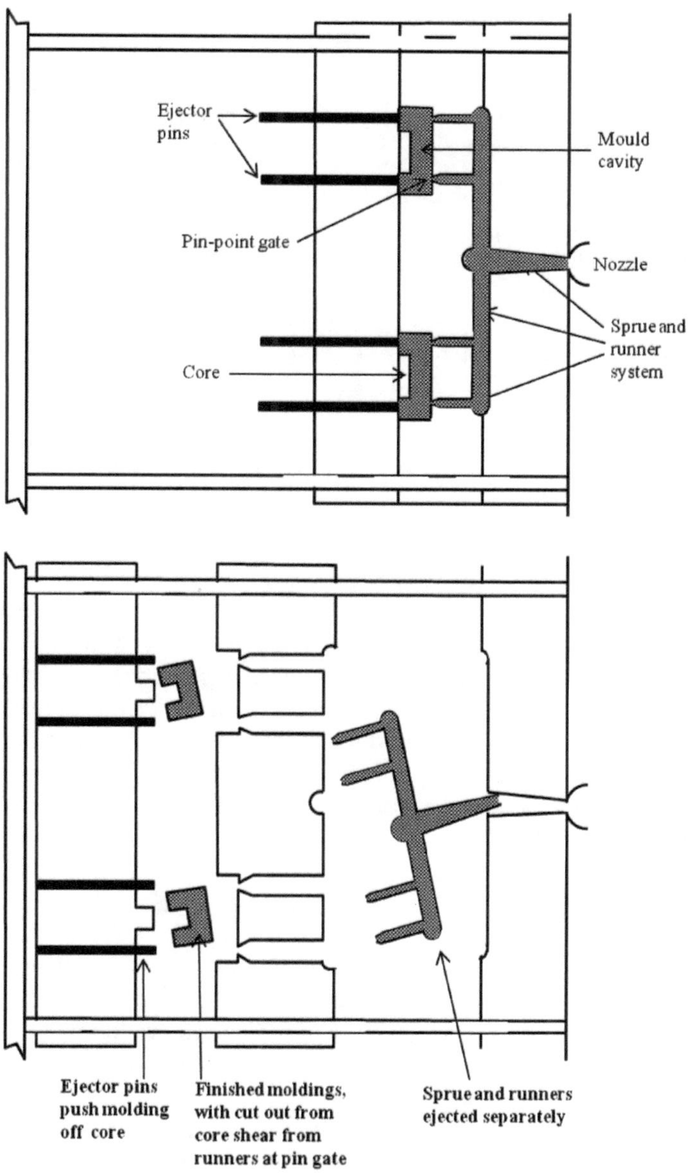

Fig. 3.8 The principle of the three-plate mold

faults may result. At the most extreme, a bubble of air can be trapped, leaving a dimple in the molding. A more common fault is burning, caused by the rapid escape of venting air; its velocity can be high enough for a temperature rise sufficient to degrade the polymer locally, and to cause characteristic burn marks on the molding.

This mold is supplied with cooling channels through which water is passed. The temperature of this water varies for different products. Chilled water gives the fastest cycle times but warmer mold temperatures are sometimes required, especially with crystalline polymers, to achieve superior properties in the finished product. The constriction at the gate has three main functions:

1. It allows rapid freezing of the polymer at the conclusion of injection. This isolates the cavity and permits withdrawal of the screw.
2. The narrow and thin solid section allows the molding to be sheared off easily after demolding, eliminating finishing trimming in most cases.
3. It increases the shear rate as the melt flows through and hence lowers the viscosity to ease rapid and complete filling of complex shapes.

Several types of gate design are used for different purposes, and some of those commonly used are illustrated in Fig. 3.9. Some of the features of these gating variations are described below.

1. Sprue gates are the simplest. There is direct feed to a single cavity from the sprue.
2. Pin gates are fed from runners. They are frequently used in three-plate molds. The small scar they leave leads to easy finishing. The narrow section gives very high shear rate, low viscosity and easy filling in thin mold Sect.
3. Side gates are the standard gate type for multi-impression tools. They feed the side of the product. Multi-impression tools should use 'balanced runners' to ensure even pressure distribution through the system. Unbalanced runners can give molding of unequal quality because the pressure and hence flow is not the same for impressions near the sprue and those remote (Fig. 3.10).
4. Ring gates are used for multi-impression molds making hollow moldings with the flow round a central core.
5. Diaphragm is similar to the ring gate but feeding directly from the sprue for single impression.
6. Fan gates cause the melt to spread fan-wise to cover large areas well
7. Film gates are also known as 'edge' or 'flash' gates, these give orientated distribution for flatness in thin flat moldings. They are much used for transparent products like polycarbonate lensing on meters, where the even flow avoids ripples
8. In tab gates, the tab eliminates 'jetting' in large plane areas by breaking the flow and making it turbulent as it enters the cavity. Jetting causes unsightly flow lines, especially in transparent materials.

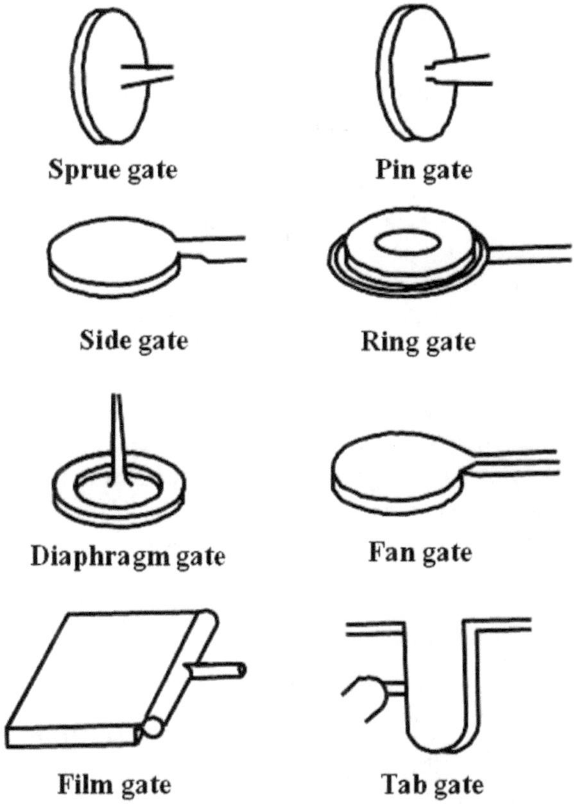

Fig. 3.9 Variations in gate design

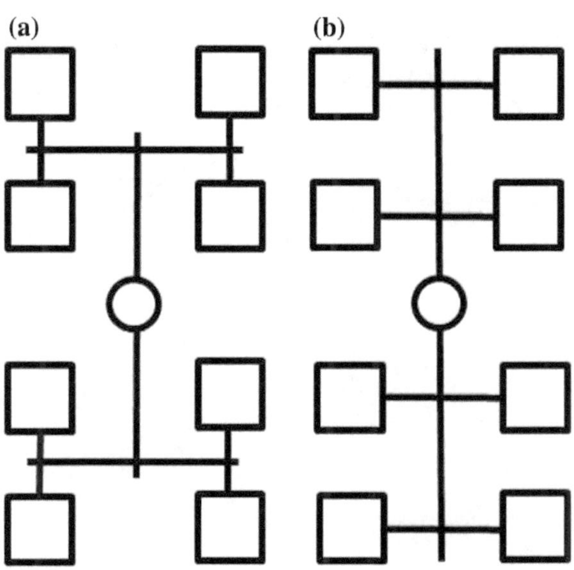

Fig. 3.10 Balanced and unbalanced runners; **a** balanced and **b** unbalanced

3.3.2 Two-Plate Mold

The simplest and most reliable mold design is the two-plate mold. This is because it normally has the fewest number of moving parts and is more straightforward to manufacture and run in production. Because of its simpler construction it is usually cheaper to manufacture than more complex designs.

Given the simplicity of its design and manufacture, mold design engineers should make sure that all possibilities of using a two-plate design have been exhausted before other more complex designs are considered. This means that the component should be examined carefully to see whether any undercut features could be designed out of the part. Screw threads are a prime example of this point. Threads normally require split tools, collapsible cores or more complex automatic unscrewing devices, but sometimes they can be jumped out of the cavity [50, 51].

A typical two-plate mold tool is shown in Fig. 3.11. Figure 3.12 shows a typical full general arrangement drawing of a two-plate mold. In practice, the general arrangement drawing would be considerably more detailed than this. However, this simplified version illustrates more clearly, the basic construction of a typical two-plate mold.

Since some of the components in this design are common to the vast majority of all injection mold designs, we will look at some of these individually.

1. Locating or register ring

This is a circular ring, screwed to the front clamping plate. It enables the tool be centered on the injection cylinder axis by locating it into a machine hole in the fixed platen. It is usually made from good-quality low-carbon steel.

2. Top plate

There are alternative names for this plate, i.e., front plate, fixed half front plate, fixed half clamping plate. It function is to allow the tool to be secured to the fixed

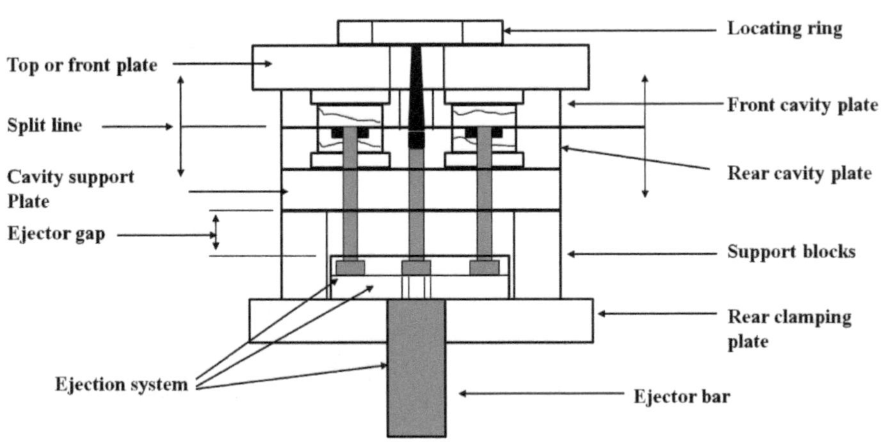

Fig. 3.11 Basic two-plate construction

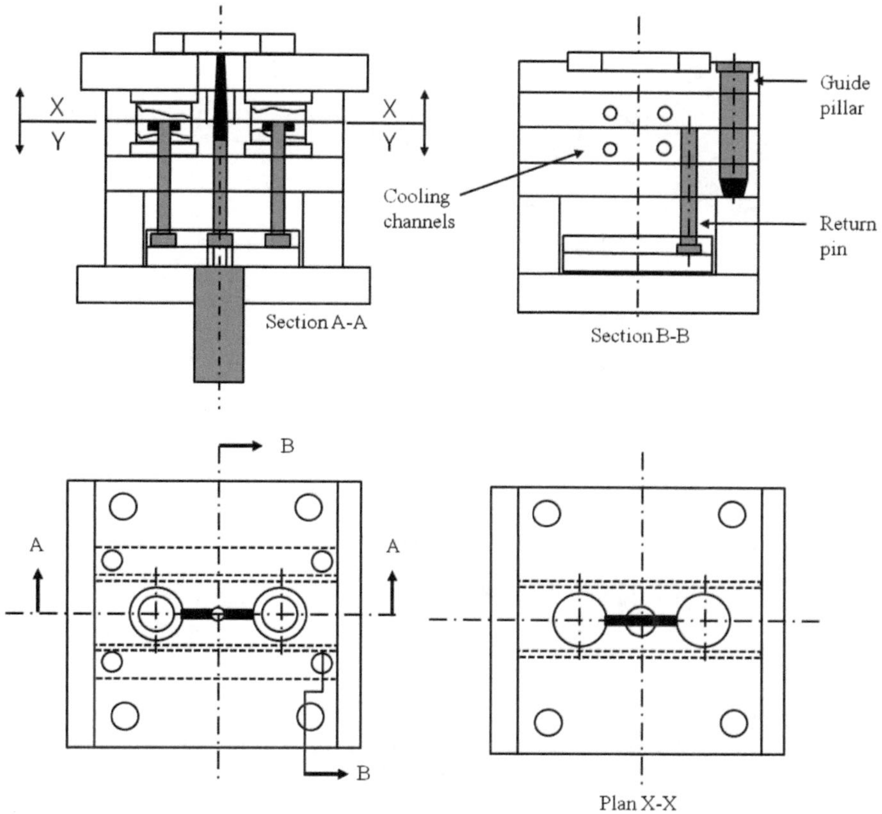

Fig. 3.12 Basic two-plate tool general arrangement

platen with cap screws to hold it in position. It is usually made from low-carbon steel or perhaps alloy tool steel for very long-running jobs. Tools are also frequently clamped to the platens with tool clamps. The rear half of the tool may be secured in a similar manner.

However, the more secure system for clamping tools on to the machine is to use direct cap screw fixing that entails providing clearance holes in the clamping plate to secure the tool to tapped holes in the platens (Fig. 3.13).

Another method for tool clamping is to provide tapped holes in the top and rear plates of the tool. Screws are passed through clearance holes in the platens and then screwed into the tool from the rear of the platens.

3. Split line

This is the plane or position at which the tool separates into two distinct parts-the fixed half and the moving or ejection half. After the tool has split at this point, space is created for the moldings to be pushed or ejected from the mold.

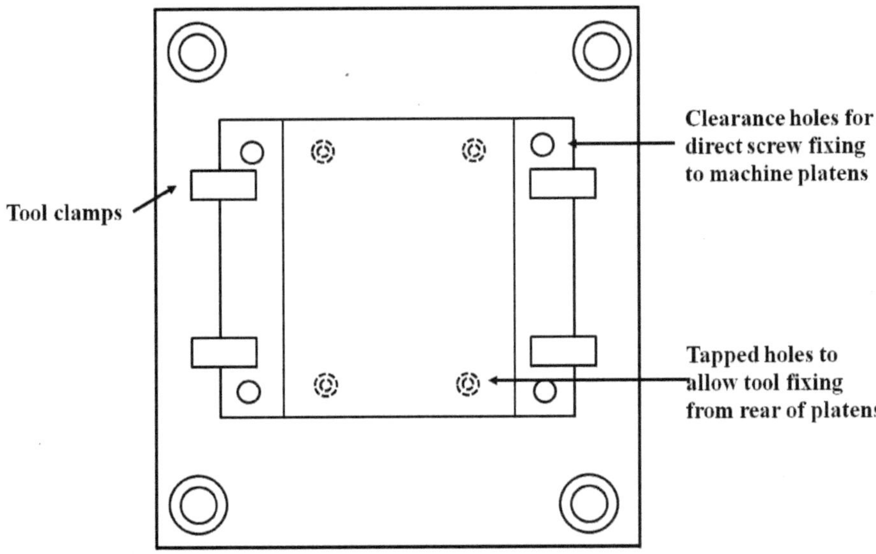

Fig. 3.13 Mold clamping arrangements

4. Cavity inserts

This is circular or rectangular piece of alloy steel that carries the form of the molding in it. Such inserts are inserted into the front and rear retaining plates as described below. Using inserts avoids machining the cavity forms directly into the cavity plates, which is more difficult. It also avoids having to harden the cavity plates, which can lead to the plates distorting.

5. Front cavity plate

Another name for this is the fixed half cavity plate. Its function is to hold the front half cavity inserts in position in the fixed half of the mold. The cavity inserts cannot move because the inserts have a shoulder on them to make them captive between the top and front half cavity plates. The plate in which they are fitted is either low-carbon steel or, for very long-running jobs, alloy steel.

6. Rear cavity plate

Rear cavity plate serves the same function as the front cavity plate. It makes the rear half inserts captive between it and the cavity support plate. The materials used are the same as the front cavity plate.

7. Cavity support plate

Cavity support plate has to withstand the force generated by the injection pressure of the melt that is exerted on the actual cavity forms. This force is given by:

$$\text{No. Imps.} \times (\text{Projected area of cavities and runner}) \\ \times \text{Injection pressure of the melt}$$

This plate must be made from an alloy steel to resist the cavity inserts being embedded into it. It must also be of sufficient depth to prevent excessive plate deflection taking place.

8. Ejection system

After the parts have been molded and have solidified sufficiently, they have to be pushed out of, or ejected from, the mold tool after it has opened at the split line. The ejector assembly carries a number of ejector pins that push or eject the parts from the cavities. The parts usually fall into a bin or onto a conveyor for packaging or for further operations to be carried out on them (Fig. 3.14).

As well as ejector pins, the ejection system may use sleeve ejectors and blade ejectors or may also operate stripper plates, double ejection systems, collapsible cores other more complex devices.

9. Ejection gap

This is the amount that the ejector system or assembly can move towards the cavity support plate form its rest position. It must be large enough to permit the whole form of the component to be fully ejected from the mold. It is essential that the parts are cleanly ejected for the tool to prevent them becoming trapped in the mold.

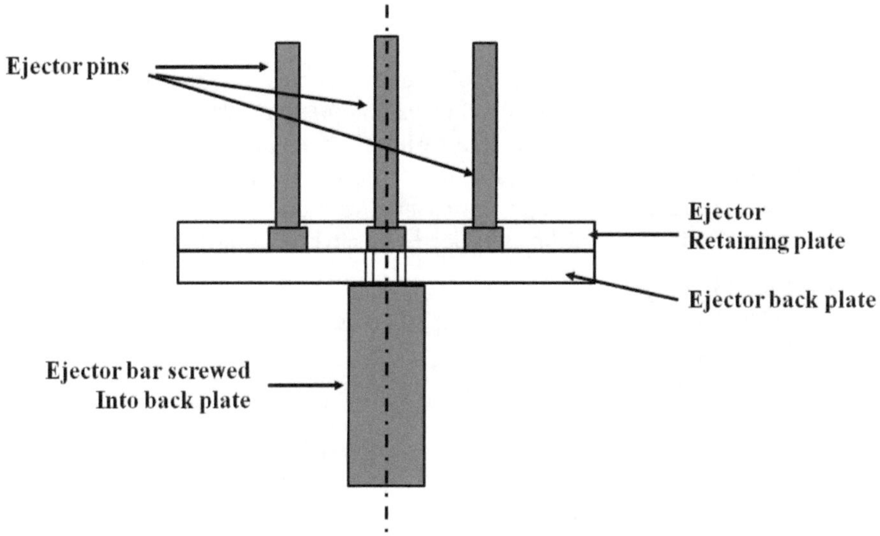

Fig. 3.14 Typical ejection system

10. Support blocks

These are also sometimes called risers. They connect the cavity support plate to the rear clamping plate or back plate. The material used is usually good-quality low-carbon steel.

11. Guide pillar

Guide pillar are used to accurately align the front and rear halves of the mold. This is necessary because the molding machine cannot be relied on to do this consistently. Unless the two mold halves are aligned accurately there could be mismatches of the cavity forms and cores in the front and rear mold halves. Guide pillars are also used in toolmaking to align and register the complete mold so that it may be stripped sown and reassembled accurately and normally made from hardened alloy steel.

12. Return pins

They are used, as the name implies, to return the ejection system to its initial or rest position when the mold closes. They stand proud of any other ejector pins and therefore as the mold closes they make contact with the front half first and thus prevent any damage to the more fragile ejector pins through contact with the front half of the mold. Return pins are also quite often called push backs for obvious reasons and made form either hardened silver steel or hardened alloy steel.

3.3.3 Mold Materials

It is very important to ensure that the correct materials are specified in all new injection mold design. Use of incorrect or inappropriate materials can lead to poor mold performance in production and to early failure of the tool [52].

In all mold materials, we are looking for certain characteristics for ease of toolmaking and good performance during production from the mold tool. Ideally we would prefer the material finally selected to have the following properties:

1. Good machining properties
2. Ease of heat treatment where hardening is required
3. Good toughness and strength
4. Polishes and accepts texturing well
5. Good resistance to heat and wear
6. Good fatigue resistance
7. High thermal conductivity for effective water cooling
8. Good corrosion resistance

Unfortunately, in practice no signal material will exhibit all these characteristics and therefore a compromise has to be reached depending on the type of tool design being employed. Among the major governing factors that should be considered are:

1. The tool life in terms of the quantity of parts required to be produced from the tool
2. The molding material being used, such as abrasive or corrosive
3. Texturing and polishing requirements
4. Whether hardening is required, such as for long running tools or for side cores and splits, etc.
5. Whether high thermal conductivity will be required
6. Exceptional requirements like the use of very high injection pressures or speeds.

3.3.3.1 Selection of the Mold Materials for the Application

There are many factors to consider when selecting a mold material. The final plastic part requirements are the most important factors to consider when selecting the proper mold material. Cosmetic requirements, resin type and filler content are the main considerations along with the expected tool life. The second category of factors is the manufacturing and servicing conditions in which the tool is expected to perform. These include production environment, expected cycle time and tool manufacturing variables: heat treatment, weldability and machinability.

For most normal applications steel is used because it has most of the properties we required. There are several other applications, however, where alternative materials may be used. The following sections discuss the materials used for most normal situations encountered. Table 3.1 gives an overview of the materials most often used.

In practice certain applications may fall outside the groupings shown owing to individual preferences or special requirements. For example, some companies use aluminium alloys for all their mold tools irrespective of whether the application is a long- or short-run job. The grouping given in the Table 3.1 is those used for most general-purpose mold tools.

3.3.3.2 Various Mold Materials

This section discusses the major properties and characteristics of the majority of the materials used for injection mold tools. The selection of mold materials in tooling can be as critical to the success of a plastics application as the selection of resin is to the end use performance requirements of the molded product. Just as resins are formulated to meet performance requirements in plastics applications, steels are alloyed to meet specific performance requirements in use.

1. Steel

Steel is the industry standard material for the manufacture of mold tools. It is considered a good all-round material embodying many of the desirable characteristics. Some applications may require a mold steel with high hardness and wear resistance for parting line durability, while others will require a mold steel with

Table 3.1 Mold materials

Application	Material used	Process
Preparation runs	Epoxy resin	Casting
Prototypes	Low-melting-point alloys, e.g., Kirksite	Casting and metal spraying
Small quantities	Mild steel or low-carbon steel	Machining
Medium production runs for straightforward parts	Mild steel Aluminium Pretoughened alloy tools steels Unhardened alloy tool steels	Machining
Medium production runs for technical or high-performance parts	Aluminium alloy Pretoughened alloy tool steels Unhardened alloy tool steel	Machining
High-speed production and close tolerance work	Hardened alloy tool steels	Machining
Large-volume production	Hardened alloy tool steels	Machining
Texturing, high polished finishes for		
Low/medium volumes	Aluminium alloy Unhardened and pretoughened alloy steels	Photo-etching Chemical ehching EDM
High volume	Pretoughened alloy steels Hardened alloy tool steels	Photo-etching Chemical EDM
High heat transfer	Beryllium copper	Casting and machining
Poor heat transfer	Ceramic titanium	Casting and machining

higher toughness for resistance to mechanical fatigue. In general, steels delivering higher hardness and wear resistance properties are those that tend to be more brittle, and in almost all cases, a steel with greater toughness will deliver some reduction in resistance to steel-to-steel wear and abrasive resistance to resin containing glass fiber or mineral fillers [52–54].

There are many different types of steel available for a wide range of general purpose applications and may specialist steels for more extreme applications. For our purposes, however, we will restrict our discussion to those steels commonly used for the manufacture of injection mold.

A number of steels are used for the construction of mold tools. All steels contain carbon and, in general, the higher the level of carbon, the tougher and stronger the steel will be. Sometimes plain or mild steel is used with low carbon content, and on other occasions higher carbon content steel is used, often alloyed with other elements to improve the properties of the steel. Carbon steels that have been alloyed with other elements to increase their performance are usually called tool steels or alloy tool steels.

For nearly all mold steels the most common alloying elements are chromium, nickel, molybdenum, tungsten, cobalt and vanadium. For most mold steels, four of these elements are normally used.

Table 3.2 Mold steel applications

Category	Uses	Comments
Low-carbon, also known as plain plates, carbon or mild steel	Backplates, Support locating rings	Used for economy in noncritical applications
Prehardened or pretoughened alloy tool steels	Cavities, punches and cores	Alloy tool steel used unhardened. Can be locally heat-treated.
Fully hardening alloy tool steels	High volumes where high quality, strength and hardness are required for cavities and cores	High-quality tool steel. First choice for long-running jobs
Stainless steel	Used for corrosion resistance to protect against corrosive materials such as PVC	Stainless steels contain up to 18% chromium and hence have worse heat transfer
High-carbon-high-chrome	Cavities and cores where high levels of hardness are required	Very hard when heat treated. Used for small parts. Prone to stress cracking
Nitriding steels	Areas of high wear. Sliding surfaces, ejector pins and sleeves	Extensively used for small sliding parts

1. Chromium: improves surface resistance to wear and corrosion
2. Nickel: Improves low-temperature toughness and increases fatigue resistance
3. Vanadium: Increases strength, hardness and impact resistance
4. Tungsten: Increases strength, toughness and higher temperature performance.

Although these alloying elements are designed to improve certain properties, it must be noted that they can also have adverse effects on other properties. For example, with increasing chromium content the thermal conductivity of alloy steel decreases.

Table 3.2 lists the different types of steel available for mold plates together with a brief account of their properties and uses.For large mold tools, plate steel from rolled or forged stock is often not available in the sizes that may be required. Even when plate steel is available in these size it is seldom economic to machine cavity forms directly into such large pieces of metal.

Deep drawn forms would also be extremely difficult to machine on components like refrigerators, wheelie bins and TV surround cases. Additionally, to machine such large forms would result in up to half of the original steel billet being machined away, making it an uneconomic process. To overcome this problem, large mold tools would be cast form suitable steel, which is usually very similar to the plate steels previously discussed for smaller tools.

The cavity form is cast in a similar manner to other metal casting techniques using a pattern in a sand mold. It is common practice to incorporate during the casting process complete water temperature control channels and cored ejector pinholes and similar features that would be difficult or expensive to machine in the finished casting. A specialist foundry would carry out such casting work.

Although the steels used for casting are similar to those used for plates, there are slight differences because the structure of cast steel makes it not quite as strong as that of rolled or forged steels. Casting-grade steels are therefore selected to give as fine a crystal structure as possible for maximum quality and strength.

An important factor is that the cast steel should easily weldable, as pits, blow-holes and shrinkage holes frequently occur in the surface of the casting. Long-term damage from thermal shock is common in cast steels and results in surface cracks appearing. Welding also normally repairs these. The sand casting process is used for producing large mold tools of up to about 4 tons per casting resulting in finished molds of up to 10 tons in weight.

Also, the type of resin is an important factor to consider when selecting a mold steel. Many of the resins-when exposed to elevated temperature-will emit a corrosive gas. Polyvinyl Chloride (PVC) resins operating at 165 °C will release hydrogen chloride when combined with humidity to form a corrosive hydrochloric acid (HCl). In addition, cellulose acetate will also create a corrosive reaction by forming ethanoic acid at elevated temperatures. Corrosion of the mold steel may also occur when processing resins with fire retardant additives. Chlorine and bromine base flame-retardant additives can become highly corrosive to the mold steel.

The type of filler is another critical factor. Typically glass-filled resins will be more abrasive to the mold steel and a higher steel hardness should be used. If the resin is corrosive and has abrasive fillers, a stainless steel grade should be selected and the hardness should be maximized to a level suitable for the design. When corrosion is a concern-either with the resin or filler-stainless steel cavity and mold base material should be considered as an alternative to plating or coating regular tool steels.

2. Aluminium alloys

Recently, the use of aluminium alloys has been significantly increased in the UK, the USA, and Europe. Compared to steel, aluminium alloys has many advantages as following; cheap cost, good machineability and thermal conductivity, low distortion by heat-treatment, weight-light, etc. [55, 56].

The service life of aluminium alloy mold tools can quite readily achieve around 200,000 shots depending on the molding conditions. There are a few notable exceptions to this where longer service lives of up to a million shots are achieved for straightforward parts, although certain mold parts may have to be replaced because of damage and wear.

There is also an increasing use of hybrid aluminium-steel mold tools in which the advantages of both materials are used to best effect-steel for high-wear areas and aluminium for less critical areas and where high rates of mold cooling are desirable. Table 3.3 shows two of the main grades of aluminium alloy used for mold tool applications.

3. Zinc alloys

Zinc alloys have poor mechanical strength and are unsuitable for production tooling. They are, however, often used for prototype tooling where only a relatively small number of parts are required for preproduction evaluation purpose [57, 58].

Table 3.3 Properties of aluminium alloy for molds

AA No	7075-T6	7029-T6
Composition	Al, Zn, Mg, Cu, Cr	Al, Zn, Mg, Cu
Treatment	Solution treated, Artificial aged	Solution treated, Artificial aged
Density (kg/dm^3)	2.80	2.74
Coefficient of thermal expansion	23.3×10^{-6}	24.7×10^{-6}
Thermal conductivity (W/(kg-m)	140	126
Ultimate tensile strength (MPa)	572	537
Yield strength (MPa)	503	469
Brinell hardness (BHN)	150	125

Table 3.4 Properties of Mazak and Kirksite

	Mazak	Kirksite
Density (g/dm^3)	6700	6700
Melting point (°C)	390	380
Shrinkage	1.1	0.7–1.2
Thermal expansion (10^{-6}/K)	27	27
Tensile strength (MPa)	220–240	226
Elongation (% in 50 mm)	1–2	3
Shear strength (MPa)	300	246

They have a low casting temperature of around 400–450 °C and are therefore well suited to casting from patterns of plaster, wood, steel and aluminium. Zinc alloys can also be cast by sand molding. The resulting castings reproduce intricate pattern detail extremely well, giving a smooth nonporous surface that ultimately produces good-quality plastic moldings.

In view of the poor mechanical strength of this material, normally only the cavity forms are made from it. The resulting casting or hobbed cavity would be inserted into a steel bolster to support it. Two of the best-known zinc alloys are Mazak and Kirksite (Table 3.4).

4. Beryllium-copper alloys

Pure copper is a very soft, ductile material and is unsuitable for use as cavities or cores. Copper is used in mold tools only as a heat exchanger, usually inside core pins. Its excellent thermal conductivity makes it ideal for conducting heat away from smaller core pins where direct water-cooling is not possible.

When copper is alloyed with beryllium, the resulting beryllium-copper alloy is a much stronger, tougher material and is used for cavity and punch applications. With increasing beryllium content, the mechanical properties improve but the thermal properties worsen. At least 1.7% beryllium content is required for direct cavity and punch applications, with around 2% being the norm. Cavities made from this material may be heat-treated to around 400 °C, giving a hardness of about 40 Rc.

The surface hardness may also be increased by about 15% by the use of ion implantation techniques.

5. Epoxy resin

This material is frequently used for making cavity forms because of the ease with which it can be done. A resin and hardener are mixed together and then a finely divided aluminium powder is added to the mixture. The aluminium is an essential additive to help with heat conduction, as the resin itself is a poor conductor of heat.

The resulting mixture is thoroughly mixed under a high vacuum to obviate the inclusion of air bubbles. Once a homogeneous mixture is achieved, the resin may be carefully cast over a master to achieve a cavity insert, which is then loaded into a bolster.

Owing to the low mechanical strength of the material and its poor thermal conductivity, its use is limited to prototype work only. However, many very intricate prototypes have been successfully molded from molds with epoxy resin inserts. Several major car manufactures use this technique to obtain good quality small plastic moldings for evaluation. It is a cheap and quick method of making cavity inserts from suitable master forms [59, 60].

6. Introduction of functionality on mold

6.1 Corrosion protection

Nickel plating or stainless steels may be needed to help prevent mold corrosion when molding in a high humidity environment. Corrosion is most likely to occur with a cold mold where condensation, then oxidation may occur, or when using a molding material that may emit a gas that is aggressive to most steels.

Nickel plated or stainless steel molds are not normally required to mold GE resins because mold temperatures should be no cooler than 140 °C and only a few injection molding grade of acrylic-styrene-acrylonitrile (ASA) resins have an aggressive PVC component. It is generally suggested that, if there may be occasions of long-term mold storage, where corrosion protection beyond preventative spraying may be necessary, nickel plating may be employed. Electroless nickel plating offers excellent chemical protection and is relatively inexpensive when compared to chrome or their techniques.

Finally, nickel plating can allow for steel selection offering higher mechanical properties such as toughness, hardness, abrasion or adhesion wear resistance, and higher thermal conductivity than stainless steels.

6.2 Heat treatment

Heat treating is a science in itself and should be left to specialists. One of the reasons the designer should stay with a minimum selection of tool steels is to avoid the many heat treatment specifications, which are usually different for each type of steel; this way both the designer and the heat treatment company both become familiar with particular steels and know what to expect.

Nearly all mold toolmakers use outside suppliers for heat treatment because it is a specialist process and also because of the investment that would otherwise be required for the necessary plant and equipment.

Sometimes steels may required significantly different heat treatment to achieve the optimum properties required form the steel. The designer need only specify the required hardness of the part, as it is not necessary for the designer to have a deep knowledge of the heat treatment process. It is desirable, however, that the basic procedures are understood so that the designer will be aware of the results that different hardening procedures give.

Where the same degree of hardness can be obtained by using alternative hardening procedures, the designer should specify which process is required since these may affect, wear, toughness, and mechanical strength. There are several different techniques available for mold tools, such as thorough-hardening, pretoughening or prehardening, carburising or case hardening, nitriding, tuftriding, ion implantation.

6.3 Through-hardening

This technique is used for most cavities, punches, core pins, side cores and other items that come into contact with the molding material. Through-hardening consist of heating the steel to a given temperature and then quenching it in air, water or oil. The usual steel selected for this purpose is AISI H13 alloy or equivalent.

The cavity, punches and core pins need to be hardened for medium- to long-run jobs to avoid them becoming bruised, scuffed or worn during repeated cycling of the tool. If these items are not hardened, the tool will quickly deteriorate, resulting in damage and possibly seizure on mating sliding parts. Unhardened tools are also prone to early signs of flashing, which can rapidly worsen with continued production.

Through hardening in mold tools is restricted to nickel-chromium, high carbon-high chromium or similar alloy steels. To through harden steel it must have at least 0.35% carbon content. Steels with a lower carbon content than this will not through-harden and other methods of hardening have to be used. Most through-hardening processes involve the work piece being heated to a high temperature followed by quenching. During this procedure the steel undergoes a high degree of thermal shock, which induces a very brittle structure in it. Subsequent tempering procedures have to be carried out to change this structure; otherwise the steel will remain too brittle and fracture easily.

3.3.4 Other Considerations for Mold Design

The feed system of an injection mold is mainly consisted of the sprue, runner, and gate. Figure 3.15 shows the basic feed system.

1. Sprue Bushes

The sprue should be tapered 3–5° inclusive angle in order for it to be pulled out of the tool more easily. It should also be highly polished in the line of draw to assist

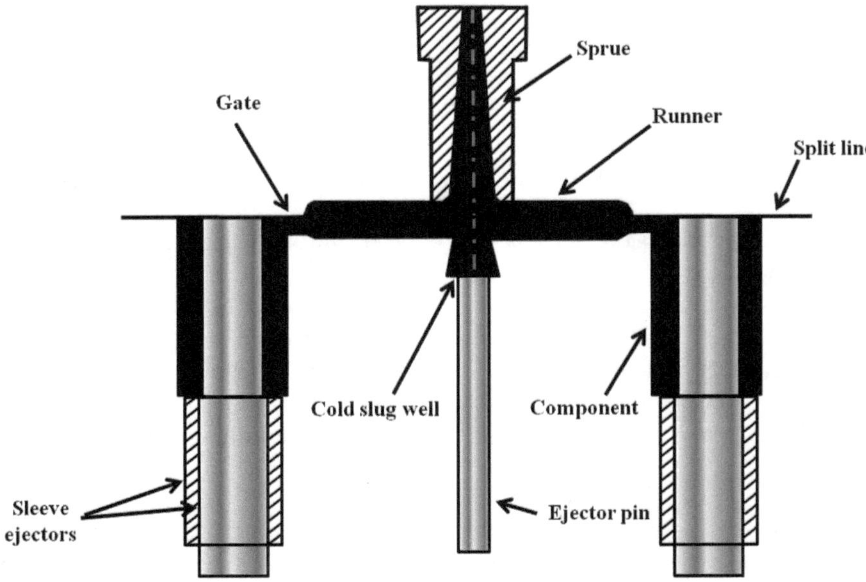

Fig. 3.15 Basic feed system

withdrawal and in the direction of flow of the melt for more efficient flow. The diameter at the narrow end should be larger than the machine cylinder nozzle opening.

Sprue bushes may be categorized as either cold sprues or heated sprues. The cold sprue bush is low in cost, easy to install and requires very little maintenance. However, this type of system generates more material that must be reground and reprocessed than the heated sprue. Should the cavity depth require a long sprue bush to accommodate the runner system, the heated sprue bush should be considered. Since the sprue is not ejected with the part, the sprue trimming and scrap are eliminated and molding time can be reduced by the amount of time normally allowed for cooling the sprue.

The internal surface of the bushing should taper from a diameter of 16–19 mm at the nozzle end to a diameter of 9.5 mm at the cavity end, and should be perfectly smooth and polished. A 450 chamfer is desirable for nozzle seal-off and seating. The orifice (gate) should be a minimum of 3.2 mm and a land length can be minimized by sinking 1.5–2.3 mm from cavity end. Counter sinking removes the gate land from the immediate vicinity of the cavity, thus preventing damage to the orifice and reducing the risk of freezing off.

The diameters of the outer surface of the bush should slip fit into the mold plate when both the bush and mold or cold. There should be a shoulder not more than 40 mm from the cavity to prevent expansion of the bush into the cavity when heat is applied. For bushes more than 200 mm long, a second shoulder, situated 75–100 mm from the cavity end, is recommended. The remainder of the outer surface of the bush should be machined to provide an insulated air gap which will minimize heat transfer to the mold plate and cavity, and also provide areas for mounting the heated bands. Two 125 W heater bands are usually satisfactory. Before the bush is installed in the mold, it should be heated to the operating temperature range of 1750–2050 °C, followed by tightening of the heater bands.

There are two methods of heating the hot sprue bush. The one most commonly used is the installation of heater bands around the outside diameter of the bush. The other method of heating the sprue is by installing heater cartridges within the wall of the sprue bush or directly in the melt flow chamber. Heated sprue bushes should be made from a good grade of steel, and be oil or air-hardened after machining.

2. Runners

The purpose of the runner is to transport the melt from the sprue to the gates. For most normal injection molding, runners with a circular (full round) section provide maximum flow. The most efficient runner section is the full round and this should be used wherever possible. The semi circular of half round runner severely restricts flow and should be avoided although it is frequently seen in production mold tools. Square section and rectangular section runner should never be used, as shown in Fig. 3.16.

The size of the runner should be based on the thickness of the molding wall section. It must be larger enough to provide adequate pressure to all the cavities. This ensures that there will be no packing pressure shortfall and thus permit adequate control over the molding conditions to achieve satisfactory moldings. An alternative method is to calculate the runner cross-section based on an appropriate pressure drop along the length of the runner. Also, main runners should have a cross sectional diameter of 6–12.7 mm. Secondary runners should have a cross sectional diameter of 4.8–8 mm. normally all main runners to a given mold should be equal in diameter, and all secondary runners in the mold should be equal in diameter, but they need not be equal in length. When the runner system is cut into only one mold face, trapezoidal runners are preferred to half-round runners of the same depth.

It is advisable to locate a cold slug well opposite the sprue, and to extend the main runners beyond each intersection with the secondary runners so that the extensions also act as cold slug wells. To help reduce friction and material hang-up,

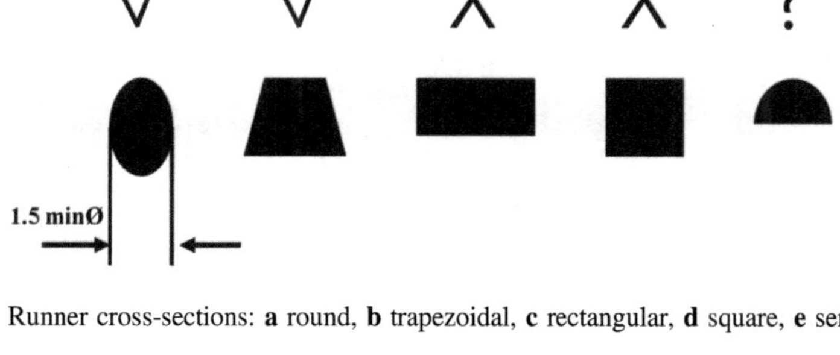

Fig. 3.16 Runner cross-sections: **a** round, **b** trapezoidal, **c** rectangular, **d** square, **e** semi-circular

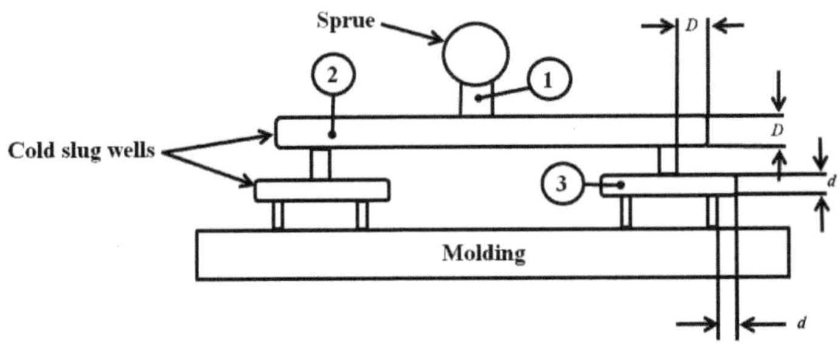

Fig. 3.17 Correct runner design

a highly polished chrome plated runner system may be used. The diameter should be a minimum of 8 mm for small parts and 9.5–11 mm for larger parts. Hot runner molds require runners of 12–19 mm diameter.

All runner intersections should have a cold slug well beneath each intersection and preferably have an ejector pin beneath them. The cold slug well helps the flow of material through the runner system and into the cavity. The length of the well is usually equal to the runner diameter. Figure 3.17 shows the correct design for including cold slug wells in runner systems. A cold slug well must also be placed at the end of each runner after it intersects with another runner. Figure 3.18a shows a frequently used incorrect rectangular layout design and Fig. 3.18b reveals the correct design.

3. Gate

A gate us a small opening through which the polymer melt enters the cavity. Gate design for a particular application includes selection of the gate type, dimensions and location. The gate design is largely classified as the part geometry, part specifications, used material and fillers, cycle time, and de-gating requirements.

Unless it is necessary to use multiple gates, a single gate is generally preferred. Multiple gates always create problems of weld and meld lines. The cross-section of the gate is typically smaller than that of the runner and the part, so that the part can be easily de-gated. Gate thickness is usually two-thirds of the part thickness.

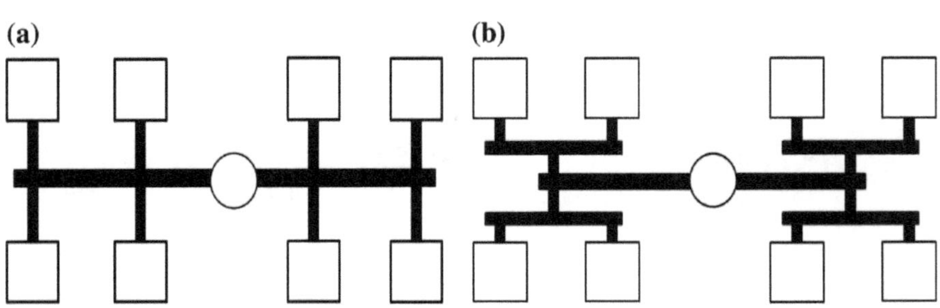

Fig. 3.18 **a** Incorrect runner layout and **b** correct runner layout

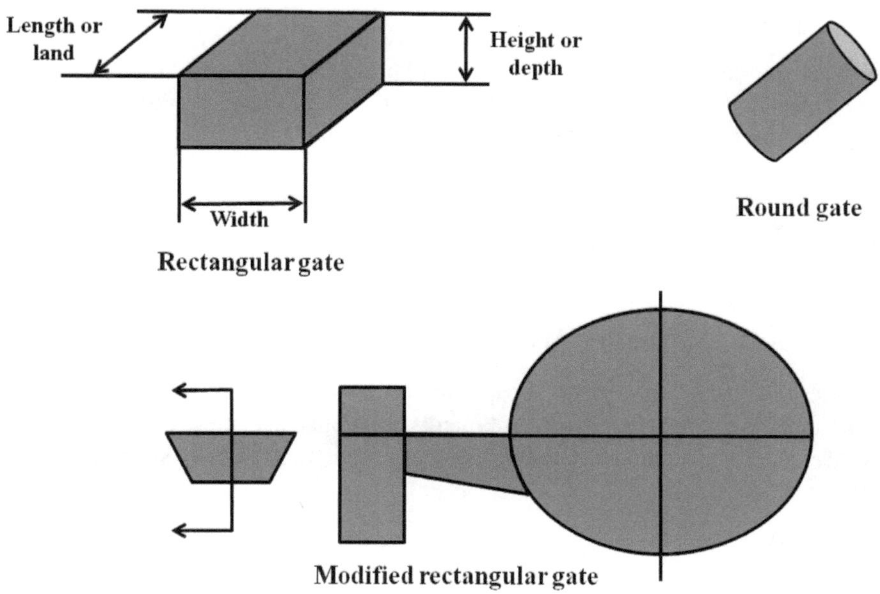

Fig. 3.19 Gate terminology

Technically speaking, the materials freezing off at the gate indicate the end of the cavity packing phase. Hence there is no point in maintaining pressure after this point is reached. This in turn means the gate must be large enough to make sure that the molding is properly filled before gate freezing occurs.

A large gate dimension will reduce viscous heating, permit lower velocities, and allow the application of high packing pressure to increase the density of the material in the cavity. If low stress is a requirement, owing to concerns of aesthetic appearance or dimensional stability, a large gate may be necessary. Figure 3.19 defines the terms used to describe gate geometry.The gate location should be selected in such a way that rapid and uniform mold filling is ensured and the weld/meld line and air vents are positioned properly. The gate should be positioned away from load-bearing areas. This is because the high melt pressure and high velocity of following material at a gate causes the area near a gate to be highly stressed

The gate land should be as short as possible to achieve lower pressure for filling and to improve de-gating by minimizing the height of gate vestige. The only reason for not using very short lands is the loss of strength of the steel at this area, which may cause the steel to bruise or fracture. The land lengths should be from 0.25 to 1.0 mm for the average molding.

Too small a gate can be recognized by blemishes at the gate and by surface imperfections. However, even before it affects the resin, it dramatically affects the available injection pressure so that the product will not be filled, and the gate will freeze off too soon. Too large a gate often results in an unsightly vestige. It will affect the mold-closed time, and requires an increase in cycle time. While too small a gate can easily be corrected by increasing the gate size, too large a gate will require a new gate insert, or even a new cavity if there was no insert.

4. Venting

Molds must be well vented to eliminate any possibility of gas or air being trapped as the part is molded. The entire circumference of each cavity should have venting to prevent gas or air burns at the parting lines. In some instances a minimum of 40% venting is recommended for intricately designed parts. Vents should normally be 0.05 mm deep. If internal vents are necessary, ejector pins, sleeves and core pins may be utilized if they are located where the trapped gas and/or air occur. Ejector pins and sleeves are preferred because the movement of ejection in the molding cycles automatically cleans the vents.

5. Considerations for part quality

Among the quality problems that can often be minimized by design there are three main factors, such as weld lines, sink marks and voids, and stress concentrations at corners, leading to product failure in service selection of the most suitable material for the product.

5.1 Weld lines

These form where polymer flows meet and they can sometimes be avoided; e.g. the tub-shaped molding in Fig. 3.7 will have weld lines. If welds are unavoidable they can often be moved to a position on the molding where they are unimportant, by control of the gate position. Once design has minimized the incidence, process control can be invoked to minimize the effect. This will entail ensuring adequate temperature and pressure for a good weld. A weld is always a potential air trap, because of the converging melt fronts, and venting may be required at this point in the mold.

5.2 Sink marks and voids

These related molding faults occur when the product section is too thick. The thick part retains heat and is drawn down by contraction forces-especially crystallization which involves a large density change. If the outer skin hardens, and so resists further sinking, internal voids form as the tensile strength of the solidifying melt is exceeded. This is essentially a design problem, to be designed out as far as possible by avoiding thick sections, but it can be helped by carful control of hold-on pressure. When thick sections are required in a molding, e.g. to confer stiffness, it is better to adopt a modified process such as foam-cored molding which avoids the problem of sinking and voiding altogether. Alternatively, a pattern of thin-section ribs may serve as shown in Fig. 3.20.

5.3 Stress concentrations

The consequence of stress concentrations in molding with sharp corners can often be failure, especially if the product is load-bearing. Sometimes there will be distortion, especially when fibre-reinforced grades of polymer are used. Figure 3.21 illustrates this and shows a few design features which can assist.

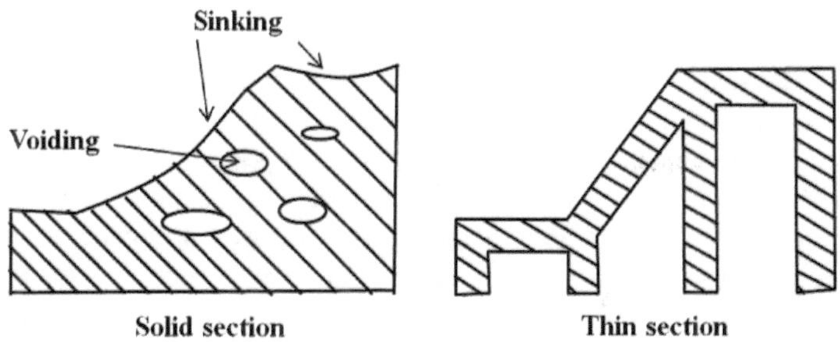

Fig. 3.20 Use of ribs instead of a solid section

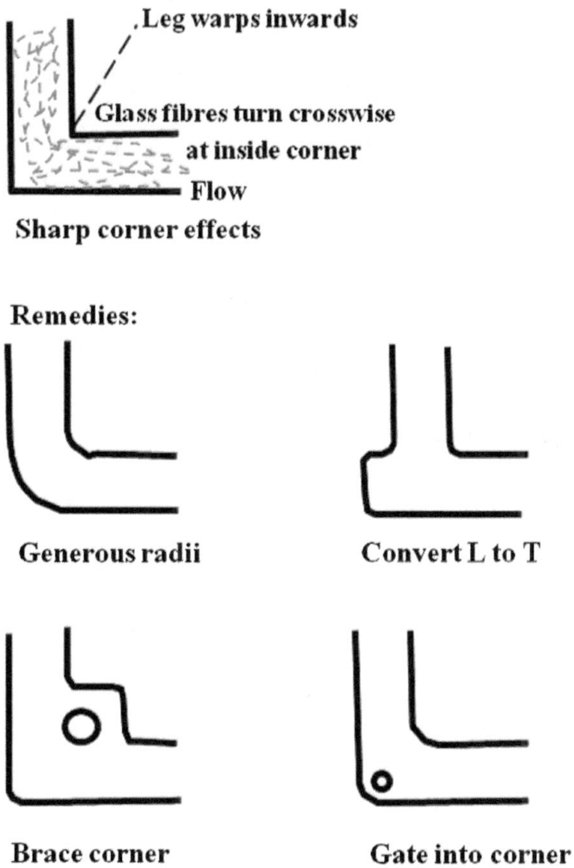

Fig. 3.21 Design features to avoid the effects of sharp corners

5.4 Polymer selection

The subject of selection of the correct polymer for a given application is a very large one. It is not really possible to draw up a comprehensive guide; much depends on individual experience and often several polymers will do the job equally well. In such cases the final choice will depend on cost and the processor's preference. However, once again the advent of cheap computer power has made possible a type of CAD approach to this problem. Databases are used which contain the general, mechanical, electrical, etc. properties of a large number of polymers; these are accessed to match the design property requirements in the product, and suitable materials are selected by the computer. An example is the 'EPOS' system, jointly produced by ICI plc and LNP Plastics Ltd.

3.3.5 Mold Design by Computer-Aided Design

The large extent of tasks to be managed by a designer in the course of the process of mold design can be seen from the many functions a finished mold has to be able to perform. The fundamental tasks of injection molds comprise the following: admission and distribution of the runner and gate-mold cavity-heating/cooling system, as mentioned prior section.

The variety of requirements placed on the shaping mold illustrates the complexity of the tasks inherent in the design of injection molds and hence the high demands placed on design and designer. If one intends to utilize the aid of computers for performing all these tasks, it becomes necessary to systematically analyze the design process and divide it into different stages with clearly defined tasks [61], as shown in Fig. 3.22.

The task of the first stage comprises the determination of the 'mold principle'. In this stage the designer decide type of mold. The second stage of the design process includes dimensioning of the functional complexes and component groups selected in the course of the first stage. The third stage comprises drawing up to documentation for the manufacture, such as drawings, part lists etc.

The tasks of the first stage are predominantly intellectual and creative. However, a well structured program will be able to assist the designer in the second state. This means that the designer advances step by step towards the solution by going through various possibilities both graphically and by means of calculation. The goal to be achieved during the second stage of mold design is the dimensioning of the mold and its component parts. Using computer for performing these tasks in cooperation with the designer is often paraphrased as computer aided engineering (CAE) [62, 63].

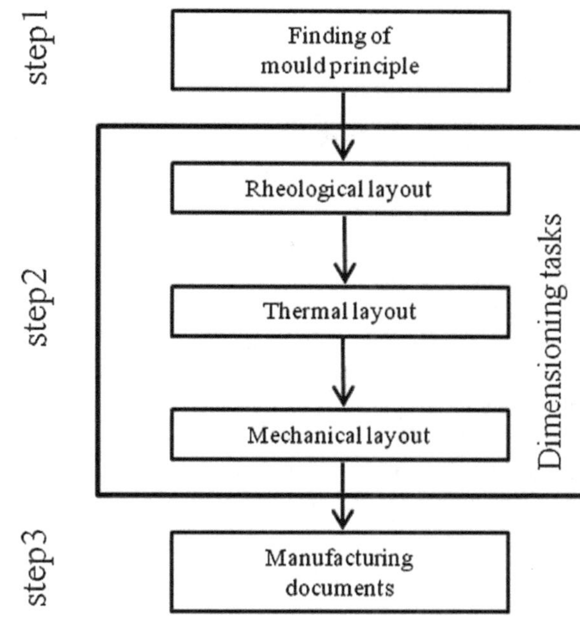

Fig. 3.22 Main steps of mold design

3.3.5.1 Computer Aided Design System for Mold Design

Design and manufacturing technology has been rapidly renovated by introducing computer support technology and tools throughout the whole product development processes. This trend is largely accelerated by the severe competition among manufacturing industry in globalized market. Always high quality, low cost and quick delivery are the decisive factors for winning the competition. For drastically improving such factors, computer support technology is indispensable, which can automate routine engineering processes and amplify human expertise for creative activities [64].

Mold technology is very important for efficient mass production of various kinds of discrete parts. And design and manufacturing of molds require a lot of technical know-how. Many existing knowledge representation and management methods can be effectively applied for supporting these processes. In addition, the mold design process is generally the critical path of a new product development. Conventionally, mold design has always been a much "mystified" art, requiring years of experience before one can be relatively proficient in it. Due to the initial difficulty in learning this art, less and less people are benefiting from the experience and knowledge of the experts in this field. To change the current situation, one way is to use a computer-aided design (CAD) system [65–67].

CAD as an everyday term has grown to a broad range of capabilities and has applications in fields ranging from education for school teaching to three-dimensional mechanical design. At the present time, most CAD systems provide only the geometric modeling functions that facilitate the drafting operations of mold design, and do not provide mold designers with the necessary knowledge to design the molds.

An important development of the past few years has been the emergence of CAD methods, pioneered by the Moldflow company. The database of the system contains rheological/temperature/pressure data on polymers of different types. A proposed mold design can be simulated by the computer and the melt flow through it analyzed. Different runner and gate size and positions can be tried and the optimum found. The technique is particularly valuable for difficult multi-cavity molds where the flow pattern may be difficult if not impossible to predict. Traditionally such tools were made with undersized channels which were then adjusted by trial and error methods on the plant, a time-consuming and expensive procedure. The Moldflow program allows the trials to be simulated and a perfect mold to be fabricated directly. The tools for which it has been used include the so-called 'family molds' in which several different components are made simultaneously. Previously, these tools, though attractive economically because they ease assembly when several components of a product can be produced together, have generally been regarded as impracticable. The new technique allows their manufacture with relative ease. An example is a family mold designed for the simultaneous molding of the three lengths of piping for an industrial vacuum cleaner manufactured by Flymo Ltd. The moldings are shown diagrammatically in Fig. 3.23.

The original design for this mold by the toolmakers had runners arranged as shown in Fig. 3.24a. It is a conventional symmetrical design, with the runner to the gates and the gates themselves having different sizes, with the shortest component to be molded at the central position. There was some unease felt that the design might give trouble and the design was subjected to a Moldflow analysis. The runner arrangement to emerge from this shown in Fig. 3.24b. It has the same sized runners and gates for all three cavities. This is the best way to ensure even filling of the three cavities, as long as the rate of supply is controlled through the runner sizes. It has long been recognized that this is the ideal, but the calculations to achieve it

Fig. 3.23 Components in a family mold

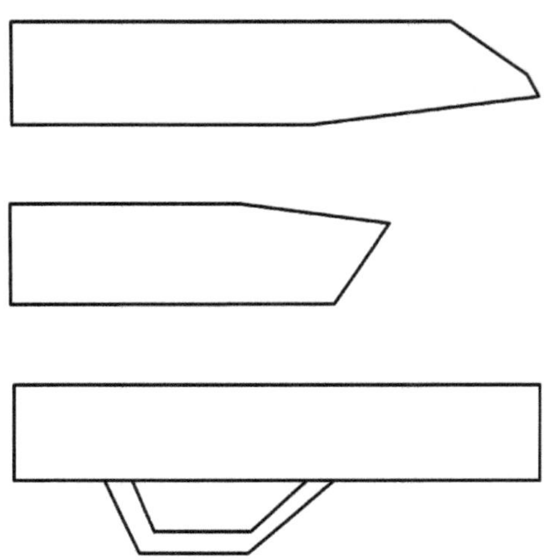

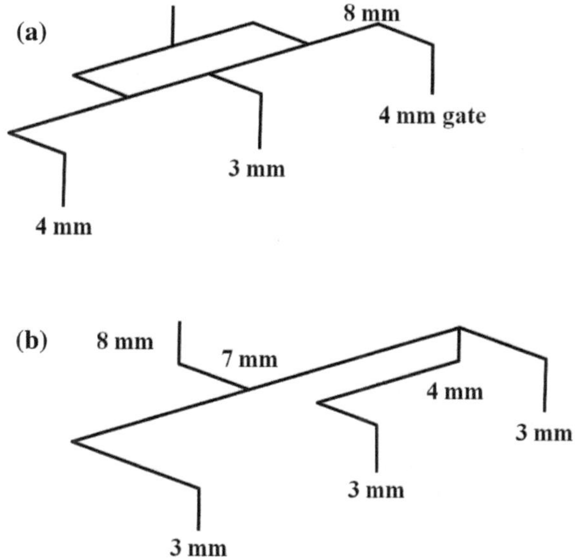

Fig. 3.24 Arrangement of runners in a family mold; **a** conventional design approach, **b** using Moldflow CAD

without CAD techniques, for non-Newtonian, temperature-dependent fluids, are impossibly difficult. The runner layout depicted in Fig. 3.24b is rather unexpected, and the designer has stated that it is a layout that would not have occurred to him in the normal way. The new design was made and put into production, where it gave perfect moldings right from the start.

3.3.5.2 Undercut

To shorten design and development lead-times, automatic and rapid design is critical. New CAD-enabled and tailor-made technology is thus needed to realize the automatic and rapid design in injection mold CAD. In mold CAD, undercut feature recognition is an important step, as it is related to the determination of parting directions and parting lines, generation of core and cavity, and the creation of secondary molding tools which include side-cores and internal pins. The determination of optimal parting direction depends on the types, number, volumes, directions and locations of undercut features [68–70].

To recognize undercut features, the topological relationships of geometrical entities in a molding need to be extracted. In tandem with this, many researchers conducted research on undercut feature recognition. The undercut features are classified into different categories based on the edge number and edge characteristics of the target surface. The undercut direction is then defined as the direction whereby the local tool can be withdrawn easily and most effectively to avoid the warpage of the molding. It is thus also the withdrawal direction of the local tool by which the undercut is molded.

Undercuts can be defined as the convex and concave portions of the molding that cannot be removed from the mold along the parting direction. If the possible undercuts cannot be molded in by cores and cavities, they will require the incorporation of a side-core or side cavity in the mold structure. In determining the optimal parting direction, all the possible undercuts should be recognized and extracted first [71–73].

The undercuts can be classified into two types: external and internal undercuts. The external undercut is the restriction region which prevents the molding from being withdrawn from the cavity, and the internal undercut is the one of that prevents the molding from being ejected from the core. The external undercuts can be further classified into inside and outside external undercut. In order to define the inside external undercuts shown in Fig. 3.25a, the target surface is investigated. It is found that the edges of the garget surface form the different edge-loops in which the edges are linked together. The largest edge-loop is the outside boundary of the target surface and is designated the external edge-loop like external edge-loop1 as shown in Fig. 3.25b. The other edge-loops lying inside the largest edge-loop are known as the internal edge-loop as internal edge-loop2 and edge-loop3.

On the other hand, from the viewpoint of surface features, the adjacent surfaces (S_A) of a target surface are classified into external and internal adjacent surface. In Fig. 3.25a the target surface is the top surface of the block, the external adjacent surfaces are four-sided surfaces. Its internal adjacent surfaces compose two undercuts lying inside the target surface. The internal adjacent surfaces can be divided into different groups in which the adjacent surfaces are linked together and form an undercut. The undercuts are classified in such a way that it is easy to set up their determining criteria. Similarly, the internal undercuts can also be classified into inside and outside internal undercuts. The definitions are similar to the above.

Undercut feature parameters are undercut feature volume and undercut feature direction. The undercut feature volume can be determined based on the geometric entities of the undercut feature, while the undercut directions can be determined by

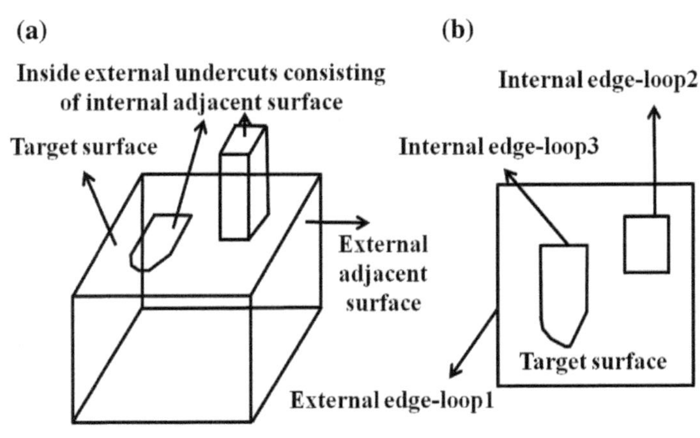

Fig. 3.25 Inside external undercuts and their edge-loops

the visibility map (V-map) of the undercut feature surfaces. The V-map of a surface is a notion to describe the surface visibility and is formed by the points on the unit sphere from which the surface is completely visible from infinity. The surface V-map is a twin motion of Gauss map, which is introduced by Gauss through mapping the surface normal of a given point in a surface onto a unit sphere to describe the surface property [74, 75].

3.4 Summary

In this chapter, we described a brief introduction of the processing methods for multi-materials and focused on introduction of the considerations for mold design including computer-aided design system.

Multi-materials are defined as the class of composite consisting of two or more different materials. For the multi-materials-based parts, there are various processing methods, such as injection molding, extrusion molding, compression molding and transfer molding. Among these methods, injection molding is most useful method and the multi-material molding, which is specialized injection molding technique, is also used for manufacturing multi-material objects.

The properties of a multi-material objects strongly depend on each material features, such as molecular orientations, crystallinity, and miscibility, and processing conditions, such as melt temperature, mold temperature, injection pressure, injection speed, and mold pressure, and the mold. In these factors, the mold design is one of the most important elements to obtain high-quality multi-material parts and determining the shape of final product.

The mold is comprised of mold base, core and cavity and the mold design process is generally the critical path of a new product development. In most cases, quality of mold is very important for the economics of total injection molding process. Therefore, mold designers are required to possess thorough and broad experience, because detailed decisions require the knowledge of the interaction among various parameters. To design the mold, the manual and computer-aided design methods are usually used. In these days, generally, computer-aided design (CAD) systems have become one of the most important methods. In addition, intelligent CAD tools that can assist in the various tasks of the mold design process are very important to the productivity of the mold-making industry.

References

1. Goodship V, Love JC (2002) Multi-material injection moulding. Rapra Technology LTD., Shawbury, UK
2. Malloy RA (1994) Plastic part design for injection molding and introduction. Hanser Gardner, Cincinnati, Ohio

3. Priyadarshi AK, Gupta SK, Gouker R, Krebs F, Shroeder M, Warth S (2007) Manufacturing multi-material articulated plastic products using in-mold assembly. Int J Adv Manuf Technol 32:350
4. Tang SH, Kong YM, Sapuan SM, Samin R, Sulaiman S (2006) Design and thermal analysis of plastic injection mould. J Mater Process Technol 171:259
5. Bozdana AT, Eyeríoğlu Ö (2002) Development of an expert system for the determination of injection moulding parameters of thermoplastic materials: EX-PIMM. J Mater Process Technol 128:113
6. Li X, Gupta SK (2004) Geometric algorithms for automated design of rotary-platen multi-shot molds. Comput Aided Des 36:1171
7. Ilinca F, Hetu JF (2003) Three-dimensional simulation of multi-material injection molding: application to gas-assisted and co-injection molding. Polym Eng Sci 43:1415
8. Banerjee AG, Li X, Fowler G, Gupta SK (2007) Incorporating manufacturability considerations during design of injection molded multi-material objects. Res Eng Design 17:207
9. Griffith K, Williams R, Knight T, Sass L, Kamath A (2012) Cradle molding device: an automated CAD/CAM molding system for manufacturing composite materials as customizable assembly units for rural application. Autom Constr 21:114
10. Sakamoto S, Katayama T, Yokogawa R, Kimura T (2001) Construction of PC-based intelligent CAD system for cold forging process design-intergration of CAD system and development of input method. J Mater Process Technol 119:58
11. Zhang LP, Fuh JYH, Nee AYC (2003) Tool path regeneration for mold design modification. Comput Aided Des 35:813
12. Kuo Y, Kamal MR (1976) The fluid mechanics and heat transfer of injection mold filling of thermoplastic materials. AIChE J 22:661
13. Park SJ, Kwon TH (1998) Thermal and design sensitivity analyses for cooling system of injection mold, Part 1: Thermal analysis. J Manuf Sci Eng 120:287
14. Chiang HH, Hieber CA, Wang KK (1991) A unified simulation of the filling and postfilling stages in injection molding. Part I Formulation Polym Eng Sci 31:116
15. Karmaker AC, Youngquist JA (1996) Injection molding of polypropylene reinforced with short jute fibers. J Appl Polym Sci 62:1147
16. Guan Q, Shen K, Ji J, Zhu J (1995) Structure and properties of self-reinforced polyethylene prepared by oscillating packing injection molding under low pressure. J Appl Polym Sci 55:1797
17. Suarez H, Barlow JW, Paul DR (1984) Mechanical properties of ABS/polycarbonate blends. J Appl Polym Sci 29:3253
18. Yalcin B, Cakmak M (2004) Superstructural hierarchy developed in coupled high shear/high thermal gradient conditions of injection molding in nylon 6 nanocomposites. Polymer 45:2691
19. Huggett R, Zissis A, Harrison A, Dennis A (1992) Dimensional accuracy and stability of acrylic resin denture bases. J Prosthet Dent 68:634
20. Ulcer Y, Cakmak M, Miao J, Hsiung CM (1996) Structural gradients developed in injection-molded syndiotactic polystyrene (sPS). J Appl Polym Sci 60:669
21. Rosato DV, Rosato DV, Rosato MG (2000) Injection molding handbook. Kluwer Academic Publishers, USA
22. Rubin II (1977) Injection molding. Introduction to polymer science and technology. Wiley, NY
23. Malloy RA (1994) Plastic part design for injection molding: an introduction. Hanser Verlag, Munich Vienna New York
24. Takashima H, Miyagai K, Hashida T, Li VC (2003) A design approach for the mechanical properties of polypropylene discontinuous fiber reinforced cementitious composites by extrusion molding. Eng. Fracture Mechanics 70:853
25. Rauwendaal C (2001) Polymer extrusion. Hanser Verlag, Munich
26. Baird DG, Collias DI (1998) Polymer processing principles and design. Wiley, New York, USA

27. Smith DE, Tortorelli DA, Tucker CL (1998) Analysis and sensitivity analysis for polymer injection and compression molding. Comput Methods Appl Mech Eng 167:325
28. Issyev AI (1987) Injection and compression molding fundermentals. Marcel Dekker, New York
29. Laza JM, Julian CA, Larrauri E, Rodriguez M, Leon LM (1999) Thermal scanning rheometer analysis of curing kinetic of an epoxy resin: 2. An amine as curing agent. Polymer 40:35
30. Dumont P, Orgéas L, Corre SL, Favier D (2003) Anisotropic viscous behavior of sheet molding compounds (SMC) during compression molding. Int J Plast 19:625
31. Odenberger PT, Andersson HM, Lundström TS (2004) Experimental flow-front visualisation in compression moulding of SMC. Compos Part A 35:1125
32. Ikegawa N, Hamada H, Maekawa Z (1996) Effect of compression process on void behavior in structural resin transfer molding. Polym Eng Sci 36:953
33. Davis BA, Gramann PJ, Rios AC (2003) Compression molding. Hanser Verlag, Munich
34. Parasnis NC, Ramani K (1998) Analysis of the effect of pressure on compression moulding of UHMWPE. J Mater Sci Mater Med 9:165
35. Um MK, Lee WI (1991) A study on the mold filling process in resin transfer molding. Polym Eng Sci 31:765
36. Lin MY, Murphy MJ, Hahn HT (2000) Resin transfer molding process optimization. Compos A 31:361
37. Peters ST (1998) Handbook of composites. Chapman & Hall, UK
38. Schmidt LR (1974) A special mold and tracer technique for studying shear and extensional flows in a mold cavity during injection molding. Polym Eng Sci 14:797
39. Chen SC, Chung YC (1992) Simulations of cyclic transient mold cavity surface temperatures in injection mold-cooling process. Int Commun Heat Mass Transfer 19:559
40. Kantz MR, Newman HD Jr, Stigale FH (1972) The skin-core morphology and structure–property relationships in injection-molded polypropylene. J Appl Polym Sci 16:1249
41. Pye RGW (1983) Injection mould design: a design manual for the thermoplastics industry. Longman Scientific & Technical
42. Austin C (1985) Developments in injection moulding. Elsevier Applied Science, Barking, UK
43. Mok CK, China KS, Lan H (2008) An internet-based intelligent design system for injection moulds. Robotics Computer-Integrated Manuf. 24:1
44. Chu CH, Song MC, Luo VCS (2006) Computer aided parametric design for 3D tire mold production. Comput Ind 57:11
45. Nakao M, Yamada S, Kuwabara M, Otubo M, Hatamura Y (2002) Decision-based process design for shortening the lead time for mold design and production. CIRP Annals-Manuf. Technol. 51:127
46. Lichius U (1983) Rechnerunterstützte Konstruktion von Werkszeugen zum Spritzgießen von thermoplastischen Kunststiffen. Diss. an der RWTH Aache
47. Jones P (2008) Mould design guide. Rapra Technology Ltd.
48. Lin SIE (2001) Near-net-shape forming of zirconia optical sleeves by ceramics injection molding. Ceram Int 27:205
49. Kim HS, Son JS, Im YT (2003) Gate location design in injection molding of an automobile junction box with integral hinges. J Mater Process Technol 140:110
50. Kazmer D (2007) Injection mold design engineering. Hanser Verlag, Munich
51. Beaumont JP, Nagel R, Sherman R (2002) Successful injection molding: process, design, and simulation. Hanser Verlag, Munich
52. Whelan A (1982) Injection moulding materials. Applied Science Publishers, Barking, UK
53. Fallböhmer P, Rodríguez CA, Özel T, Altan T (2000) High-speed machining of cast iron and alloy steels for die and mold manufacturing. J Mater Process Technol 98:104
54. Moritz T, Richter HJ (2006) Ceramic bodies with complex geometries and ceramic shells by freeze casting using ice as mold material. J Am Ceram Soc 89:2394
55. Pehlke RD, Jayarajan A, Wada H (1992) Summary of thermal properties for casting alloys and mold materials. University of Michigan, Ann Arbor, MTIS-PB83-211003

56. Perry MJ, Xu J, Ma Y, Lee LJ, Liou MJ (1992) Analysis of thermal characteristics of alternative mold materials for resin transfer molding. Soc Aero Mater Process Eng 23:20
57. Santos GA, de Neto CM, Osório WR, Garcia A (2007) Design of mechanical properties of a $Zn_{27}Al$ alloy based on microstructure dendritic array spacing. Mater Des 28:2425
58. Kaufman JG, Rooy EL (2004) Aluminium alloy castings: properties, processes, and applications. ASM International
59. Cooper AG, Kang S, Kietzman JW, Prinz FB, Lombardi JL, Weiss LE (1999) Automated fabrication of complex molded parts using mold shape deposition manufacturing. Mater Des 20:83
60. Young WB (2009) Thermal behaviors of the resin and mold in the process of resin transfer molding. J Reinf Plast Compos 28:833
61. Manzione LT (1987) Applications of computer aided engineering in injection molding in computer aided engineering for polymer processing. In Bernhardt EC (ed) Hanser Publishers, Munich-Vienna-New York
62. Kim SW, Turng LS (2004) Developments of three-dimensional computer-aided engineering simulation for injection moulding. Model Simul Mater Sci Eng 12:S151
63. Lee RS, Chen YM, Cheng HY, Kuo MD (1998) A framework of a concurrent process planning system for mold manufacturing. Comput Integr Manuf Syst 11:171
64. Kimural F, Ariyoshi H, Ishikawa H, Naruko Y, Yamato H (2004) Capturing expert knowledge for supporting design and manufacturing of injection molds. CIRP Ann Manuf Technol 53:147
65. Nee AYC, Fu MW, Fuh JYH, Lee KS, Zhang YF (1998) Automatic determination of 3-D parting lines and surfaces in plastic injection mould design. Ann CIRP 47:95
66. Nakao M, Yamada S, Kuwabara M, Otubo M, Hatamura Y (2002) Decision-based process design for shortening the lead time for mold design and production. Ann CIRP 51:127
67. Kong L, Fuh JYH, Lee KS, Liu XL, Ling LS, Zhang YF, Nee AYC (2003) A Windows-native 3D plastic injection mold design system. J Mater Process Technol 139:81
68. Ran JQ, Fu MW (2010) Design of internal pins in injection mold CAD via the automatic recognition of undercut features. Comput Aided Des 42:582
69. Fu MW, Fuh JYH, Nee AYC (1999) Undercut feature recognition in an injection mold design system. Comput Aided Des 31:777
70. Nee AYC, Fu MW, Fuh JYH, Lee KS, Zhang YF (1997) Determination of optimal parting direction in plastic injection mould design. Ann CIRP 46:429
71. Majhi J, Gupta P, Ravi J (1999) Computing a flattest, undercut-free parting line for a convex polyhedron, with application to mold design. Comput Geom 13:229
72. Ye XG, Fuh JYH, Lee KS (2001) A hybrid method for recognition of undercut features form moulded parts. Comput Aided Des 33:1023
73. Chen LL, Chou SY, Woo TC (1993) Parting directions for mould and die design. Comput Aided Des 25:762
74. Gan JG, Woo TC, Tang K (1994) Spherical maps: their construction, properties, and approximation. ASME, J Mech Des 116:357
75. Hilbert D, Cohn-Vossen S (1983) Geometry and the imagination. Chelsea, New York

Chapter 4
Injection Molding for Multicomponent Materials

Soo-Jin Park and Seul-Yi Lee

4.1 Introduction

4.1.1 Basics of Injection Molding

Injection molding is one of the most valuable and important manufacturing processes capable of mass-producing delicate plastic parts in net shape with excellent dimensional allowance. Injection molding is used to create many things such as wire spools, packaging, bottle caps, automotive dashboards, pocket combs, some musical instruments (and parts of them), one-piece chairs and small tables, storage containers, mechanical parts (including gears), and most other plastic products available today. For many years, injection molding process and quality control has been an active research area, as part quality and yield requirements become more rigid.

In addition, the injection molding is the most well-known process used to manufacture plastic parts. It is ideal for producing high volumes of the same object. A typically injection cycle is composed of three fundamental phases: (a) filling, (b) packing, and (c) cooling [1–4]. In the beginning, the polymer is granulated the pellet-shaped form, it arrives the molten state by heating and shearing (screw) in a sleeve. At the extremity of the sleeve, a nozzle makes possible the polymer injection in the mold cavity. While the first phase, once the mold is closed, a melted polymer fills the cavity part, moving by means of the spure, runners, and gates. In the packing phase, extra melt polymer enters the cavity to balance the possible shrinkage caused by cooling. The cooling phase takes place concurrently with the filling and packing phases, and considers polymer plasticization and the additional time required to obtain a more than 80% solidified product [5]. This is necessary for the ejection of the part, which together with mold opening completes the process.

S.-J. Park (✉) · S.-Y. Lee
Department of Chemistry, Inha University, 253, Nam-gu,
Incheon 402-751, South Korea
e-mail: sjpark@inha.ac.kr

J.K. Kim et al. (eds.), *Multicomponent Polymeric Materials*,
Springer Series in Materials Science 223, DOI 10.1007/978-94-017-7324-9_4

While these phases, interactions between material properties, machine parameters, and process variables make fabrication complex. The quality of the final products, which may be characterized in terms of dimensional stability, appearance, and mechanical properties, depends directly on these factors [5, 6].

Recently trends are seen in the direction by obtaining optimum modeling conditions, develop and present a methodology for molding condition optimization for the practical application. In this chapter, we will contribute the classification of multi-materials injection molding technique, their application, and prospective.

4.1.2 Multi-materials Injection Molding

With standard injection molding becoming more and more established in practical manufacturing, special variants are attracting increasing attention. Injection molding is the most commercially important of all plastic processing methods. It makes production of the great variety of manufactured goods with intricateness and refinedness. Also, it allows the high precision, three-dimensional parts at high production rates. The different types of polymer materials are commercially made to available for injection molding process. In many applications, traditional materials such as metal, glass, silicon, wood can be substituted for plastics.

In recent, actually almost every industry (for example, automotive, consumer goods, toys, electronics, power tools, appliances) that makes use of traditional single-material injection molding process is replaced to use multi-material molding processes. Some common applications include multicolor objects, skin-core arrangements, in-mold assembled objects, soft-touch components (with rigid substrate parts) and selective compliance objects. Typical examples of each class of application are shown in Fig. 4.1 [7, 8].

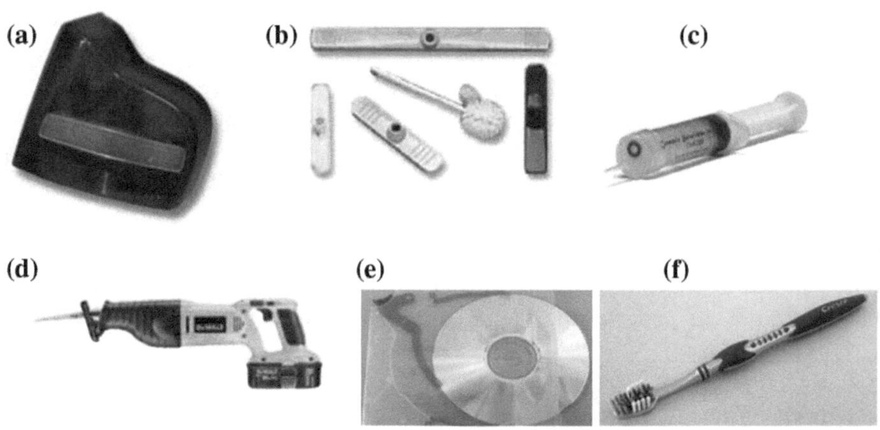

Fig. 4.1 Examples of multi-materials injection molding objects. **a** Taillight (multi-color). **b** Assorted brushes (skin/core arrangements). **c** One-piece syringe (in-mold assembly). **d** Cordless saw housing (soft-touch grip). **e** e-slimcase by ejector. **f** Crest toothbrush

Up to now, the variety of materials for macro-, micro-, and nano technology applications has been dominated by silicon, polymers, certain metals, and so on. In particular, the approaches on multi-material injection molding are worth mentioning. Over the past few decades, various multi-materials injection molding processes have emerged for manufacturing multi-materials objects. These multi-materials objects are attributable by the different portions are made of different materials, which allows significant decrease in assembly operations and production cycle times, owing to the fabrication and assembly steps being performed inside the molds. In addition, the quality of product can be improved; meanwhile, the possibility of manufacturing defects and total manufacturing costs can be decreased.

In multi-materials injection molding, different materials are sequentially injected into a multi-stage mold. The divisions of the molding body that are not to be filled during a molding stage are temporally blocked. Once the first injected materials are arranged, and then one or more blocked portions of the mold are opened and the next prepared materials is injected. This process continues until the required multi-materials objects are obtained.

4.1.3 Multi-materials Injection Molding and Quality Control

In practice, once a multi-materials injection mold is built and mounted on a machine, a molding engineer (or setup person) has to set up the machine parameters based on information from the material supplier, mold designer, and, to a great extent, his or her own experience. Then the engineer has to go through an extensive "mold trial" procedure, which entails producing some parts and adjusting machine parameters by trial and error until the molded parts meet the design specifications consistently. Such an exercise is not only time consuming and costly, but also has to be repeated for each specific material/mold/machine configuration.

Furthermore, even when the machine is properly set up, periodic inspection of part quality is still necessary as a measure of quality assurance. There is a good possibility that the part quality may wander outside the quality limits due to a number of unpredictable reasons, which include variation in material properties (particularly when reground resins are used), change in the ambient environment (e.g., humidity or temperature in the shop), and machine characteristics (particularly those using hydraulic power). If that happens, the process conditions have to be readjusted in order to bring the part quality back within the tolerance limits. To overcome the aforementioned difficulties and to achieve and maintain first-class product quality in injection molding, considerable research has been conducted over the past two decades on suitable process control systems with specific control technologies either online or off-line for the purpose of automatic and adaptive quality control. Given the numerous variables involved in the complex injection molding process, it is impossible to develop a feasible control strategy without a thorough understanding of the relationship and dependency among those variables.

Table 4.1 Three-level variables in injection molding

Level	Conditions	
Level 1 Machine variables (independently controllable)	Temperature	Barrel temperature (in several zones) Nozzle temperature Coolant temperature
	Pressure	Pack/hold pressure Back (recovery) pressure Maximum injection pressure Clamp/fill/pack/hold/recovery/eject switch over point
	Sequence and motion	Injection (ram) speed Screw (rotation) speed Shot volume and cushion (via screw displacement)
Level 2 Process (dependent) variables	Melt temperature (in the nozzle, runner, or mold cavity) Melt pressure (in the nozzle, cavity) Melt-front advancement Maximum shear stress Rate of heat dissipation and cooling	
Level 3 Quality definitions (final response)	Part weight and part thickness Shrinkage and warpage Sink marks Appearance and strength at the weld lines Other aesthetic defects: burn marks, gate blushes, surface texture, etc.	

By examining closely all of the important variables involved, it was suggested that they could be categorized into three distinctive levels; namely, level 1-machine variables, level 2-process variables, and level 3-quality variables, as tabulated in Table 4.1 [9, 10]. These variables have a direct impact on the final part quality and process economics. The further information is presented in [11].

4.2 Multi-materials Injection Molding

Multi-materials injection molding can be considered a relatively young manufacturing technology, many modified versions of the basic process have been developed, for example, injection-compression, gas-assisted, water injection, co-injection, and over-molding.

For a clarifying explanation, an overview of the multi-materials molding process is presented in Fig. 4.2. A number of processes fall into this molding category. Various terminologies are named in multi-materials injection molding to clearly outline specific process.

There are fundamentally three different types of multi-material molding processes here; multi-component, multi-shot, and over (insert) molding.

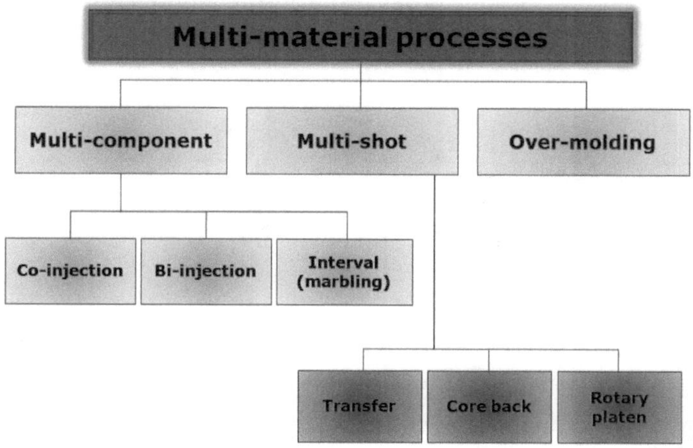

Fig. 4.2 Process options for multi-materials molding

4.2.1 Multi-component Injection Molding

Multi-component injection molding might be considered the simplest and most common method of multi-materials injection molding. It is simultaneous or sequential injection of two (or more) different materials (either the same or different gate locations) in a single mold.

As shown in Fig. 4.3, the process starts with the injection of component A. At a certain point in the filling process, a second materials B follows, completing the filling. The first injected materials, A, cools at the mold surface and builds up the outer layer of the molding. The second injected materials, B, in the core layer are still liquid, pushing the first materials to the mold wall and to the end of the flow path. In the finished molding, the skin layer is formed by the materials injected first and the core layer is formed by the second materials [12–14].

Moreover, it is again divided to co-injection molding, bi-injection molding, and interval (marbling) molding.

4.2.1.1 Co-injection Molding

The co-injection molding is known as dual injection or sandwich molding. It involves making sequential injection into the same mold with one material as the core and another as the skin. It is a process that creates a skin and core materials arrangement in a molded part. At first, the skin materials are injected into the cavity in molding body, and are directly followed by core materials. The core is fully encapsulated to produce a sandwich like molding. The co-injection molding has advantages of; (1) the foamed core for reduced weight and noise transmission, (2) glass filled cores for improved physical properties, (3) low cost core for cost

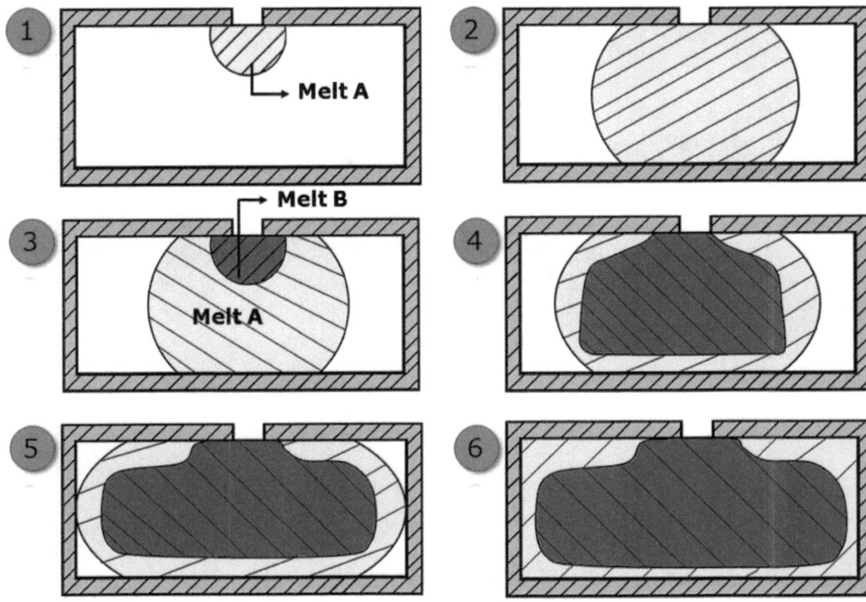

Fig. 4.3 Scheme of multi-materials injection moldings

savings, (4) high gloss skin materials over structural core material for combination of aesthetic and structural properties (for example, fiber-reinforced materials, which may have undesirable surface properties), (5) post consumer recycled materials in core (environmental friendly), (6) post industrial recycled material in core, and (7) reground painted parts recycled into core. In particular, the co-injection molding utilizes a property of liquid polymers called "fountain flow" to keep the core material inside the skin material as they flow and subsequently harden in the mold. Figure 4.4 presents a sequential injection process in powder co-injection molding [15].

Figure 4.5 presents a typical A-B-A co-injection sequence to obtain a simple skin/core arrangement component. As shown in Fig. 4.5, in the step 1, the valve to barrel A is open during the valve to barrel B is closed. This makes some skin material to be injected, incompletely filling the cavity. Next, in the step 2, the valve to barrel A is closed during the valve to barrel B is opened. This makes the core materials to be injected into the cavity, penetrating the initial skin layer. The two materials are not mixed completely. And the core will not puncture the skin because of laminar flow. And then, in the optional step 3, more skin materials are injected as in the step 1. This ensures complete encapsulation of the core material. Finally, the finished part then cools and hardens so it can be expelled from the molding body [16] (Figs. 4.6 and 4.7).

The process itself can be performed with simultaneous injection of both materials or just with sequential injection [15, 16]. With sequential injection, the melt front may stop for a short time at the switch from component A to B. This may lead

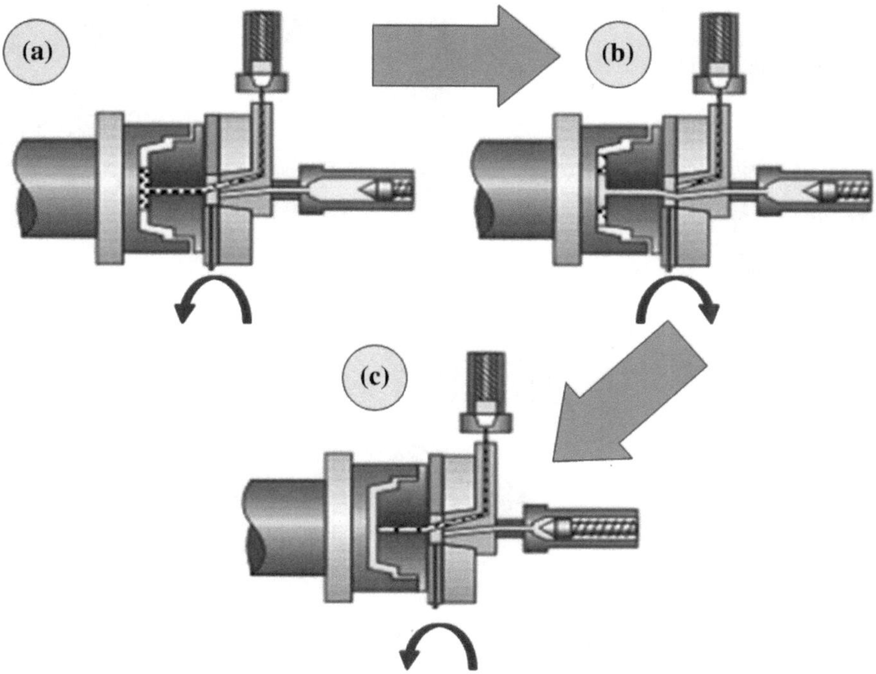

Fig. 4.4 Sequential injection process in powder co-injection moulding; **a** injection of the skin, **b** injection of the core, and **c** injection of a short shot to complete the skin

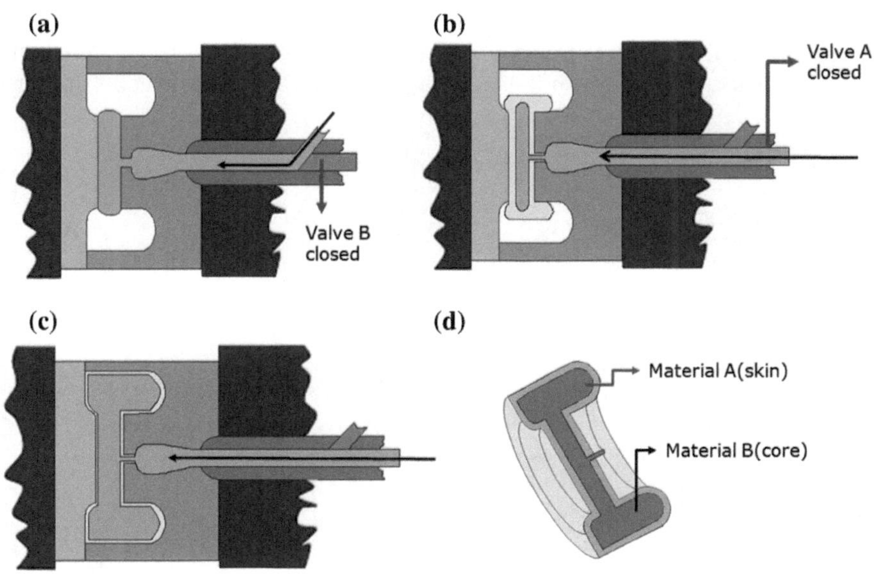

Fig. 4.5 Schematized co-injection molding process. **a** Step 1: shot A. **b** Step 2: shot b. **c** Step 3: final filling. **d** *x*-section of complete part

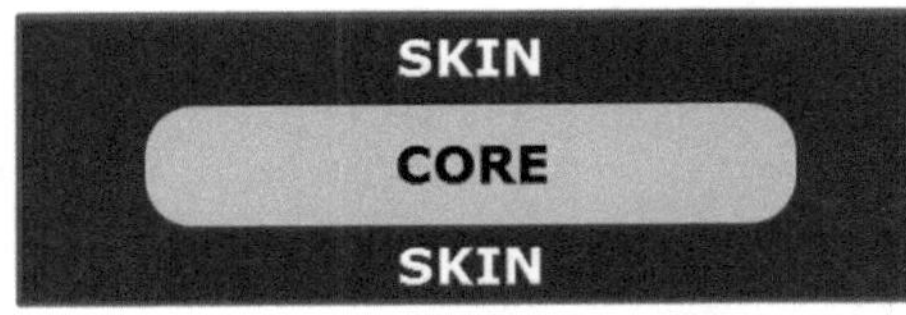

Fig. 4.6 Cross section of sandwich structure of a co-injection molding

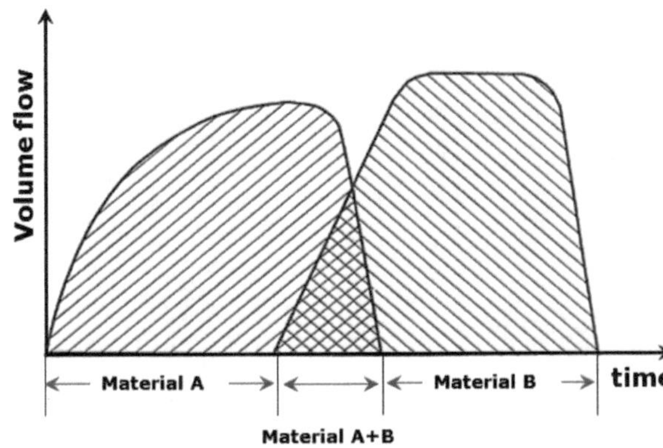

Fig. 4.7 Two-component injection molding with simultaneous injection phase

to visible marks on the molding. Injecting materials simultaneously instead of switching helps prevent this (Table 4.2).

4.2.1.2 Bi-Injection Molding

The bi-injection molding is a process in which two different resins are simultaneously injected at different gates in the same molding body. As the materials injected

Table 4.2 Current applications for co-injection moldings

Materials combination	Properties	Application
Soft feel skin/hard core	High strength core with soft feel skin	Door handles, gear lever [17]
Unfilled skin/core with conductive filler	Electromagnetic interference shielding (EMI)	Computer housings
Virgin skin/recycled core [18]	Environmentally friendly production, cost saving	Garden furniture, automotive bumpers and fascias [19]
Unfilled skin/reinforced core	High surface finish, structural performance	Automotive door handles [20]
In mold paint/variable core	No finishing of product required after molding	Wheel trims [21]
Unfilled skin/foamed core	Good surface finish, low density, high rigidity	Automotive body panels [22]
Pigmented skin/uncolored core (or reverse)	Reduced pigment cost, aesthetics	Yogurt pot

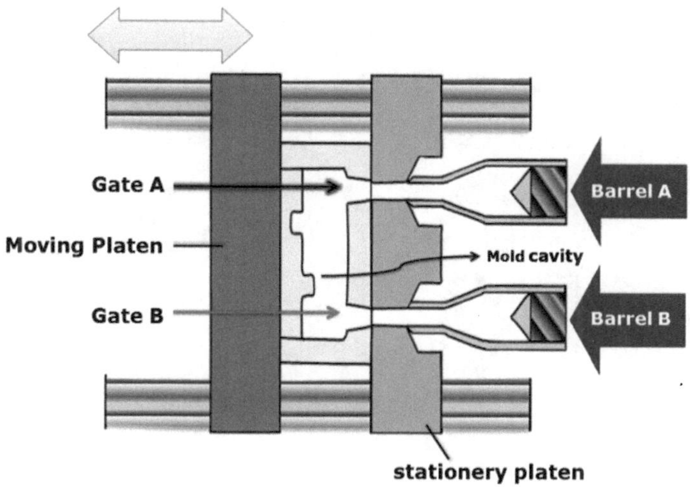

Fig. 4.8 Schematized bi-injection molding machine

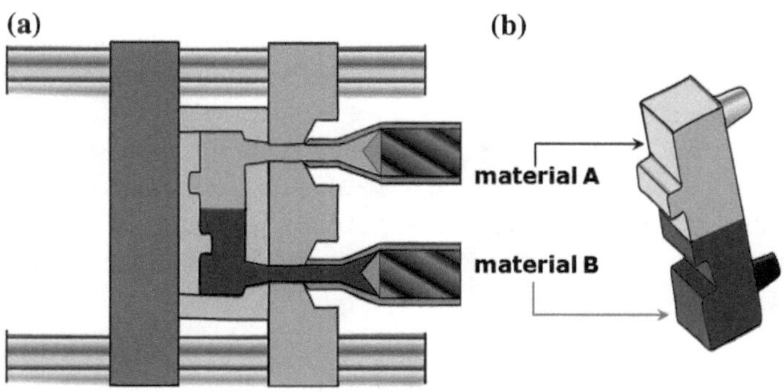

Fig. 4.9 Schematized bi-injection molding process. **a** After injection. **b** Finished part

from the different gates flow into the single part, they meet along a common interface and cross-polymerize. The bi-injection molding [23] is relatively simple and is only used to produce simple, low-tolerance parts. The interfaces are naturally formed when the separate resins meet are usually very simple planar surfaces. Figure 4.8 schematically illustrates a bi-injection molding machine and Fig. 4.9 illustrates the process.

4.2.1.3 Interval Injection Molding

The interval injection molding produces marbling type effects and is generally, though not always, confined to moldings of various colored materials of the same type. Two plastic components are injected alternately into the cavities during

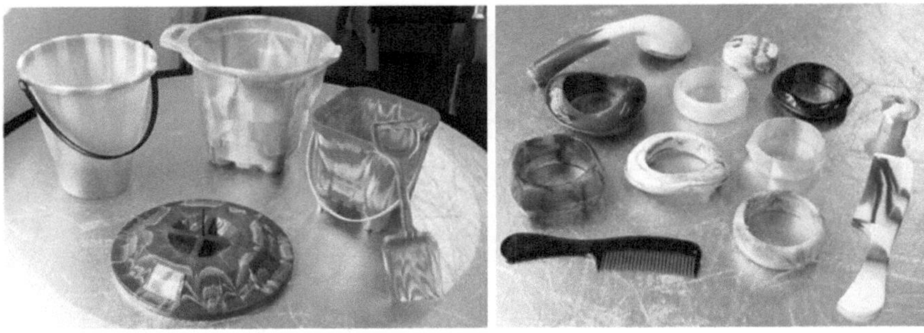

Fig. 4.10 Examples of multi-materials interval (marbling) molding products

interval injection molding. This is similar to methods used to produce co-injection molding. However, the injection is not used to produce a skin/core configuration. It has also the color effects created by the two plastics flowing into each other in the case of interval injection molding technology in comparison with co-injection molding. It is intended that the two colors blend together by this procedure (no clear color separation, this is, a random or a regular color distribution pattern). The two materials are injected into the mold through this special nozzle, which results in the typical color effect. The coloration is influenced by the cycle sequence, the design of the molded part, the position of the sprue and the flow properties of the injection molding materials. The decisive influencing para-meters are simultaneous or alternating injection, the control of the interval cycle rates and modifying the speed of the injection. In the method marketed by Arburg [24], 'the two injection units are coupled together using a special interval unit, inside which the mixing nozzle is located' (Fig. 4.10).

4.2.2 Multi-shot Injection Molding

Multi-shot injection molding is the most complex and versatile of the multi-materials injection molding processes. In a specialized sequence, it makes to inpour the different materials into the appointed mold, where the mold cavity geometry may partially or completely change between sequences. That is, multi-shot injection molding technique joins two or more plastics into one solid, integrated component. This is also known as "two-shot molding". To achieve this, it is performed that two resins are directly injected into one mold held in a rotary platen press.

Figure 4.11 is illustrated the multi-shot injection molding process. At first, two unique resins are flowed in the cavities, which are from two independently programmed barrels. The first injection is usually the hard materials. After the first shot of plastic is performed, the mold opens and rotates, and is carefully placed so a second shot can fill on top of the first. Finished parts are taken away the mold. Empty cavities are then ready for the first shot of the next part. Finally, while every

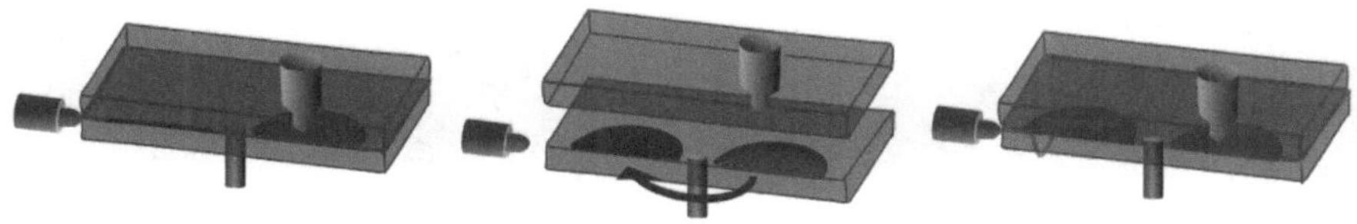

Fig. 4.11 Multi-shot injection molding process

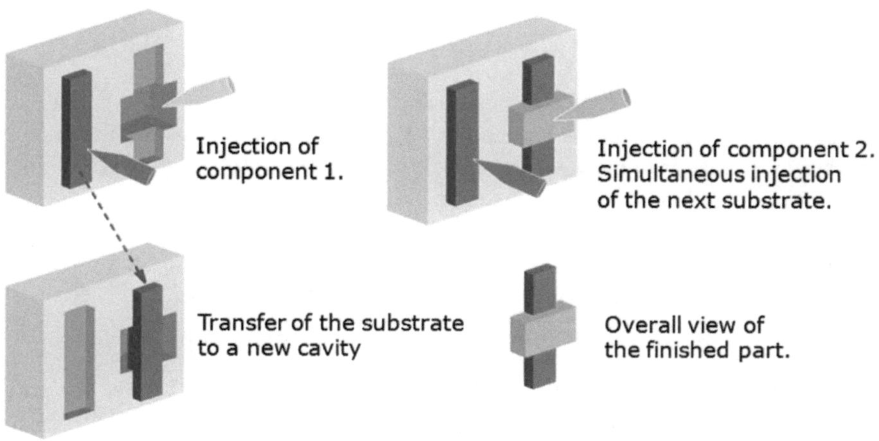

Fig. 4.12 Scheme of universal transfer process

molding cycle, one part accommodates the first shot during the other gets the second. Proper and right bonding is come from the proficient process control, accurate timing, and the selection of suitable materials.

4.2.2.1 Transfer Multi-shot Molding

In the transfer multi-shot molding process, the substrate is injected and then transferred by a robot to a new cavity or to a different injection molding machine. This method is the most flexible multi-component method. Such parts necessitate complete removal of the substrate from the first cavity. Transfer from one machine to the other is common for thermoplastic/rubber composite parts as it is a simple way to thermally separate the two components (Fig. 4.12).

4.2.2.2 Core Back Multi-shot Molding

The principle of multi component is that two different (or more) materials (or two forms of the same material) should be used for the core and the skin. That is, the expensive high performance materials are used for the skin and the low-cost

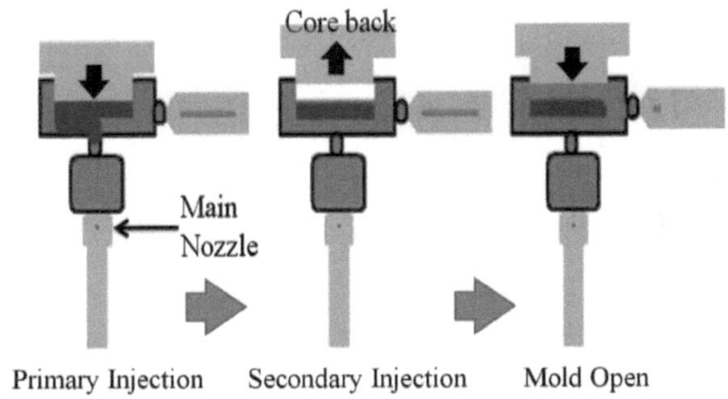

Fig. 4.13 Simultaneous injection molding with core back and co-injection

commodity or recycled plastic is used for the core. However, in-molding technique needs time and labor to insert skin material (cover stock) into mold [25–27]. When molding multilayer products with a conventional injection technique, core materials are firstly injected to cavity. Then, the mold is turned around for skin materials to be injected. This process requires a complicated injection machine. In addition, in the phase of molding skin materials, the interface between two layers of melt occasionally becomes instable [27, 28].

Park et al. [28] reported the new method that could produce multi-component products with the core material being injected sufficiently into the cavity. Then, the core moves back and the skin materials are injected into the secondary cavity formed by moving back of core. After a period of time, the core moves forward for packing and gluing materials, as shown in Fig. 4.13.

In comparison with the conventional injection molding process, the new method of sequential core-back technique makes a perfect boundary between two materials. It also requires fewer manufacturing steps and less labor. Moreover, by means of moving core-back for forming skin cavity, and also moving core forward for packing, the quality of part is high and the new method requires low pressure for filling and packing phase [28].

4.2.2.3 Rotary Platen Multi-shot Molding

Rotary platen multi-shot molding is the simplest and most common incarnation of multi-shot molding. The basic idea is that the moving half of the mold contains as many cavities as there are shots, and it rotates these cavities to the injection position before each shot [29].

Figure 4.14 illustrated a simplified schematic of rotary platen multi-shot injection molding machine. The key feature to note is the rotary platen on the left side of the diagram. The rotary platen is attached to the core plate which contains two identical cores mirrored across the centerline of the platen with coincides with the axis of rotation. The cavity plate attached to the stationary platen contains two corresponding cavities with differing geometries. In essence, the rotary platen accomplishes the task of switching the partially completed component between

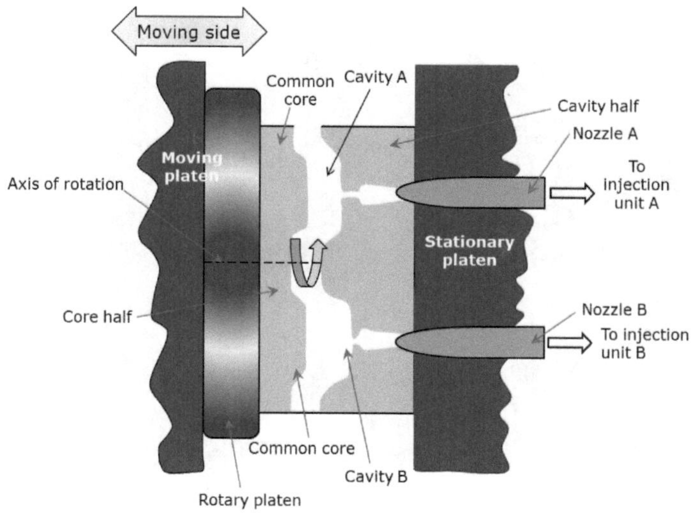

Fig. 4.14 Schematized rotary platen multi-shot injection molding machine

molds for each stage. This eliminates the need for manually changing molds via hand or robot. For this type of multi-shot injection moldings, the core for both material stages is exactly the same while the cavity is different [23].

The rotary platen molding process, as illustrated in Fig. 4.15, is described as follows [23]:

1. The mold must first reach steady state operation, where at least one complete component AB has been produced. This ensures the system is ready for the subsequent ("(n + 1)th") cycle. The partially-complete nth component is in cavity B (Fig. 4.15a).
2. The (n + 1)th shots of materials A and B are simultaneously injected into their respective cavities and allowed to cool. This produces the nth complete component AB in cavity B and the (n + 1)th partially-complete component AB in cavity A (Fig. 4.15b).
3. The mold opens and the nth complete component AB is ejected (Fig. 4.15c). A cross section of a complete component AB is shown in Fig. 4.15d.
4. The rotary platen rotates 180° and the mold closes (not shown). The cycle is now ready to repeat.

Although Figs. 4.14 and 4.15 only show a two-material rotary platen machine, it is possible to accommodate more materials. Three-shot and four-shot injection molding machines are also available to manufacture three-material and four-material objects, respectively. Normally, depending on how many different materials there will be, the rotary platen can be rotated by 90°, 120°, or 180°. Special molding presses are required to provide the rotation needed for the core side. This can drive up the mold cost significantly [23].

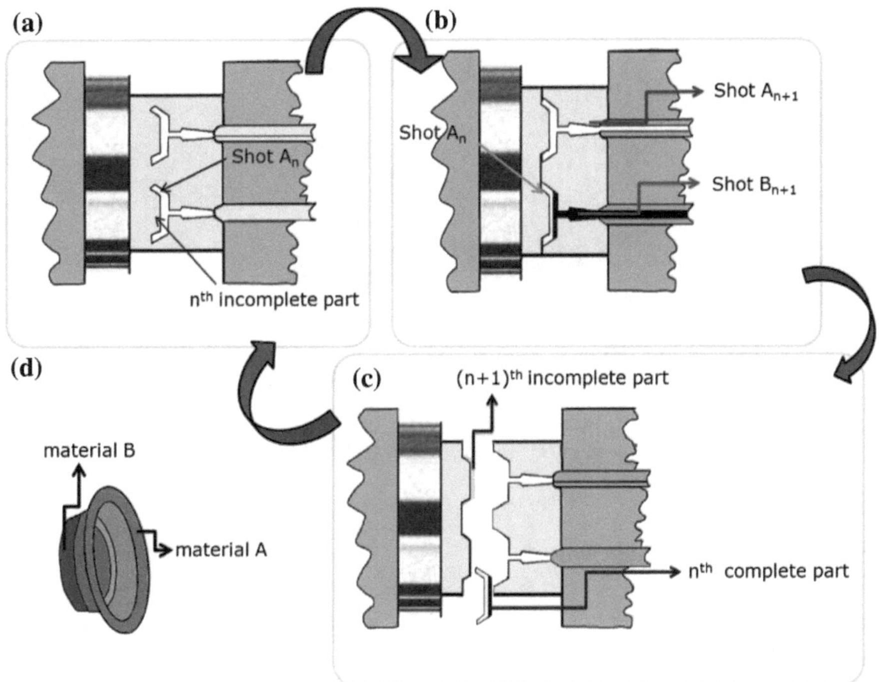

Fig. 4.15 Schematized rotary platen molding process. **a** After rotation. **b** Injection of (n + 1)th shots. **c** Ejection of nth completed part. **d** complete part

Figure 4.16 illustrates this condition by depicting the two molding stages in rotary platen multi-shot molding. Although the shape of the core remains identical in both the stages, the cavity shape changes and already molded component A acts as an additional "mold piece" in the second shot. Moreover, the first stage part that acts as plastic "mold piece" is not separated from the final assembly. These forces is to avoid applying some of the traditional molding design rules on certain portions of the gross shape of the overall object also referred as gross object. By gross object, we mean the solid object created by the regularized union of the two components. That is why, simply ensuring that the first stage part and the gross shape are moldable do not solve this problem either [2].

Multi-shot injection molding techniques are well established, their growth being pushed by development of thermoplastics elastomer materials, enabling rigid and flexible material combinations to be employed. These can also be seen in a variety of application from automotive seals to bras [30].

A number of molding methods can be employed to produce multi-shot injection moldings but those produced from the same materials in multi-color multi-shot enjoy the highest market share. Multi-shot techniques produce both multi-color and multi-materials moldings. Some applications for multi-shot injection molding are listed in Table 4.3.

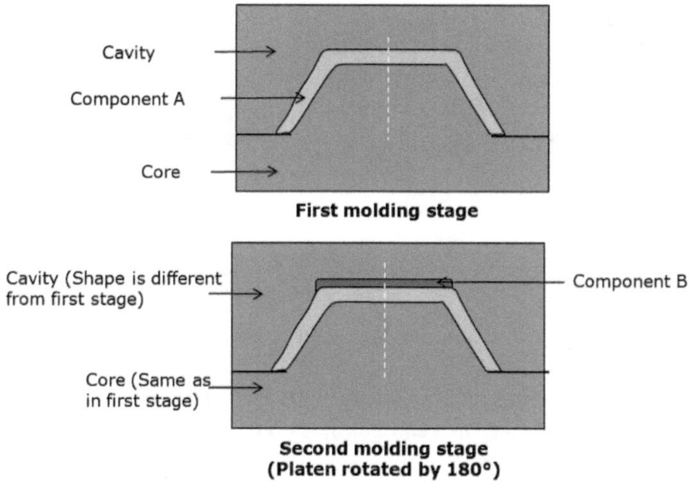

Fig. 4.16 Two different stages in rotary platen multi-shot molding process

Table 4.3 Applications for multi-shot moldings

Materials combination	Properties	Application
Same materials/different colors	Aesthetics	Buttons, mobile phone casings [31], toothbrushes, automotive light casings [32]
Soft feel/hard feel [33]	Increased customer appeal through 'soft touch' properties	Various handles, toothbrushes [34], cameras [35], screwdrivers, kettle handles [36], vacuum cleaners, lawn mowers and electric toothbrushes [34]
Transparent/colored	Viewing panel incorporated into molding	Light casings [32], battery casings [37]
Very flexible TPE, hard substrate	Sealing properties	Lids with mold on seals
Liquid silicone rubber, hard substrate	Sealing properties	Seals

All of the multi-shot injection molding processes requires specialized equipment not found in standard single-materials injection molding. In terms of the mold side of the equipment, molding presses designed specifically for each multi-shot injection molding process must be used to produce multi-shot components. Multi-shot molding also requires careful control of the mold temperature at all times so that any moving or rotating components can properly operate. For example, if a brass slide or core lifter is incorporated into a steel mold, the temperatures have to be controlled in order that the slide/lifter will not lock up or jam owing to different coefficients of thermal expansion between the two metals.

Multi-shot injection molding has the advantage of the cost efficiency owing to the elimination of a process, the improved craftsmanship (clean and smooth transition of materials), the improved durability over two piece designs, the complexity reduction by integrating seals, the commercially available mold-in-color elastomers, and high gloss or textured seals available for styling.

4.2.3 Over (Insert) Injection Molding

Over injection molding is a process, which uses multiple molds to exhibit a multi-material component. Intrinsically, the first materials are injected into a mold through standard single materials molding techniques and then moved to a different mold where the second materials can be injected to combine with the first material. This has been schematically shown in Fig. 4.17.

Over injection molding is usually applied to combine various colors of the same or different materials in one molding, without supplementary operations like assembly, bonding, or welding [38–40]. A typical molding obtained by this process is a typewriter key or multicolor car taillight. Figure 4.18 provides a multicolor process injection molding by over injection molding technique.

Also, this technique is a molding process by a resin around a previously made injection-molded plastic part. In particular, the only difference compared with other technique is that a preform, either metal or plastic, is placed into the mold before the resin is injected. Over injection molding is related to manufacture a preform, sequentially transferring it to another mold after it has at least fractionally cured. And then, the second materials are shot into the new mold with the inside of preform. The preform is usually made in-house on a separate injection unit. And, it

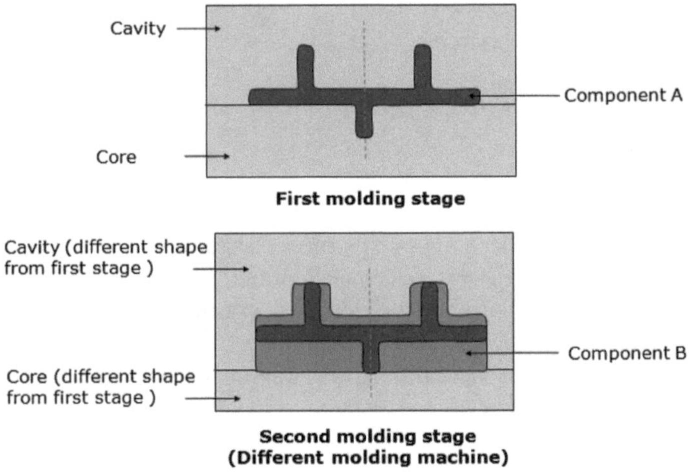

Fig. 4.17 Two different stages in over-molding process

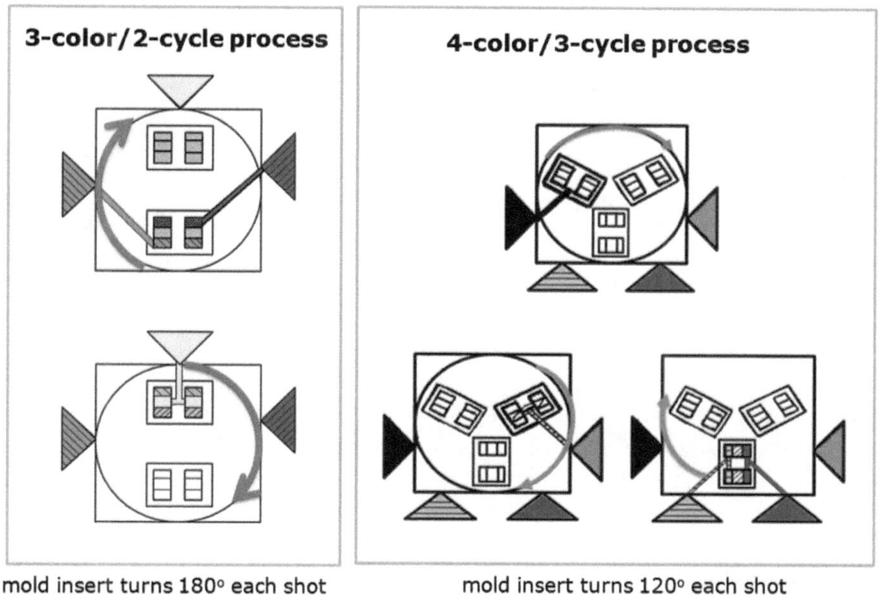

Fig. 4.18 Multi-color process over injection molding

is known that, once the resin is injected into the mold, it flows over, under, and around the preform and hardens, locking the preform inside of it.

And its advantage is that it is highly integrated. Meanwhile, a disadvantage is the necessity for a more complicated mold and a unique injection molding machine, which must have at least two plasticating units and a special control system.

The insert injection molding is really just a special modified technique of over injection molding. The insert injection molding with plastics is a two-step process. The first preform component is placed into the open mold cavity. And then, the injection proceeds as with traditional injection molding methods with injection of a molten plastic onto the preform [41]. This is presented in Fig. 4.18. The preform for insert injection molding is made for a while before the injection cycle. This means that the preform can be outsourced to a supplier. The preform is usually composed with metal, and also can be made by any standard injection molding or metal-forming or machining process. Generally, the preform is knurled, roughened, or treated with an adhesive in order to the better interaction between the resin and its surface. In addition, mechanical interlocking is often utilized to hold the preforms to the injected resin.

This process is not confined to two materials components and the obtained molding can be transferred in this way until the required number of layers is accomplished. Inserts can be performed by hand (as shown in Fig. 4.19) or by the use of robots. It must be accurate in not only their dimension but also their placement into the over molding tool in order to prevent tool damage and provide accurate registration of one materials on another.

Fig. 4.19 Insert molding

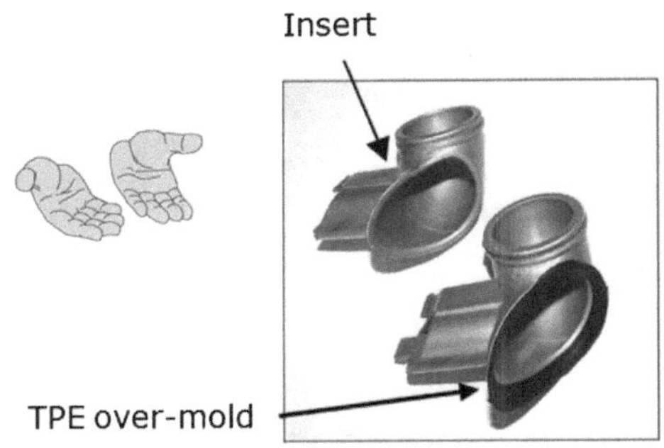

The over (insert) molding is that the perform has to be inserted, either manually or robotically, into the mold before each shot, which is only real difference in comparison with single materials injection molding. There are not required the special equipment requirements on the mold or clamp side of the machine, due to the demand of the only one injection unit per molding machine. Over molding is one process capable of producing a variety of products, including the in-mold assemblies. In many cases, over molding is applied to provide soft-touch user-interface sections on rigid product housings.

Generally, many typical products are manufactured by the over molding technique; for example, power tools, hand tools, telephone, pagers, computer peripheral devices, small appliances, writing instruments, personal care products (e.g., hairbrushes and toothbrushes), and toys. Additionally, over molding is used extensively for producing electronic cabling. Figure 4.20 shows some typical components manufactured using over molding. Meanwhile, typical insert molding applications are shown in Fig. 4.21.

As mentioned above, each of the three classes of multi-materials injection molding is considerably different and unique. Each specific multi-materials injection molding process is positively necessary its own sequences of specialized equipment of machine. But, there are certain equipment requirements that are customarily similar to all types of multi-materials injection molding.

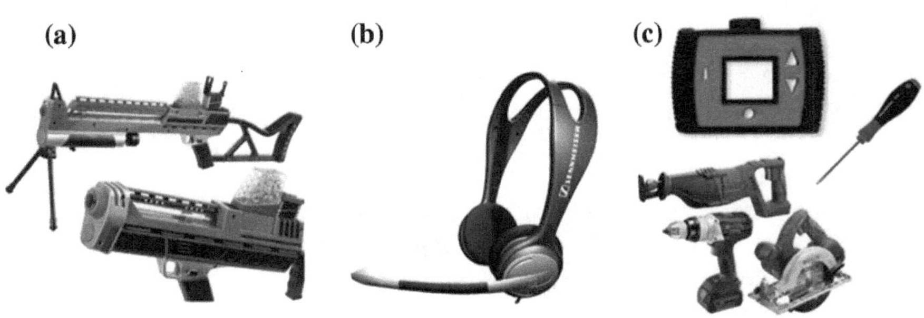

Fig. 4.20 Some typical components manufactured using over molding. **a** Paint gun. **b** Head set. **c** Various appliances' housing

(a) (b)

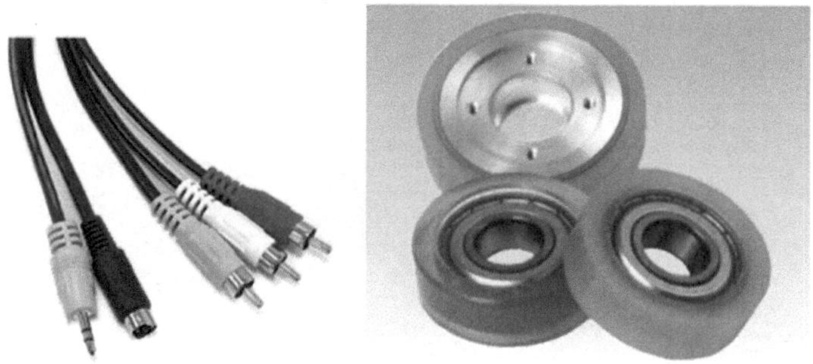

Fig. 4.21 Some typical components manufactured using insert molding. **a** Digital cables. **b** Urethane rollers

4.2.4 *Others*

Recent, based on the same principle above mentioned, Yang et al. [42] proposes the multi-fluid multi-injection molding. It main feature is that the mold cavity is partially filled with the polymer melt that is used as the first fluid, and then the second fluid penetrates the first one and drives it to fill out the entire cavity. A schematic representation of multi-fluid multi-injection molding is shown in Fig. 4.22.

Multi-fluid multi-injection molding, an innovative molding method, has been set up to allow multiple resins to be injected sequentially into one mold to make a single part with multi-layered structure. The practical design in industry of multi-fluid multi-injection molding is known as co-injection molding, or sandwich injection molding, whose feature is that the polymer melts in two different injection units are injected sequentially into the cavity to form an encapsulated sandwich-structure product. To be specific, one polymer melt forming the skin layer is injected first and then the other melt injected afterwards will penetrate the former to be the core layer. Economically, recycled plastics can be used as the core layer while a high quality virgin materials form the exterior regions of the parts. At the same time, it offers the advantage of combining two (or more) material properties in

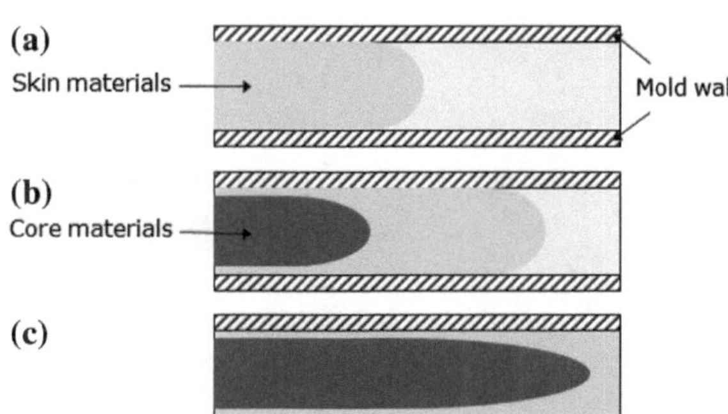

Fig. 4.22 Schematic representation of multi-fluide multi injection molding: **a** skin materials injection, **b** core materials injection, and **c** melt packing stage

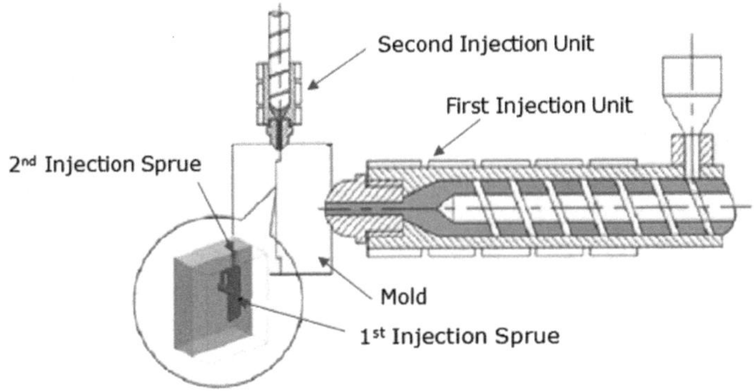

Fig. 4.23 Schematic principle of multi-fluid multi-injection molding setup utilized from Yang's group [42]

specific applications [43], such as, using the parts with unfilled skin/core with conductive filler in electromagnetic interference shielding for computer housings. Figure 4.23 illustrated a schematic representation of the multi-fluid multi-injection molding apparatus used in Yang's study [42].

Meanwhile, one typical kind of multi-fluid multi-injection molding is gas-assisted injection molding where gas serves as a second penetration fluid.

As in the two-component injection process, the gas-assisted injection molding process includes two phases of filling (Fig. 4.24). First the cavity is filled with polymer to a certain point. Instead of a second polymer being pushed into the cavity, a gas, usually nitrogen, is then pushed into the cavity. The polymer is pushed farther forward to complete the filling [14, 44–48]. The polymer is pressed against the mold wall and solidifies immediately. The gas advances through the core layer, forming a hollow part. The gas also transfers holding pressure to the molding. An advantage is that there is almost no pressure drop within the gas

Fig. 4.24 Gas-assisted injection molding

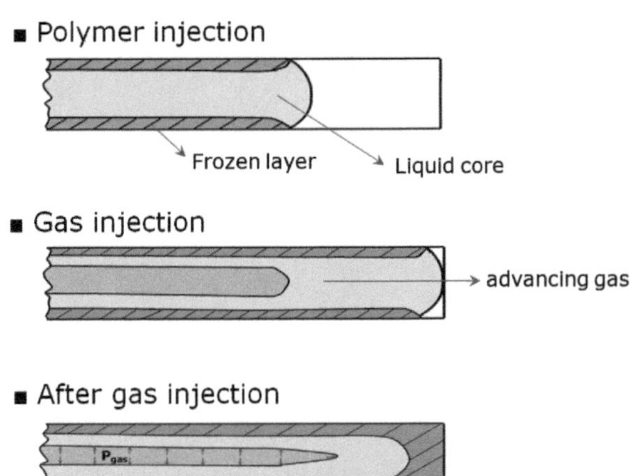

channel, and the holding pressure acts uniformly all over the molding. Moreover, there is no sealing time (as in conventional injection molding) during which holding pressure is not effective.

The gas-assisted injection molding process has several attractive features; (1) The presence of the gas channels leads to a reduction of materials and weight, (2) Parts of the same weight can be stiffer, because of the greater distance of the materials from the neutral axis, (3) Cycle times can be shorter because there is no hot melt in the core to be cooled, (4) Thick-walled parts without sink marks and air bubbles can be produced economically, and (5) A lower clamping force is needed because the gas pressure usually is lower than the holding pressure in an equivalent conventional injection process [49].

A gas injection unit is necessary in addition to an injection molding machine (Fig. 4.25). Nitrogen from a conventional tank fills a pressure transformer, which may build up a pressure of up to 400 bar. Then the gas can be injected through the machine nozzle or, if a gas injection needle is used, directly into the cavity [50].

The process also has disadvantages; (1) The machinery costs more because of the gas injection unit, (2) Gas channels must be taken into account during part design, and (3) There may be surface marks on the molding, because the polymer flow stops when the process is switched to gas injection.

The surface marks can be avoided if a modified gas injection process is used [51]. In this case the cavity is filled completely by the polymer, and then a gas needle in the mold pushes the still flowable materials in the core section back into the plasticating unit. In this process variation, not only are surface marks avoided, but also the polymer wall thickness can be varied within wide limits (Fig. 4.26).

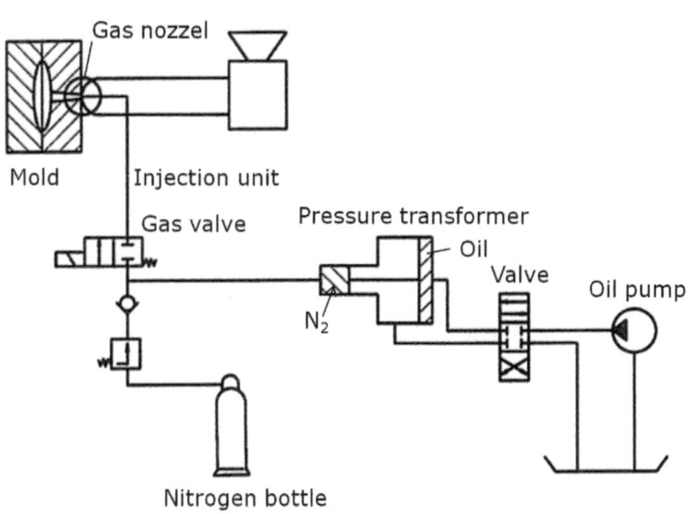

Fig. 4.25 Machine configuration for gas-assisted injection molding

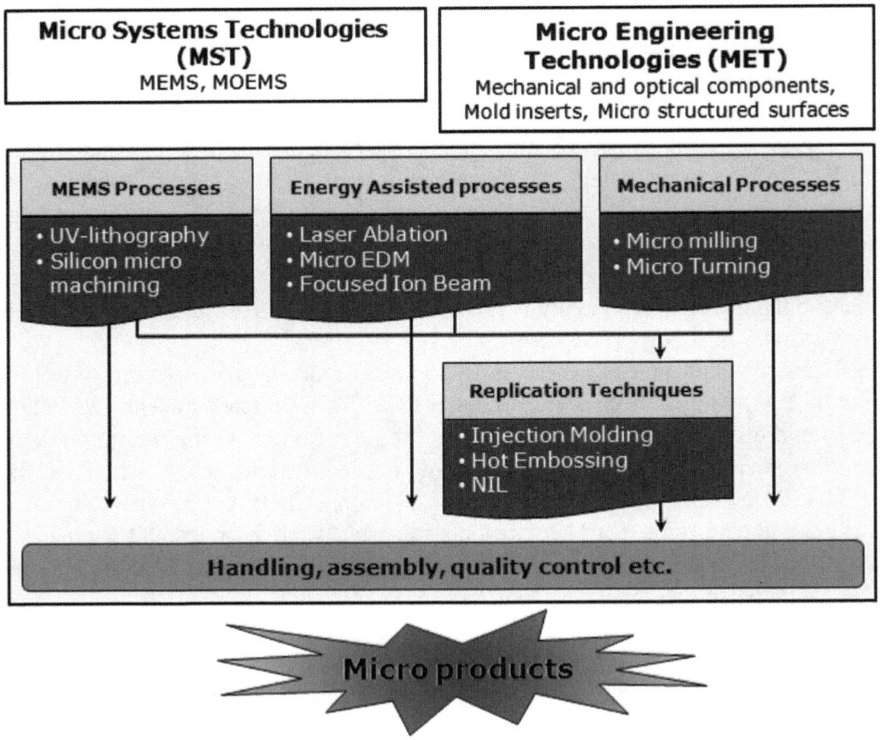

Fig. 4.26 Process technologies for micro-manufacturing [53]

4.3 Prospects on Multi-materials Injection Molding

4.3.1 Micro-powder Injection Molding

Miniaturization is one of the most important technological trends, which will gain further importance in the future. It offers the decisive advantage of a higher functional integration on smaller room and thus the possibility to develop new product concepts. This opens a wide range of new application fields associated with handy, portable products [52].

In response to the required performance, micro-powder injection molding is a fast-developing micro-manufacturing technique for the production of metal and ceramic components. Shape complexity, dimensional accuracy, replication fidelity, materials variety combined with high-volume capabilities are some of the key advantages of the technology [54]. In this field, 'micro-components', i.e., components have to make belong to one or more of the following classes [54–56]; micro-parts (parts with a maximum size below 10 mm and features in the micron rage [57]), microstructured parts (parts with dimensions between several millimetres and several centimetres with three-dimensional microstructures located on one

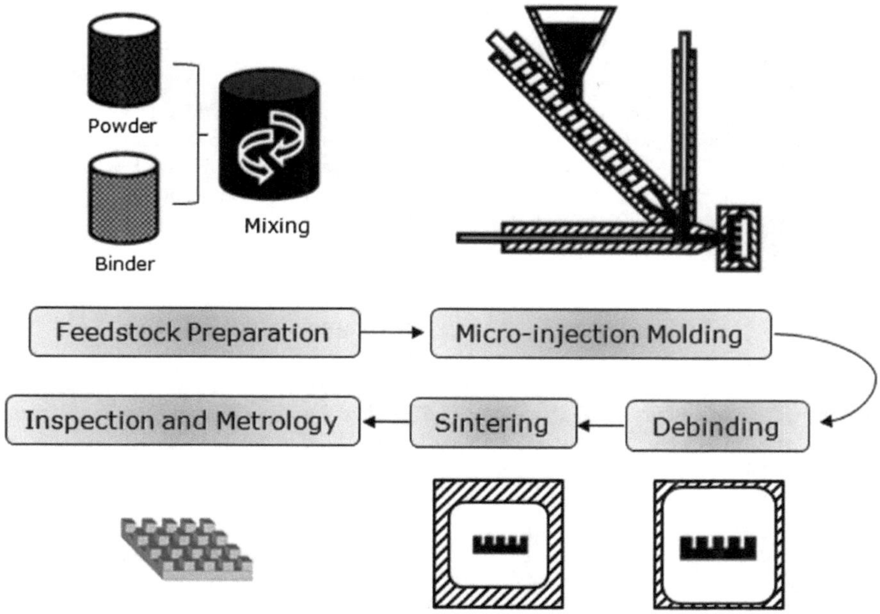

Fig. 4.27 Process for micro-powder injection molding

or more surface areas [57, 58], and micro-precision parts (parts of unlimited size, but with tolerances in the micron range or smaller [59]).

As shown in Fig. 4.27, micro-powder injection molding generally follows a specific process chain in which a metal of ceramic powder is mixed with a binder, the mix molded, the binder removed and the powder then sintered [54].

Micro-powder injection molding is done where the state of the art in these areas. This is predominantly the case for micro-power injection molding as a near-net-shape manufacturing technique and as a technique based on polymer flow. Thus micro-powder injection molding has already been used in multiple industrial application areas. These can be classified into five main groupings, as shown in Table 4.4 [54, 60].

As can be observed from the table, industrial application areas cover all the categories of the micro-components, i.e., micro-parts, micro-structured components, and micro-precision components, which indicate the wide capabilities of the micro-power injection molding. Table 4.4 shows also that typical application of micro-powder injection molding components focus on niche areas were other micro-fabrication processes are not feasible in terms of functional requirements and cost.

Reports suggest that micro-powder injection molding accesses about 10% of the micro-system market, because of its competitive costs and ability to form complex shapes [57]. But, considering that materials consumption for micro-components is typically small, a case for investment in new feedstock production requires that low volumes of feedstock sales can return that investment.

Table 4.4 Application area of micro-powder injection molding: applications that are actually on the market are marked with asterisks [54]

Application	Examples
Micro-mechanical components: custom-designed micromechanical components or replacements for plastic parts to make use of properties of metals or ceramics (such as mechanical strength, corrosion resistance or high temperature performance)	Micro-gearbox and stepped-gear wheel structures [61–65]. Micromold inserts [66–69]. Electrodes for micro-EDM [66]. Divertors for power applications [70]. *Micro-components for cameras [60], e.g. by Taisei Kogyo (www.taisei-kogyo.com) in Japan. *Clock parts, such as operating cams and weights [71] produced by e.g. Citizen (www.citizenwatch.com). *Printer components [71]
Micro-system technologies: this market segment is dominated by silicon technologies, but there seems to be a portion of this market available for micro-powder injection molding products as well	Nozzle structures using ceramic materials [59, 62]. Micro-optical benches for communication systems [59, 62]. Opto-electronic systems [57]. Data-storage devices [57] Communication systems [57]. *Automotive systems [57], such as airbag components [71]
Micro-fluidics and micro-reaction technologies	High-temperature gas phase reactors or reactors for highly corrosive media [60]. Microfluidic devices for chemical and medical applications, such as micro-mixers and micro-heat exchangers [61, 66, 72–74]
Medical technology: there is request for biocompatible materials (such as ceramics, stainless steel or titanium and reliable). Manufacturing technologies to produce complex shaped medical components.	Small replacement bones [57, 60]. *Micro-needles for selective nerve simulation [63, 64]. *Dental instruments and medical biopsy tools [57, 75, 76]. Minimally invasive surgical tools [57, 72]
Biosensors	The micro-structured surfaces of micro-power injection molding components are functionalized to react with specific biomolecules [60]

In addition, micro-powder injection molding has a lot of advantages as a micro-manufacturing technique; (a) Shape complexity (3D features relative to 2½D features typical to silicon techniques) [57] and accurate replication of microstructures [61], (b) Net-shape or near net-shape forming. PIM in general offers high material utilization, where probably 97% of the powder is delivered in final components [77–80], (c) micro-powder injection molding is applicable to several material functional classes. These include materials for magnetic, mechanical or electric properties, as well as for high wear resistance [55, 65, 81–85], (d) micro-powder injection molding increases the range of materials that can be used for microsystem applications, in comparison with other techniques, such as micro-cutting, laser ablation, silicon etching or LIGA (Lithography, Electroplating, and Molding) technique (Fig. 4.28), which are often limited in respect to processible materials and/or to series production [59, 86], (e) μPIM is cost-effective for mass production relative to other micro-manufacturing techniques, such as

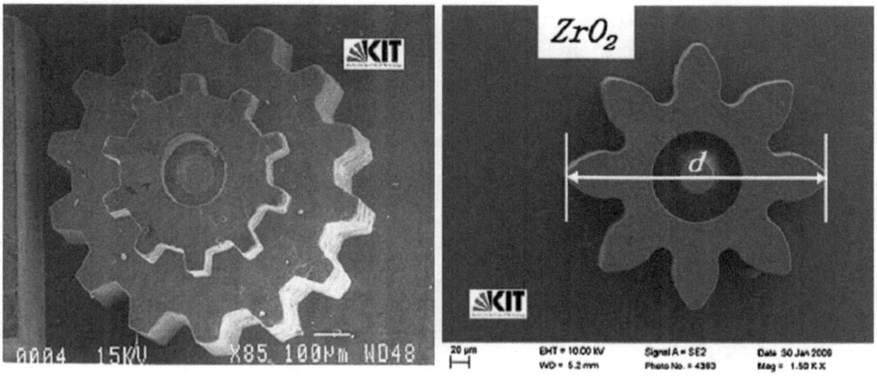

Fig. 4.28 Two ceramic micro gear wheels injection-molded using LIGA (lithography, electroplating, and molding) mold inserts, state of the art 1999 (*left*) and nowadays (*right*) [79]

microcutting, laser ablation, silicon etching and LIGA [60, 61, 64, 66], (f) For small series and prototyping, low-pressure powder injection molding can be implemented, where pressure values are limited to 0.1-1 MPa, and low-viscosity paraffin or wax is used instead of a high-viscosity polymeric binder [87–89], and (g) A considerable amount of know-how is already available from established PIM techniques [58, 81].

Therefore, it is expected that the capabilities of micro-powder injection molding by investigating variant processes would be important for meeting increasing demands of the Microsystems market in terms of functional integration and structural complexity. To searching new micro-system applications for micro-powder injection molding would create a market demand for the technology helping it to evolve into a well-established, high-volume micro-manufacturing method.

4.4 Summary

In this chapter, we have presented that a variety of multi-materials injection molding and its variations exist. It covers the basics of injection molding technique, multi-materials injection molding, their some application, and vision in the industry.

For the first part an overall view about the basics of injection molding and multi-materials injection molding was presented, which include their injection process and mold design, thereby quality control of multi-materials injection molding.

The second part exhibits the classification of multi-materials injection molding. These have been split into four categories in this chapter; multi-component, multi-shot, over (insert), and others molding technique. Each category is departmentalized in detail in the each section, which is described the definition, process, application, and so on.

In the last part, one of the most vital versatile researches, micro-power injection molding technique is presented. It is likely considered that in the future, commercially successful multi-materials injection molding technique will require the abundant understanding and adaptability to utilize a wide range of these process options. It is also likely that developments will continue to extend beyond the common material that is required to create the new one or hybrid multi-materials of many different materials including plastics, metals, ceramics, and so on.

References

1. Han KH, Im YT (1997) Compressible flow analysis of filling and postfilling injection molding with phase-change effect. Compo Struct 38:179
2. Holm EJ, Petter LH (1999) A unified finite element method model for the injection molding process. Comput Met Appl Mech Eng 178:413
3. Kumar A, Ghoshdastidar PS, Muju MK (2002) Computer simulation of transport process during injection mold-filling and optimization of the molding conditions. J Mater Proc Tech 120:438
4. Courbebaisse G (2005) Numerical simulation of injection molding process and the pre-modeling concept. Comput Mater Sci 34:397
5. Galantucci LM, Spina R (2003) Evaluation of filling conditions of injection molding by integrating numerical simulations and experimental tests. J Mater Process Tech 141:266
6. Hill D (1996) Further studies of the injection molding process. Appl. Math Model 20:719
7. Banerjee AG, Li X, Fowler G, Gupta SK (2007) Incorporating manufacturability considerations during design of injection molded multi-material objects. Res Eng Design 17:207
8. Gouker RM, Gupta SK, Bruck HA, Holzschuh T (2006) Manufacturing of multi-material compliant mechanisms using multi-material molding. Int J Adv Manuf Technol 30:1049
9. Wang KK, Zhou J, Sakurai Y (1999) In ANTEC'99 in plastics engineering, p 611
10. Wang KK, Hieber CA, Wang N, Zhou J, Lee MC, Sakurai Y (200) In Cornell injection molding program (CIMP) progress report, No. 21. Cornell University, Ithaca, New York, Ch. XI
11. Chen Z, Turng LS (2005) Review of current developments in process and quality control for injection molding. Adv Polym Technol 24:165
12. Eckardt H (1985) Mehrkomponenten-Spritzgießen ermöglicht das Herstellen abgeschirmter Gehäusemit gutter Oberfläche in einem Arbeitsgang. Kunststoffe 75:145
13. Johannaber F (1985) Spritzgießen-Neue Prozeßvarianten. Kunststoffe 75:560
14. Pötsch G, Michaeli W (1995) Injection molding: an introduction. Hanser/Gardner Publications, Inc.
15. Alcock J Stephenson D (1996) The powder injection moulding process. Materials World
16. Young SS, White JL, Clar ES, Oyanagy Y (1980) A basic experimental study of sandwich injection molding with sequential injection. Polym Eng Sic 20:798
17. Garcia-Etxabe R, Campo JJ, Manchado JC, Clemente R (2002) Application of co-injection technology to handles for the gear lever. In: Injection moulding. Barcelona, Spain, pp 18–19
18. Eckardt H (1987) Production of high-quality mouldings from recycled materials by means of the multi-component injection moulding process. Plaste Kautsch 34:267
19. Smith M, Valentage R (2000) Coinjection moulded to fascia production. In: SPE automotive TPO global conference, p 275
20. Toensmeier PA, Hilpold L (1999) New coinjection designs add applications potential. Mod Plastics Int 29:77

21. Love JC, Smith GF, Pharoah MW (2002) In mould painting using granular injected paint technology. In: Injection moulding. Barcelona, Spain
22. Atkinson P, Bagdatlioglu I (1994) Foam sandwich for automotive body panels. In: International polypropylene conference. London
23. Fowler GT (2004) Cost and performance evaluation models for comparing multishot and traditional injection molding, Master of Science's thesis in the Faculty of the Graduate School of the University of Maryland
24. Arburg (1999) Technical information. Multi component injection moulding, 522771-GB-08
25. Heim HP, Haber H (2003) Specialized molding techniques: application, design, materials and processing. Plast Des Libr
26. Crawford RJ (1998) Plastic engineering, 3rd ed.. Butterworth-Heinemann
27. Baird DG, Collias DI (1998) Polymer processing: principles and design. Wiley-Interscience Publication
28. Park HS, Anh DBH (2010) A new approach for molding multi component parts. Int J Precision Eng Manuf 11:291
29. Li X, Gupta SK (2003) Manufacturability analysis of multi-material objects molded by rotary platen multi-shot molding process, In: International mechanical engineering congress exposure (IMECE2003). Washington DC, USA
30. Higgs R (2000) Plastics seeking to usurp the underwire. Plast News (USA) 40:9
31. Smith C (2000) Turning a profit. Plast Rubber Wkly
32. Sonnery (1992) Multi-material moulding opens doors to design and development. Br Plast Rubber 3:4
33. Ford Motor Co. (2000) Ford reduces costs with eight santoprene tpe parts. Mod Plast Int (USA) 30:19
34. Vink D (2000) Electrifying productivity. Euro Plast News 27:49
35. Maniscalco M (2001) Overmoulding winner merges design and process. Injection Molding 9:52
36. Sifam (1996) Making multi-shot work. Plast Rubber Wkly 1625:21
37. Making it with multi-shot (1995) Sifam European community; European unionUK; Western Europe. Plast Rubber Wkly 1583:13
38. Thoma H (1988) Rationalisieren durch Mehrkomponenten-Spritzgieben. Kunststoffe 78:665
39. Reker H, Ullmann R (1989) Konstruktion and fertigung eines rasierergehäuses in "Hart-Weich-Technik". Kunststoffe 79:164
40. Schultheis SM (1991) Ventildeckel mit dichtung in einem arbeitsgang herstellen. Kunststoffe 81:876
41. Arning M (2000) Two-shot injection moulding high performance and conventional elastomers. In: High performance elastomers
42. Zhang K, Liu Z, Yang B, Yang W, Lu Y, Wang L, Sun N, Yang M (2011) Cylindritic structures of high-density polyethylene molded by multi-melt multi-injection molding. Polymer 52:3871
43. Seldén RJ (1997) J. Injec. Mold. Techno. 1:189
44. Meridies R (1981) Verfahren zum herstellen von sandwichspritzgußteilen mit einum gas als kernkomoponente. Kunststoffe 71:420
45. Klamm M, Feldmann F (1988) Gasinnendruckverfahren beim spritzgießen. Kunststoffe 78:767
46. Klotz B, Bürkle E (1989) Neue mőglichkeiten beim spritzgießen durch das gasinnendruckverfahren. Kunststkffe 79:1102
47. Pearson T (1986) Fomteilherstellung nach dem cinpres-verfahren. Kunststoffe 76:667
48. Leyrer KH (1990) Spritzgießen mit dem gasinnendruckprozeß. Plastverarbeiter 41:39
49. Eyerer P, Bürkle E (1991) Spritzgießen mit reduzierter schließkraft. Kunststoffe 81:851
50. Anon Durch (1990) Injektortechnik unabhängig vom schmelzanguß. Plastverarbeiter 41:34
51. Jaroschek C (1990) Gasinnendruck zum ausblasen überschüssiger schmelze. Kunststoffe 41:34

52. Fleischer J, Kieckmann AM (2006) Automation of the powder injection molding process. Microsyst Technol 12:702
53. Brousseau EB, Dimov SS, Pham DT (2010) Some recent advances in multi-material micro- and nano-manufacturing. Int J Adv Manuf Technol 47:161
54. Attia UM, Alcock JR (2011) A review of micro-powder injection moulding as a microfabrication technique. J Micromesh Microeng 21:043001
55. Tay BY, Liu L, Loh NH, Tor SB, Murakoshi Y, Maeda R (2005) Surface roughness of microstructured component fabricated by μMIM. Mater Sci Eng A 396:311
56. Attia UM, Marson S, Alcock JR (2009) Micro-injection moulding of polymer microfluidic devices. Microfluid Nanofluid 7:1
57. German RM (2009) Medical and dental applications for microminiature powder injection moulding (microPIM)-a roadmap for growth. PIM Int. 3:21
58. Liu L, Loh NH, Tay BY, Tor SB, Murakoshi Y, Maeda R (2007) Effects of thermal debinding on surface roughness in micro powder injection molding. Mater Lett 61:809
59. Ruprecht R, Gietzelt T, Műller K, Piotter V, Haußelt J (2002) Injection molding of microstructured components from plastics, metals and ceramics. Microsyst Technol 8:351
60. Petzoldt F (2008) Micro powder injection moulding-challenges and opportunities. PIM Int 2:37
61. Yin H, Jia C, Qu X (2008) Micro powder injection molding-large scale production technology for micro-sized components. Sci China Ser E: Technol Sci 51:121
62. Piotter V, Gietzelt T, Plewa K, Ruprecht R, Hausselt J (2001) Tiny parts made by micro powder injection molding. Proc Adv Powder Metall Part Mater 13–17:728
63. Li SG, Fu G, Reading I, Tor SB, Loh NH, Chaturvedi P, Yoon SF, Youcef-Toumi K (2007) Dimensional variation in production of high-aspect-ratio micro-pillars array by micro powder injection molding. Appl Phys A 89:721
64. Fu G, Loh NH, Tor SB, Tay BY, Murakoshi Y, Maeda R (2005) Injection molding, debinding and sintering of 316L stainless steel microstructures. Appl Phys A 81:495
65. Loh NH, Tor SB, Tay BY, Murakoshi Y, Maeda R (2008) Fabrication of micro gear by micro powder injection molding. Microsyst Technol 14:43
66. Rota A (2002) New features in material issues for metallic micro components by MIM. In: Proceedings of PM2TEC, 10/49-/57
67. Fu G, Tor SB, Loh NH, Hardt DE (2010) Fabrication of robust tooling for mass production of polymeric microfluidic devices. J Micromech Microeng 20:085019
68. Rota A, Duong TV, Hartwig T (2002) Wear resistant tools for reproduction technologies produced by micro powder metallurgy. Microsyst Technol 7:225
69. Gietzelt T, Piotter V, Jacobi O, Ruprecht R, Hausselt J (2003) Fabrication of micromolds for gearwheels by micro powder injection molding. Adv Eng Mater 5:139
70. Zeep B, Norajitra P, Piotter V, Boehm J, Ruprecht R, Hausselt J (2007) Net shaping of tungsten components by micro powder injection moulding. Fusion Eng Des 82:2660
71. Yoshikawa K, Ohmori H (2001) Outstanding features of powder injection molding for micro parts manufacturing. RIKEN Rev 34:13
72. Laddha SG, Wu C, Park SJ, Lee S, Ahn S, German RM, Atre SV (2010) Characterization and simulation of macroscale mold-filling defects in microminiature powder injection molding. Int J Powder Metall 46:49
73. Wang MW (2010) Microceramic injection molding of a multilayer micropatterned micropart. Int J Adv Manuf Technol 51:145
74. Meng J, Loh NH, Fu G, Tor SB, Tay BY (2010) Replication and characterization of 316L stainless steel micro-mixer by micro powder injection molding. J Alloy Compd 496:293
75. German RM (2010) Materials for microminiature powder injection molded medical and dental devices. Int J Powder Metall 46:15
76. Thomas P, Levenfeld B, Várez A, Cervera A (2009) Production of alumina microparts by powder injection molding. Int J Appl Ceram Technol 8:617
77. German RM (1999) Rationalization of injection moulding of stainless steel based on component features. Met Powder Rep 54:38

78. Rota A, Duong TV, Hartwig T (2002) Micro powder metallurgy for the replicative production of metallic microstructures. Microsyst Technol 8:323
79. Okubo K, Tanaka S, Ito H (2009) The effects of metal particle size and distributions on dimensional accuracy for micro parts in micro metal injection molding. In: Proceedings of annual technical conference (ANTEC)
80. Piotter V, Beck M, Plewa K, Ritzhaupt-Kleissl H, Ruh A, Hausselt J (2009) Micro PIM moves into the zone of industrial possibility. Met Powder Rep 64:35
81. Piotter V (1999) PIM looks for role in the micro world. Met Powder Rep 54:36
82. Urval R, Lee S, Atre SV, Park SJ, German RM (2008) Optimisation of process conditions in powder injection moulding of microsystem components using a robust design method: part I. Primary design parameters. Powder Metall 51:133
83. Mohamad NNH, Muhamad N, Ismail MH, Jamaludin KR, Ahmad S, Ibrahim MHI (2009) Flow behaviour to determine the defects of green part in metal injection molding. Int J Mech Mater Eng 4:70
84. Zauner R (2006) Micro powder injection moulding. Microelectron Eng 83:1442
85. Piotter V, Finnah K, Plewa R, Ruprecht R, Hausselt J (2005) High-resistive micro components produced by high-pressure powder injection moulding. In: Proceedings of 4 m international conference. Elsevier, Amsterdam
86. Ruprecht R (1997) Various replication techniques for manufacturing three-dimensional metal microstructures. Microsyst Technol 4:28
87. Rak ZS (1999) New trends in powder injection moulding. Powder Metall Met Ceram 38:126
88. Piotter V, Hanemann T, Heldele R, Mueller M, Mueller T, Plewa K, Ruh A (2010) Metal and ceramic parts fabricated by microminiature powder injection molding. Int J Powder Metall 46:21
89. Piotter V, Bauer W, Knitter R, Mueller M, Mueller T, Plewa K (2011) Powder injection moulding of metallic and ceramic micro parts. Microsyst Technol 17:251

Chapter 5
Extrusion of Multicomponent Product

Aloke Kumar Ghosh

Abstract Extrusion is a process used to create objects of a fixed cross-sectional profile. A material is pushed through a die of the desired cross-section. The two main advantages of this process over other manufacturing processes are its ability to create very complex cross-sections, and to work materials that are brittle, because the material only encounters compressive and shear stresses.

5.1 Introduction

Polymer are extruded through dies with diversified morphological shapes and design to obtain both monolayer and multilayer products for applications in packaging materials, medical devices, and modern engineered structures. Advanced coextrusion in which two or more polymeric matrices are extruded to fabricate complex multilayer products in a single structure. Such technique allows end-users to manipulate and combine the several desirable properties of multiple polymeric matrices into one structure with improved performance. Multilayer sheet, blown films, cast film, wire, tube like structure are extensively manufactured by extrusion process. This chapter will present the details of technical aspect of extrusion includes materials design, rheological parameters with subsequent case studies.

5.2 Extrusion

Extrusion is a process used to create objects of a fixed cross-sectional profile. A material is pushed through a die of the desired cross-section. The two main advantages of this process over other manufacturing processes are its ability to

A.K. Ghosh (✉)
Department of Inorganic and Physical Chemistry, Indian Institute of Science, C V Raman Ave., Bengaluru 560012, Karnataka, India
e-mail: akghosh1982@gmail.com

J.K. Kim et al. (eds.), *Multicomponent Polymeric Materials*,
Springer Series in Materials Science 223, DOI 10.1007/978-94-017-7324-9_5

Fig. 5.1 Extruded aluminium with several hollow cavities; T slots allow bars to be joined with special connectors

create very complex cross-sections, and to work materials that are brittle, because the material only encounters compressive and shear stresses. It also forms parts with an excellent surface finish [1].

Extrusion may be continuous (theoretically producing indefinitely long material) or semi-continuous (producing many pieces). The extrusion process can be done with the material hot or cold. Commonly extruded materials include metals, polymers, ceramics, concrete, play dough, and foodstuffs. The products of extrusion are generally called "extrudates". Drawing metal is the main way to produce wire and sheet, and bar and tube are also often drawn.

Hollow cavities within extruded material cannot be produced using a simple flat extrusion die, because there would be no way to support the center barrier of the die. Instead, the die assumes the shape of a block with depth, beginning first with a shape profile that supports the center section. The die shape then internally changes along its length into the final shape, with the suspended center pieces supported from the back of the die. The extrusion process in metals may also increase the strength of the material (Fig. 5.1).

In 1797, Joseph Bramah patented the first extrusion process for making lead pipe. It involved preheating the metal and then forcing it through a die via a hand-driven plunger. The process was not further developed until 1820 when Thomas Burr constructed the first hydraulic powered press. At this time the process was called "squirting". In 1894, Alexander Dick expanded the extrusion process to copper and brass alloys [2].

5.3 Process and types of extrusion

The process begins by heating the stock material (for hot or warm extrusion). It is then loaded into the container in the press. A dummy block is placed behind it where the ram then presses on the material to push it out of the die. Afterward the extrusion is stretched in order to straighten it. If better properties are required then it may be heat treated or cold worked [2].

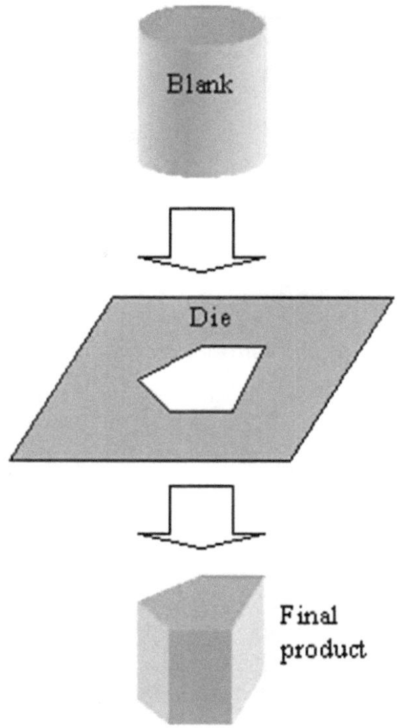

Fig. 5.2 Extrusion of a round blank through a die

The extrusion ratio is defined as the starting cross-sectional area divided by the cross-sectional area of the final extrusion. One of the main advantages of the extrusion process is that this ratio can be very large while still producing quality parts (Fig. 5.2).

5.3.1 Hot Extrusion

Hot extrusion is a hot working process, which means it is done above the material's recrystallization temperature to keep the material from work hardening and to make it easier to push the material through the die. Most hot extrusions are done on horizontal hydraulic presses that range from 230 to 11,000 metric tons (250–12,130 short tons). Pressures range from 30 to 700 MPa (4400–101,500 psi), therefore lubrication is required, which can be oil or graphite for lower temperature extrusions, or glass powder for higher temperature extrusions. The biggest disadvantage of this process is its cost for machinery and its upkeep [1] (Table 5.1).

The extrusion process is generally economical when producing between several kilograms (pounds) and many tons, depending on the material being extruded. There is a crossover point where roll forming becomes more economical. For instance, some steels become more economical to roll if producing more than 20,000 kg (50,000 lb) [2].

Table 5.1 Hot extrusion temperature for various metals [1]

Material	Temperature [°C (°F)]
Magnesium	350–450 (650–850)
Aluminium	350–500 (650–900)
Copper	600–1100 (1200–2000)
Steel	1200–1300 (2200–2400)
Titanium	700–1200 (1300–2100)
Nickel	1000–1200 (1900–2200)
Refractory alloys	Up to 2000 (4000)

- Aluminium hot extrusion die

Front side of a four family die. For reference, the die is 228 mm (9.0 in) in diameter.

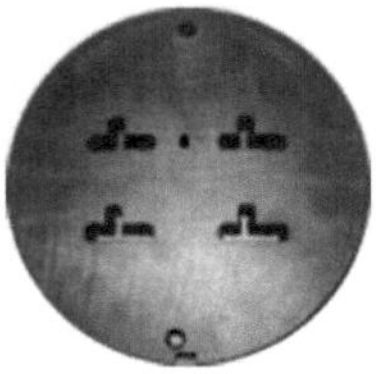

Close up of the shape cut into the die. Notice that the walls are drafted and that the back wall thickness varies.

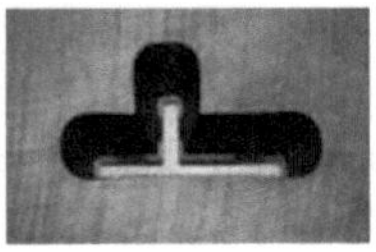

Back side of die. The wall thickness of the extrusion is 3 mm (0.12 in).

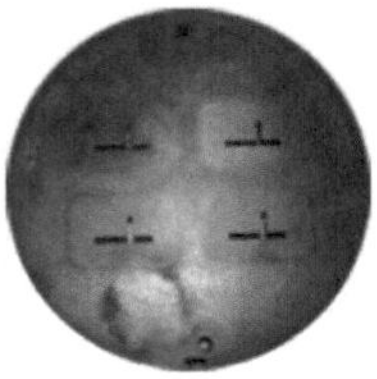

5.3.2 *Cold Extrusion*

Cold extrusion is done at room temperature or near room temperature. The advantages of this over hot extrusion are the lack of oxidation, higher strength due to cold working, closer tolerances, better surface finish, and fast extrusion speeds if the material is subject to hot shortness [1].

Materials that are commonly cold extruded include: lead, tin, aluminum, copper, zirconium, titanium, molybdenum, beryllium, vanadium, niobium, and steel. Examples of products produced by this process are: collapsible tubes, fire extinguisher cases, shock absorber cylinders and gear blanks.

5.3.3 Warm Extrusion

Warm extrusion is done above room temperature, but below the recrystallization temperature of the material the temperatures ranges from 800 to 1800 °F (424–975 °C). It is usually used to achieve the proper balance of required forces, ductility and final extrusion properties [3].

5.4 Extrusion Defects

Surface cracking occurs when the surface of an extrusion splits. This is often caused by the extrusion temperature, friction, or speed being too high. It can also happen at lower temperatures if the extruded product temporarily sticks to the die.

- Pipe—A flow pattern that draws the surface oxides and impurities to the center of the product. Such a pattern is often caused by high friction or cooling of the outer regions of the billet.
- Internal cracking—When the center of the extrusion develops cracks or voids. These cracks are attributed to a state of hydrostatic tensile stress at the centerline in the deformation zone in the die (a similar situation to the necked region in a tensile stress specimen).
- Surface lines—When there are lines visible on the surface of the extruded profile. This depends heavily on the quality of the die production and how well the die is maintained, as some residues of the material extruded can stick to the die surface and produce the embossed lines.

5.5 Equipment

There are many different variations of extrusion equipment. They vary by four major characteristics [1]:

1. Movement of the extrusion with relation to the ram. If the die is held stationary and the ram moves towards it then it is called "direct extrusion". If the ram is held stationary and the die moves towards the ram it is called "indirect extrusion".
2. The position of the press, either vertical or horizontal.

Fig. 5.3 A horizontal hydraulic press for hot aluminum extrusion (loose dies and scrap visible in foreground)

3. The type of drive, either hydraulic or mechanical.
4. The type of load applied, either conventional (variable) or hydrostatic.

A single or twin screw auger, powered by an electric motor, or a ram, driven by hydraulic pressure (often used for steel and titanium alloys), oil pressure (for aluminium), or in other specialized processes such as rollers inside a perforated drum for the production of many simultaneous streams of material.

Typical extrusion presses cost more than $100,000, whereas dies can cost up to $2000 (Fig. 5.3).

5.5.1 Forming Internal Cavities

Two-piece aluminum extrusion die set (parts shown separated). The male part (at right) is for forming the internal cavity in the resulting round tube extrusion.

There are several methods for forming internal cavities in extrusions. One way is to use a hollow billet and then use a fixed or floating mandrel. A fixed mandrel, also known as a German type, means it is integrated into the dummy block and stem. A floating mandrel, also known as a French type, floats in slots in the dummy block and aligns itself in the die when extruding. If a solid billet is used as the feed material then it must first be pierced by the mandrel before extruding through the die. A special press is used in order to control the mandrel independently from the ram [1]. The solid billet could also be used with a spider die, porthole die or bridge die. All of these types of dies incorporate the mandrel in the die and have “legs” that hold the mandrel in place. During extrusion the metal divides, flows around the legs, then merges, leaving weld lines in the final product [2] (Fig. 5.4).

Fig. 5.4 Two-piece aluminum extrusion die set (parts shown separated). The male part (at *right*) is for forming the internal cavity in the resulting round tube extrusion

5.5.2 Direct Extrusion

Direct extrusion, also known as forward extrusion, is the most common extrusion process. It works by placing the billet in a heavy walled container. The billet is pushed through the die by a ram or screw. There is a reusable dummy block between the ram and the billet to keep them separated. The major disadvantage of this process is that the force required to extrude the billet is greater than that needed in the indirect extrusion process because of the frictional forces introduced by the need for the billet to travel the entire length of the container. Because of this the greatest force required is at the beginning of process and slowly decreases as the billet is used up. At the end of the billet the force greatly increases because the billet is thin and the material must flow radially to exit the die. The end of the billet (called the butt end) is not used for this reason [2] (Fig. 5.5).

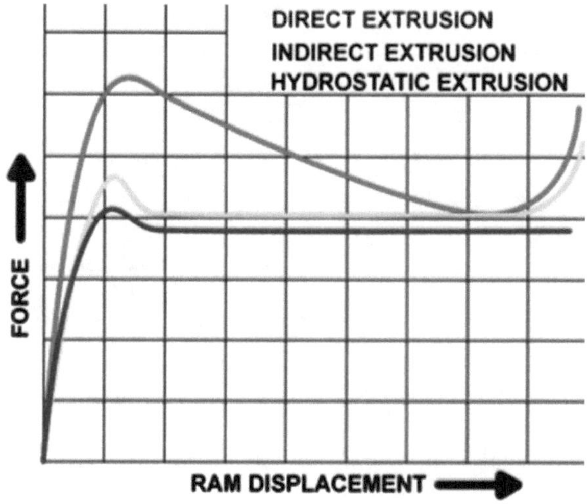

Fig. 5.5 Plot of forces required by various extrusion processes

5.5.3 Indirect Extrusion

In indirect extrusion, also known as backwards extrusion, the billet and container move together while the die is stationary. The die is held in place by a "stem" which has to be longer than the container length. The maximum length of the extrusion is ultimately dictated by the column strength of the stem. Because the billet moves with the container the frictional forces are eliminated. This leads to the following advantages [2]:

- A 25–30 % reduction of friction, which allows for extruding larger billets, increasing speed, and an increased ability to extrude smaller cross-sections.
- There is less of a tendency for extrusions to crack because there is no heat formed from friction.
- The container liner will last longer due to less wear.
- The billet is used more uniformly so extrusion defects and coarse grained peripherals zones are less likely.

The disadvantages are [2]:

- Impurities and defects on the surface of the billet affect the surface of the extrusion. These defects ruin the piece if it needs to be anodized or the aesthetics are important. In order to get around this the billets may be wire brushed, machined or chemically cleaned before being used.
- This process isn't as versatile as direct extrusions because the cross-sectional area is limited by the maximum size of the stem.

5.5.4 Hydrostatic Extrusion

In the hydrostatic extrusion process the billet is completely surrounded by a pressurized liquid, except where the billet contacts the die. This process can be done hot, warm, or cold, however the temperature is limited by the stability of the fluid used. The process must be carried out in a sealed cylinder to contain the hydrostatic medium. The fluid can be pressurized two ways [2]:

1. *Constant-rate extrusion*: A ram or plunger is used to pressurize the fluid inside the container.
2. *Constant-pressure extrusion*: A pump is used, possibly with a pressure intensifier, to pressurize the fluid, which is then pumped to the container.

The advantages of this process include [2]:

- No friction between the container and the billet reduces force requirements. This ultimately allows for faster speeds, higher reduction ratios, and lower billet temperatures.
- Usually the ductility of the material increases when high pressures are applied.

- An even flow of material.
- Large billets and large cross-sections can be extruded.
- No billet residue is left on the container walls.

The disadvantages are [2]:

- The billets must be prepared by tapering one end to match the die entry angle. This is needed to form a seal at the beginning of the cycle. Usually the entire billet needs to be machined to remove any surface defects.
- Containing the fluid under high pressures can be difficult.

5.5.5 Drives

Most modern direct or indirect extrusion presses are hydraulically driven, but there are some small mechanical presses still used. Of the hydraulic presses there are two types: direct-drive oil presses and accumulator water drives.

Direct-drive oil presses are the most common because they are reliable and robust. They can deliver over 35 MPa (5000 psi). They supply a constant pressure throughout the whole billet. The disadvantage is that they are slow, between 50 and 200 mm/s (2–8 ips) [2].

Accumulator water drives are more expensive and larger than direct-drive oil presses, and they lose about 10 % of their pressure over the stroke, but they are much faster, up to 380 mm/s (15 ips). Because of this they are used when extruding steel. They are also used on materials that must be heated to very hot temperatures for safety reasons [2].

Hydrostatic extrusion presses usually use castor oil at pressure up to 1400 MPa (200 ksi). Castor oil is used because it has good lubricity and high pressure properties [2].

5.5.6 Die Design

The design of an extrusion profile has a large impact on how readily it can be extruded. The maximum size for an extrusion is determined by finding the smallest circle that will fit around the cross-section, this is called the *circumscribing circle*. This diameter, in turn, controls the size of the die required, which ultimately determines if the part will fit in a given press. For example, a larger press can handle 60 cm (24 in) diameter circumscribing circles for aluminium and 55 cm (22 in). Diameter circles for steel and titanium [1].

The complexity of an extruded profile can be roughly quantified by calculating the *shape factor*, which is the amount of surface area generated per unit mass of extrusion. This affects the cost of tooling as well as the rate of production [4].

Table 5.2 Minimum cross-section and thickness for various materials

Material	Minimum cross-section [cm^2 (in^2)]	Minimum thickness [mm (in)]
Carbon steels	2.5 (0.40)	3.00 (0.120)
Stainless steel	3.0–4.5 (0.45–0.70)	3.00–4.75 (0.120–0.187)
Titanium	3.0 (0.50)	3.80 (0.150)
Aluminium	<2.5 (0.40)	1.00 (0.040)
Magnesium	<2.5 (0.40)	1.00 (0.040)

Thicker sections generally need an increased section size. In order for the material to flow properly legs should not be more than ten times longer than their thickness. If the cross-section is asymmetrical, adjacent sections should be as close to the same size as possible. Sharp corners should be avoided; for aluminium and magnesium the minimum radius should be 0.4 mm (1/64 in) and for steel corners should be 0.75 mm (0.030 in) and fillets should be 3 mm (0.12 in). The following Table 5.2 lists the minimum cross-section and thickness for various materials [1].

5.5.6.1 Die Forming (Plastics)

A die in polymer processing is a metal restrictor or channel capable of providing a constant cross sectional profile to a stream of liquid polymer. This allows for continuous processing of shapes such as sheets, films, pipes, rods, and other more complex profiles. This is a continuous process, allowing for constant production (assuming constant supply of polymer melt), as opposed to a sequential (non-constant) process such as injection molding.

5.5.7 Process

Die forming typically occurs immediately after polymer melt has exited an extruder. The most basic process involves guiding the stream of molten polymer under pressure through a die, which three distinct regions: manifold, approach, and lip. The 'manifold' serves to channel the polymer melt from its initial extrusion point to a near-net-shape of the final product. The 'approach' region further guides the melt into the final shape, and begins to correct for any non-uniform flow. Finally, the 'lip' forms the melt into the final desired cross section and compensates for any remaining flow asymmetry. After exiting the lip of the die, the polymer melt will undergo die swell before curing. Die swell is an expansion of the melt as the pressure is released, and is dependent on polymer chemistry and die design. After curing, the solid, continuous part is drawn onto a take-up roller or cut into transportable lengths, depending on the type of part. This process may vary significantly depending on the type of die and extrusion process [5].

5.6 Sheet/Film Extrusion

There are two major types of dies used in flat sheet extrusion: T-shaped and coat-hanger. A T-shaped die consists of two arms extending at right angles from the initial extrusion channel; these arms have a small slit along their length to allow the polymer melt to flow through. The melt is then further thinned by a short, flat approach before being pushed through the lips of the die. This setup can cause non-uniform flow across the width of the extruded sheet, with the melt at the center flowing faster than the melt at the edges of the die, resulting in buckling and other defects after exiting the die [6].

A more modern design is the coat-hanger die. This die differs from the T-shaped die in that the arms are not at right angles to the input direction; instead, the arms are at a shallower angle and are often curved. The arms also have a variable diameter, tapering down to a smaller radius further from the input channel. The approach portion of coat-hanger dies are longer than their T-shaped counterparts, further reducing any flow nonuniformity. Finally, the melt is extruded through lips as in the T-shaped die.

For products such as plastic sheets or films, cooling is achieved by pulling through a set of cooling rolls (also known as calender or chill rolls), usually 3 or 4 in number. In sheet extrusion, these rolls not only deliver the necessary cooling but also help determine sheet thickness and surface texture (in case of structured rolls; i.e. smooth, levant, hair cell, etc.). A common processing defect known as nerve may occur when contact time between the rollers and extrudate is too brief, resulting in insufficient cooling time.

Coextrusion is common in sheet and film extrusion, allowing for speedy production of multi-layered parts. This is accomplished by joining multiple polymer melts in either the manifold or approach stage. Layers of different thicknesses may be formed by introducing melts at different flow rates or different manifold sizes.

5.6.1 Blown Film Extrusion

The manufacture of plastic film for products such as shopping bags and continuous sheeting is achieved using a blown film line [7]. Polymer melt from an extruder is fed through an upright die with an annular opening. There are several types of dies that can be used, depending on final requirements of film quality and characteristics of the polymer melt: spider, crosshead, and spiral dies.

A spider die consists of an internal mandrel connected to the outer die wall by several "legs", and is a moderately complex design. The resulting film will feature weld lines wherever legs were present. These weld lines are weaker than the surrounding polymer, and may also have different optical characteristics, such as haze. This weakness is caused by incomplete healing of the polymer molecular matrix. Furthermore, a pressure gradient produced by the spider legs will cause nonuniform die swell, resulting in nonuniform film thickness.

A crosshead die splits the melt flow in two at the manifold inlet, recombining them at the opposite side of a cylindrical center mandrel. This relatively simple design results in non-symmetrical flow, as molecules take longer to reach the opposite side than the close side of the mandrel. As such, the resulting film will not be of uniform thickness. To reduce this nonuniformity, inlet diameters can be varied, and various inserts can be added to minimize stagnant regions.

A spiral die is the most complex of the three major blown film die types. The polymer melt is evenly distributed into several feed tubes, which wind around a central mandrel. Each of these feed tubes is connected to the space between the mandrel and outer die walls; the feed tubes gradually diminish in diameter as they spiral around the mandrel. At the same time, the space between the mandrel and outer die walls is increased. This allows the polymer melt to layer and blend, resulting in a uniform melt profile free of weld lines. This die design produces the most uniform films, but is also the most expensive.

Air pressure is introduced through the extrusion die so that after the polymer melt leaves the lip of the die, it expands in circumference. The tubing is also drawn along its length faster than it is being extruded. This leads to thinning of the film as it is expanded in both the draw (and machine) direction, and in the transverse (or hoop) direction. The ratio of the blown diameter to the extruded diameter is known as the blow-up ratio, and affects the resulting physical properties of the film, such as stiffness and strength. Film thickness and blow-up ratio can be varied by altering the take-up rate of the rollers, the internal pressure in the blown tube, and the melt extrusion rate.

As the film is drawn upwards, it is cooled by a ring of air blowers so that the melt first becomes an amorphous viscoelastic solid, and then a semicrystalline solid, at what is known as the frost line. After solidification, the blown film tube continues to be cooled as it is pulled up by several sets of rollers, deflating the film to form lay-flat tubing. The flat film is then wound on a spool before further processing or shipping. The height of the film line is often 10 times the diameter of the blown tube or more; film lines in excess of 30 m are possible.

Once the film tube is completely cooled, it is taken up by several nip rollers. The width of the resulting doubled-over flat film is equal to half of the blown tube's circumference. The film is then either spooled as a flattened tube, or immediately split into two separate pieces. At this point, the film is ready for further processing, such as printing or cutting into final shape (Fig. 5.6).

5.7 Over Jacketing

Over jacketing extrusion is a coating process, in which individual bare wires or bundles of pre-coated wires are coated with a layer of insulating polymer. A wide variety of materials may be used, depending on the specific application. For many applications, such as insulated cables, the polymer should be a good insulator, flexible, and wear resistant [8].

Fig. 5.6 Blow extrusion of plastic film [7]

In this process, a wire (or bundle of wires) is preheated to above the glass transition or melting temperature of the polymer coating that is to be applied. This is to ensure adhesion of the new coating. Next, this preheated bare wire is pulled very rapidly through a die which places a thin coating of polymer around the wire. Due to the geometry of the dies used, relatively high extrusion rates are possible while still avoiding melt fracture. The newly coated wire is then drawn through an air or gas flame to smooth the surface of the coating, and finally a water bath to fully cool the coated wire. Coated wires are now spooled to prepare for further processing, if desired.

There are two major types of dies used in over jacketing extrusion, both based on an overall crosshead design. Regardless of die type used, the polymer melt is often extruded at a rate less than the speed of the bare wire that is drawn through the die, typically on the order of 1–4 times the speed of the melt. This causes the polymer jacket to extend, thin, and tighten around the central wire, increasing adhesion of the new layer.

The first dye type is an annular, or tubing/jacketing, die that extrudes a tube of polymer that is initially *not* touching the bare wire. A vacuum is then applied to the still-molten polymer tube, causing it to be drawn onto and bond to the surface of the bare wire. This type of die is typically used to coat very thin wires with polymer jacketing that is highly viscous.

The second die type, known as a pressure type die, relies on contact between the jacketing polymer and bare wire inside the die. In this die type, a ring of polymer melt under pressure is forced around the bare wire. Due to the applied pressure of the melt, the opening around the inlet for the bare wire must be very small, on the

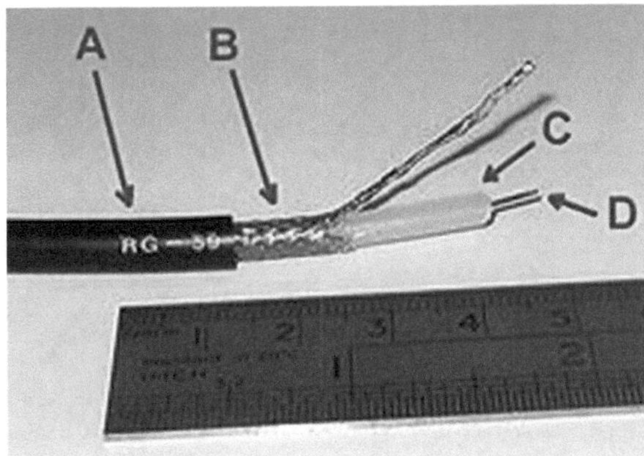

Fig. 5.7 Coaxial cable, one example of a jacketed wire [8]

order of 0.05 mm. The size of the exit opening controls the thickness of the resulting coating. This type of die results in more intimate contact between the outer coating and the bare wire than the jacketing die (Fig. 5.7).

5.8 Fiber Drawing of Polymers

Fiber drawing is a hybrid process, in which gravity or another force is used to geometrically and mechanically alter the extruded fibers. This process not only reduces the cross section of the polymer fiber, but also increases the strength of the fibers by aligning the individual polymer molecules.

Before drawing, polymer melt is pushed through a die with a large number of small holes, known as a spinneret. Typically, the fibers are air cooled without any need for curing. If curing is needed, two methods are available: dry and wet spinning. In wet spinning, the polymer is dissolved and extruded through a spinneret into a chemical bath. In dry spinning, a solvent is allowed to evaporate as the fibers are cooled.

Typically, fiber drawing occurs immediately after spinning. Application of an external force, either from gravity or take up rollers, causes the fibers to contract laterally or lengthen. This orients the individual polymer molecules along the length of the fiber, increasing strength. The radius of the fibers have been shown to decrease hyperbolically as they lengthen. Once the fibers solidify, they may begin to crystallize, with each grain initially randomly oriented. Further drawing will cause the crystal grains to elongate and reorient in the direction of the tensile axis, further strengthening the fibers.

5.8.1 Spinning Stability

In practice, not all polymers are suitable for fiber spinning or drawing. This is particularly an issue in extensional-thinning polymers, where capillary failure or necking can cause separation of the melt before solidification.

Draw resonance is the most common issue that can occur during drawing of the polymer melt, regardless of polymer suitability. Resonance occurs when the rate of mass flow is not constant between the spinneret and fiber take up roller, despite being constant at each of those individual components. When the mass flow rate is not constant, the diameter of the fiber will vary to accommodate the variation. Once started, this resonance may not correct itself, requiring a complete shutdown of the extrusion line.

It has been shown that draw resonance occurs once a critical drawdown ratio is exceeded; this ratio is dependent on the flow behavior (i.e. Newtonian, shear thinning) and viscoelastic behavior of the fluid. Draw resonance has not been found to be a function of the flow rate, however. A polymer melt approaching a Newtonian fluid such as PET can have a drawdown ratio of around 20, whereas highly shear thinning and viscoelastic polymer melts such as polyethylene, polystyrene, and polypropylene may have critical drawdown ratios as low as 3.

5.8.2 Tube Forming

Tube forming dies allow for continuous extrusion of thick walled (relative to blown film extrusion) tubes and pipes [9]. The dies themselves are almost identical to those used in blown film extrusion; the only major difference is the gap between the inner mandrel and outer die wall. Once the polymer melt is extruded from the die, it is pulled away by take-up rollers. Cooling is accomplished through the use of water baths, or a large number of cooling fans. After cooling, the tube is either wound onto large spools (if flexible), or cut into pre-set lengths and stacked (if stiff).

Tubing with multiple lumens (holes) must be made for specialty applications. For these applications, the tooling is made by placing more than one mandrel in the center of the die, to produce the number of lumens necessary. In most cases, these mandrels are supplied with air pressure from different sources. In this way, the individual lumen sizes can be adjusted by adjusting the pressure to the individual mandrels.

5.8.3 Profile Extrusion

Profile extrusion, the extrusion of complex shapes such as rain gutters, structural supports, and other components, brings with it some of the most complex die designs of any extrusion process [10]. This difficulty is due to two primary concerns: producing the initial, still molten profile, and then controlling for asymmetrical shrinkage and die swell due to varying wall thicknesses.

Unlike in blown film, pipe, and sheet extrusion, the dies used in profile extrusion are rarely round or completely flat. Whereas a round (or flat) profile has uniform flow rates along all edges, this is not the case for more complex shapes. Take, for instance, the example of a simple, solid, square profile. The velocity of the melt is highest at the center of the die, and slowest at the edges and corners due to friction between the melt and die walls. When moving from the center of the die to the midpoint of one of the edges, the velocity gradient is high, especially near the outer die wall. However, when moving from the center to one of the corners, the velocity gradient is more gradual. As a result, the extruded square profile will experience more die swell at the edges than the corners, causing the once square profile to become more circular. This can be compensated for by bowing in the sides of the die so it approximates the shape of a four-pointed star; the sides of the polymer melt will now swell to the intended dimensions.

As the desired profile becomes more complex, the die in turn becomes more complex. Care must be taken to minimize weld lines, as well as ensuring complete filling of the die to prevent bubbles and other defects in the finished extruded profile. After the initial extrusion is complete, the molten polymer profile is cooled slightly before being run through a sizing die. This die ensures that the extruded profile meets specifications, and can correct the shape to fit those specifications. After sizing is complete, the profile is cooled before any further processing.

5.9 Coextrusion

In practice, many films, sheets, and other extruded parts are multilayered; this allows for optimization of a wide range of properties, such as oxygen permeability, strength, and stiffness. The primary difficulty of coextrusion is bridging the gap in properties between each layer [11]. Adding a thin "compatibility" layer is a common solution to alleviating viscosity or stiffness incompatibilities [12].

There are two major die types for coextrusion: single manifold and multi manifold. Both types rely on a separate extruder for each polymer chemistry. In multi manifold dies, each layer is extruded separately and only combined just before the die lips. This die type is expensive due to the complex tooling required, but can alleviate vast differences in rheological behavior between the various layers. Single manifold dies form the multiple layers into a single layer, allowing contact between the polymer layers for a longer period of time. This ensures optimal bonding, but comes at the consequence of needing higher compatibility polymers.

There are two types of processing defects that can occur during coextrusion. The first defect is interface instability, causing unintended interface shapes. This can cause "encapsulation" of the higher viscosity melt by the lower viscosity melt, leading to poor final performance of the extruded part. The severity of this type of

Fig. 5.8 5:5 Layer Co-extrusion of cosmetic "Squeeze" tube [11, 12]

defect is proportional to the difference in viscosities between the two polymer melts. The other type of defect forms from oscillations in the melt flow, causing small wavelike patterns on the surface of the melt and reducing optical transparency (Fig. 5.8).

5.10 Case Study I

5.10.1 Blown Film Extrusion

Blown film extrusion is a technology that is the most common method to make plastic films, especially for the packaging industry [13]. The process involves extruding a tube of molten polymer through a die and inflating to several times its initial diameter to form a thin film bubble. This bubble is then collapsed and used as a lay-flat film or can be made into bags. Usually polyethylene is used with this process, and other materials can be used as blends with these polymers [13]. A diagram of a polyethylene chain is shown in Fig. 5.9 to the right.

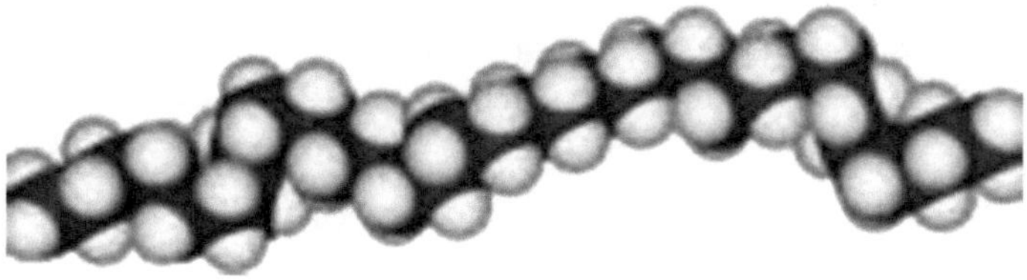

Fig. 5.9 Model of polyethylene chain from polyethylene [14]

5.10.2 Background Theory on Polymers

In the cooling step of blown film extrusion, the amorphous, transparent melt crystallizes to form a translucent, hazy, or opaque film. The point where opacity begins in the bubble is known as the frost line.

The frost line height is controlled by several parameters: the air flow, film speed, and temperature difference between the film and the surroundings [14].

Properties of the film, such as tensile strength, flexural strength, toughness, and optical properties, drastically change depending on the orientation of the molecules [14]. As the transverse or hoop direction properties increase, the machine or longitudinal direction properties decrease. For instance, if all the molecules were aligned in the machine direction, it would be easy to tear the film in that direction, and very difficult in the transverse direction.

5.10.3 The Film Blowing Process

Typically, blown film extrusion is carried out vertically upwards, however horizontal and downward extrusion processes are now becoming more common [14, 15]. The blown film extrusion procedure consists of four main steps:

1. The polymer material starts in a pellet form, which are successively compacted and melted to form a continuous, viscous liquid [16]. This molten plastic is then forced, or extruded, through an annular die.
2. Air is injected through a hole in the center of the die, and the pressure causes the extruded melt to expand into a bubble. The air entering the bubble replaces air leaving it, so that even and constant pressure is maintained to ensure uniform thickness of the film [20].
3. The bubble is pulled continually upwards from the die and a cooling ring blows air onto the film. The film can also be cooled from the inside using internal bubble cooling. This reduces the temperature inside the bubble, while maintaining the bubble diameter [19].
4. After solidification at the frost line, the film moves into a set of nip rollers which collapse the bubble and flatten it into two flat film layers. The puller rolls pull the film onto windup rollers. The film passes through idler rolls during this process to ensure that there is uniform tension in the film. Between the nip rollers and the windup rollers, the film may pass through a treatment centre, depending on the application. During this stage, the film may be slit to form one or two films, or surface treated [19].

5.10.4 Advantages

Blown film generally has a better balance of mechanical properties than cast or extruded films because it is drawn in both the transverse and machine directions. Mechanical properties of the thin film include tensile and flexural strength, and toughness. The nearly uniform properties in both directions allow for maximum toughness in the film [13, 17]. Blown film extrusion can be used to make one large film, two smaller ones, or tubes that can be made into bags. Also, one die can make many different widths and sizes without significant trimming. This high level of flexibility in the process leads to less scrap material and higher productivity. Blown films also require lower melting temperatures than cast extrusion. Measured at the die opening, the temperature of cast film is about 220 °C [18], where as the temperature of blown film is about 135 °C [19]. Furthermore, the cost of the equipment is approximately 50 % of a cast line [19].

5.10.5 Disadvantages

Blown film has a less effective cooling process than flat film. Flat film cooling is done by means of chill rolls or water [17], which have significantly higher specific heat capacities than the air that is used in the blown film cooling process. The higher specific heat capacity allows the substance to absorb more heat with less change in the substance temperature. Compared to cast film, blown film has a more complicated and less accurate method to control film thickness; cast film has a thickness variation of 1–2 % versus the 3–4 % for blown film [19]. The resins used for casting typically have a lower melt flow index [14], which is the amount of polymer that can be forced through a standard die in 10 min according to a standard procedure [20]. The melt flow index for cast film is about 5.0 g/10 min [21] where as for blown film it is about 1.0 g/10 min [22]. Consequently, the production rates for cast film are higher: cast film lines can reach production rates of up to 300 m/min where as blown film lines are usually less than half this value [23]. And finally, cast film has better optical properties, including transparency, haze, and gloss.

5.10.6 Common Problems

Air entrapment between film layers and rollers—this may cause film scratching or wrinkling, or processing problems when winding up the film due to reduced friction. Possible solutions to this is using a vacuum to remove entrapped air or by using winding rolls with a diamond shaped groove in the rubber cover to increase surface area and decrease amount of entrapped air in the film [14].

- Large output fluctuations from the die—this causes thickness variations, and can be prevented by keeping the extruder clean and by using more consistently shaped pellets in the extruder [24].
- Melt fractures—these appear as roughness or wavy lines on the film surface, and can be eliminated by lowering the viscosity of the polymer melt. This can be done by increasing the melting temperature or by adding an internal lubricant to the material composition [24].
- Thickness variations in the film—this can be avoided by centering the die in the extrusion line before every run, adjusting the air speed of the cooling system, or by using heated die lips [24].
- Die lines on the surface of the film—this defect reduces the aesthetic appeal of the film, reduces optical properties, and weakens mechanical properties such as tear strength. This can usually be avoided by routinely cleaning the inner surfaces of the die and by refinishing scratched or roughened flow surfaces [24].
- Gels—these defects are small, hard globules encapsulated in the film or stuck on the film surface and reduce the aesthetic appeal of the film and cause stress concentration points which may result in premature failure. These are caused by overheating to the point of polymer degradation in the die, and can therefore be avoided by cleaning the inner surfaces of the die on a regular basis [24].

5.11 Summary

Coextrusion: One way to improve the line efficiency of blown film extrusion is to implement coextrusion. This is the process of extruding two or more materials simultaneously through a single die. The orifices in the die are arranged such that the layers merge together before cooling [14]. This process saves time because it extrudes two or more layers at the same time, and it provides a method with fewer steps to produce multilayer films. The production rate for a coextruded multilayer film of three layers is about 65 m/min [25], and the production rate for a single layer of blown film is about 130 m/min [23]. Thus, in order to produce 10,000 m of a three layer multilayer film, it would take almost 4 h using a single layer blown film process, and only 2 and a half hours using the coextrusion process. Furthermore, the film produced from the single layer process would require an extra step to glue the layers together using some sort of adhesive. Coextrusion is the least expensive means of producing layered films and the coextrusion system is capable of quick changeovers to minimize production line down time [26].

Minimizing the Melt Temperature: The efficiency of blown film extrusion can be improved by minimizing the temperature of the polymer melt. Reduction of the melt temperature causes the melt to require less heating in the extruder. Normal extrusion conditions have a melting temperature at about 190 °C [27] despite the fact that the temperature of the melt only needs to be about 135 °C [19]. However,

Fig. 5.10 Consumer food wrap from plastic wrap

it is not always practical to decrease the melting temperature by that much. By decreasing the melt temperature 2–20 °C, the motor load can be decreased by about 1–10 % [28]. Furthermore, reduction of the melt temperature causes less need for cooling, so there is a reduced use of the cooling system. Moreover, removing heat from the bubble is usually the rate-limiting factor in this extrusion process, so by having less heat in the polymer to remove, the rate of the process can be increased, thus yielding higher productivity. A way to maintain the melt temperature at a minimum is to choose an extruder that is matched to the specific processing conditions, such as the material of the melt, pressure, and throughput [24].

Heated Extrusion Die Lips: Typically, solutions to melt fractures involve decreasing the output or increasing the melt temperature to decrease the shear stress in the extruder. Both of these methods are not ideal because they both reduce the efficiency of the blown film line. Heated extrusion die lips can solve this problem. This targeted heating method allows for film extruders to be run at higher production rates with narrower die gaps while eliminating melt fractures [29]. Direct heat is applied to the surface of the polymer melt as it exits the die so that viscosity is reduced. Therefore, melt fractures, which are caused when trying to extrude too much of the polymer at one time, will no longer act as a limiting factor to increasing the production rate [29]. Furthermore, heated die lips use less energy than increasing the melting temperature because only the surface of the melt is heated and not the bulk of the liquid. Another benefit of using heated die lips is that thickness variations can be controlled by adding heat to certain areas along the die circumference to make the film at that position thinner. This would ensure that no excess material is used [30] (Fig. 5.10).

Applications:

- Agricultural film
- Bags
- Industry packaging, shrink film, stretch film
- Consumer packaging, food wrap, transport packaging
- Laminating film

- Barrier film
- Multilayer film
- Research has been done to explore the incorporation of blown film extrusion into the large-scale manufacturing of carbon nanotube and nanowire films [31, 32].

References

1. Oberg E, Jones FD, Horton HL, Ryffel HH (2000) Machinery's handbook, 26th edn. Industrial Press, New York. ISBN 0-8311-2635-3
2. Drozda T, Wick C, Bakerjian R, Veilleux RF, Petro L (1984) Tool and manufacturing engineers handbook: forming, SME, vol 2. ISBN 0-87263-135-4
3. Avitzur B (1987) "Metal forming", encyclopedia of physical science and technology, vol 8. Academic Press Inc., San Diego, pp 80–109
4. Aluminum Extrusion Tooling—MakeItFrom
5. Tadmor Z, Gogos CG (2006) Principles of polymer processing. Wiley, New York. ISBN 978-0-471-38770-1
6. Patent No. 5395231 A
7. How to solve blown film problems (PDF) (2012) Lyondell Chemical Company. Retrieved 31 Aug 2012
8. Crawford FJ (1998) Plastics engineering. Elsevier. ISBN 978-0-7506-3764-0
9. Vogler J (1984) Small scale recycling of plastics. Intermediate Technology Publication
10. Giles HF, Wagner JR, Mount EM (2005) Extrusion: the definitive processing guide and handbook. William Andrew. ISBN 978-0-8155-1473-2
11. Rosato MG (2000) Concise encyclopedia of plastics. Springer. ISBN 978-0-7923-8496-0
12. Brydson JA (1999) Plastics materials. Elsevier. ISBN 978-0-7506-4132-6
13. Plastics Wiki (2008) Blown film extrusion, 6 Nov 2008
14. Giles HF Jr, John RW Jr, Eldridge MM III (2005) Extrusion: the definitive processing guide and handbook. William Andrew Publishing, New York
15. Chanda M, Roy SK (2007) Plastics technology handbook, 4th edn. CRC Press, Florida
16. Callister WD Jr (2003) Materials science and engineering an introduction, 6th edn. Wiley, Massachusetts
17. Rosato DV (2000) Concise encyclopedia of plastics. Kluwer Academic Publishers, Massachusetts
18. van der Molen TJ (1988) Process for the production of cast film from high-density polyethylene. European Patent EP0278569, August 1988
19. van der Molen TJ (1991) Process for the preparation of blown film and oriented film. European Patent EP0156130, September 1991
20. Chui QSH et al (2007) An interlaboratory comparison of the melt flow index: relevant aspects for the participant laboratories. Polym Testing 26(5):576–586
21. Ebnesajjad S (2003) Melt processible fluoropolymers. William Andrew Inc, New York
22. Chang AC et al (2002) Mechanisms of ductile tear in blown film from blends of polyethylene and high melt strength polypropylene. Polymer 43(24):6515–6526
23. Dyson RW (1990) Engineering polymers. Chapman and Hall, New York
24. Cantor K (2006) Blown film extrusion: an introduction. Hanser Publishers, Ohio
25. Carneiro OS, Reis R, Covas JA (2008) Small-scale production of co-extruded biaxially oriented blown film. Polym Testing 27(4):527–537
26. Cook WR (1994) Co-extrusion apparatus. United States Patent 5324187, June 1994
27. Dowd LE (1986) Blown film extrusion. United States Patent 4632801, December 1986
28. Shannon PC (2006) Polyethylene films. United States Patent 7101629, September 2006

29. Bentivoglio A (2001) Heated die lips system. Canada Patent CA 2204548, June 2001
30. Moriarity GJ (2001) Heated die lips for controlling extruded polymer film thicknesses. United States Patent 6273701, August 2001
31. Dalton A, Jurewicz I (2007) Nanotechnology meets bubbleology. Nat Nanotechnol 2:339–340
32. Yu G, Cao A, Lieber CM (2007) Large-area blown bubble films of aligned nanowires and carbon nanotubes. Nat Nanotechnol 2:372–377

Chapter 6
Compression for Multiphase Products

Ganesh Chandra Nayak and G. Hatui

6.1 Introduction to Compression Molding

Manufacturing of plastic parts is the most versatile part of plastic industry and involves many steps like shaping, materials removal and joining. With increasing demand of plastic materials the processing technology also evolved to meet the product standard. Selection of processing technique is a critical part of plastic industry and depends on the targeted product and raw material. Among many processing techniques, compression molding is the oldest and less complicated processing method for end used plastic products. Compression moulding is a high-volume, high-pressure plastic moulding method that is suitable for moulding complex, high-strength objects [1]. And with its short cycle time and high production rate, many organizations in the automotive industry have chosen compression moulding to produce parts. Compression molding is used primarily to process thermosetting systems and difficult-to-process thermoplastics, such as fiber-filled systems or thermoplastic elastomers. It is commonly used for manufacturing electrical parts, flatware, gears, buttons, buckles, knobs, handles, electronic device cases, appliance housing, and large container [2]. In this technique the resin material is heated under severe pressure within a closed mold cavity until the complete curing of thermosetting resin. The pressure liquefied the resin and distribute the liquid resin uniformly inside the mold cavity and under the influence of temperature they cross-linked and hardens to the shape of mold cavity. The product is removed after cooling down the mold. Compression molding is used for processing thermoplastics, which do not flow readily, such as highly filled systems and granular materials that

G.C. Nayak (✉) · G. Hatui
Department of Applied Chemistry, Indian School of Mines, Dhanbad 826004, India
e-mail: nayak.g.ac@ismdhanbad.ac.in; ganeshnayak2006@gmail.com

J.K. Kim et al. (eds.), *Multicomponent Polymeric Materials*,
Springer Series in Materials Science 223, DOI 10.1007/978-94-017-7324-9_6

don't really melt but only fuse under pressure [3]. The whole process involves no crosslinking reaction. Different types of thermosets and thermoplastics materials can be used for compression molding process. For example: Epoxies, Urea formaldehyde (UF), Melamine formaldehyde (MF), Phenolics (PF), Polyester, Polyamide (PI), Polyamide-imide (PAI), Polyphenylenesulfide (PPS), Polyetheretherketone (PEEK), Torlon, and Vespel etc [4]. Compression molded products are used in many fields like Automobile, aerospace, defense, chemical processing industry, electrical engineering, electronics, machine shops medical, oil and gas, seal manufacturers, semiconductor, water and wastewater treatment etc [4–9].

6.2 Design of Compression Molding

From the design point view compression molding machine consists of three major parts: (1) Mold to give shape, (2) Hydraulic to exert pressure and (3) Heating element to exert temperature. Figure 6.1 shows the process of compression molding. The mold is consists of two parts. One is bolted to the stationary end of the top platen and the other part of mold is fixed to the moving platen of the molding press. The hydraulic is attached to the moving platen of the molding press which can be designed either at the top or at the bottom to exert pressure on the preform. Depending upon the material to be molded, the mold halves are pre heated to about 150 °C or less and a thermoplastic or partially polymerized thermosetting polymer is placed in the open bottom mold cavity. During the compression molding initially the compression force increases rapidly and the preform is squeezed and heated. Subsequently the polymer is transferred into the molten state and under the influence of pressure forced to flow into the cavity and fill it. After the filling process the crosslinking reaction starts and any volume contraction during this cross linking process is compensated by the increasing mold pressure to achieve a dense and rigid product. The product is held under this pressure and temperature for a certain time, depending on the polymer, to complete the cross linking reaction. Then the mold is opened, the part is ejected, and the cyclestarts again. A slight excess of material is used to ensure proper mold filling and the excess material is squeezed out form the

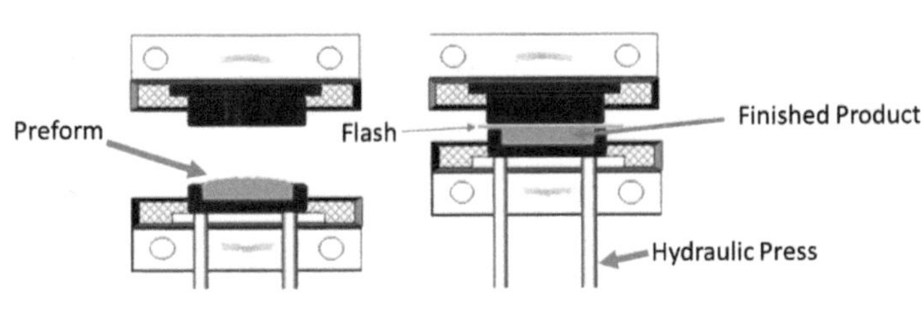

Fig. 6.1 Compression molding process

mold and known as flash material which cut out from the molded parts to get finished product. This process wastes very little material and can be used to produce large parts. However, it is difficult to produce parts with very close tolerances because the final size of the compression molded article depends on the exact amount of the preform. The advantages of compression molding includes low cost process, low maintenance cost, negligible scrap and no residual stress in the molded part [7]. Its major limitation is the time of compression molding is slightly higher than that of other similar processes so it gives slightly more cycle time.

6.3 Hydraulic System and Mold

The compression molding machine consists of a heavy metal base onto which two rods are attached which guides the top and bottom plate to move up and down during the compression process. On the bottom and top plate two heated plates are attached on which two halves of mold are fixed. The heated plates transfer the heat to the mold during compression molding. The temperature of the mold can be controlled by controlling the temperature of the heater plates which is controlled and monitored by a separate panel. The heating also accomplished by the use of circulating steam, hot oil or open flame. Hot oil is a popular heating media because it can be heated to high temperature and maintain uniform temperature distribution while electrical heating is a cleaner and greener medium. The compression pressure is controlled by an attached hydraulic. The operation of applying pressure in a compression molding machine can be executed with two types of actuations: (a) upstroke of bottom plate (upstroke type hydraulic press) and (b) down stroke of top plate (down stroke type hydraulic press) (as shown in Fig. 6.2). In the upstroke type hydraulic system the

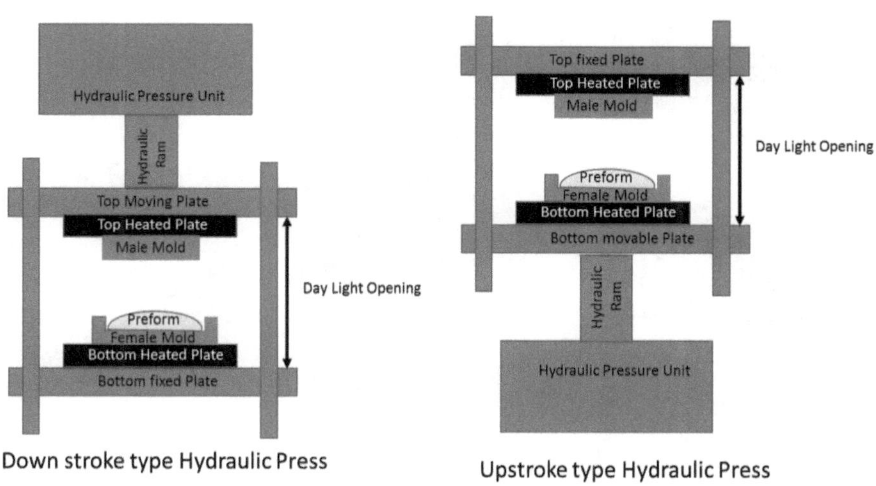

Fig. 6.2 Types of hydraulic system in compression molding

hydraulic ram moves the movable plate or bolster upward to close the mold. The attached sliding rods keep the upper and lower parts of plates and hence the mold aligned so that the mold mismatch can be avoided [7–12]. The disadvantage of this type of hydraulic system is mold opening, after the completion of compression cycle, is very slow and driven only by gravity. This limitation can be overcome by integrating a push back ram to the molding machine to speed up the mold opening. In contrast to the upstroke type hydraulic, the downstroke type hydraulic has a fixed bottom plate and hydraulic ram is fixed to the top movable plate which exert a downward force during mold closing. This type of hydraulic system is used for molding very large components which require a longer stroke.

The hydraulic system can be fully manual where a manual pump attached to the hydraulic will exert the pressure or semiautomatic where a pressure valve is activated by operator controlled switch or it can be fully automatic where the pressure valve is activated within a predefined and programmable time. The distance between the top and bottom plate is called as day light opening which is a critical parameter for selecting a compression molding machine. The day light opening should be large enough to accommodate the heating plate, complete mold and leaves space for charging the preform and taking out the molded parts. Most of the hydraulic press are sold according to the press or platen rating. For molding a product of 25 mm thickness a force of 20 MPa is required and for each additional 25 mm an additional 5 MPa force is required, approximately (depends on material property).

Apart from the two mold half of a mold set, a third component is used in the mold system to eject out molded part after the completion of one compression cycle. This component is called the ejector which consists of an ejector plate and ejector pin as shown in Fig. 6.3. The mold halves are manufactures in such a manner that they can be attached to the heating plate with the help of locking

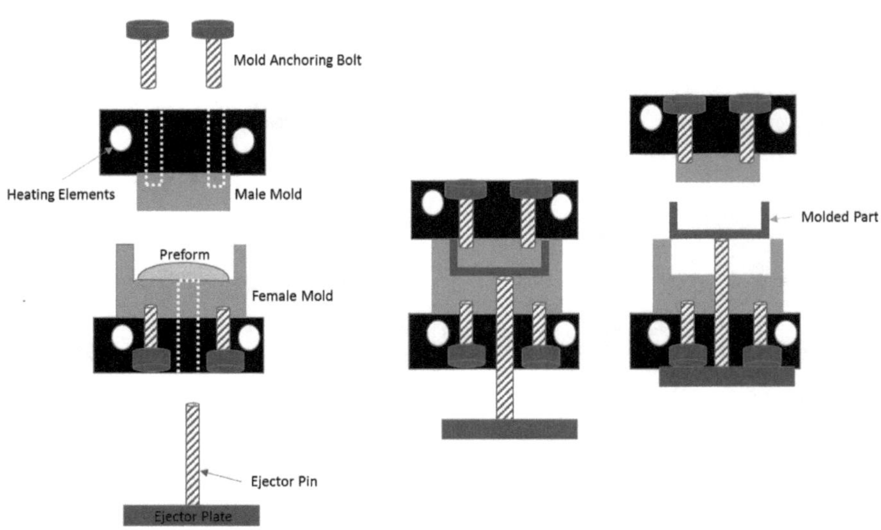

Fig. 6.3 Working mechanism of mold set

mechanism or by simple use of screws. The advantages of this arrangement is that the same compression molding machine can be used for the production of different products by just switching the molds. The heating plates integrated within the heating coils which are heated by passing the current though it and the temperature can be controlled by controlling the current.

Once the preform is charged in the mold cavity the compression cycle starts and the preform takes up the shape of mold cavity. After the completion of the compression cycle, the mold opens up and the ejector system is activated which pushed the molded parts out of mold cavity and the mold is ready for the next cycle. This ejection system can be activated by either a small hydraulic piston or can be integrated with the molding cycle, where the movement of the compression assembly away from the base will automatically trigger the ejection system.

Capacity of hydraulic system (press capacity or machine rating or machine size) must be accessed before selecting a compression molding system for the production of plastic article by compression molding [7]. The press capacity (N) can be calculated by multiplying area of ram (m^2) and hydraulic pressure (Pa). To select a correct hydraulic press for molding a particular type of plastic material, first the force required to mold the part is calculated and then compared with the press capacity of the hydraulic system. The required mold force is dependent on many factors including the specific resin, resin form (granules or preforms), resin viscosity, fillers, additives, reinforcements, molding temperature, part thickness, and complexity of part design. Different resins have different viscosities and flow characteristics. With increase in viscosity, the force required to completely fill the mold cavities also increased due to decreased in flow of material inside the mold. Similarly, if the area (projected area) of the mold or targeted product increases then the force required is also increased to fill the mold cavities in the predefined molding time. The additives and fillers in the resin system also affects the viscosity and flow characteristic of the resin and hence affect the molding force. Overall, there are many factors which affect the mold force and it is very difficult to draw a direct mathematical relation to calculate the mold force and hence the actual molding force is best set through experience or experimentation. If the mold force required is lesser than the press capacity, the hydraulic system can be used for molding that material at that temperature and that size of the product. If any of the variable i.e. molding temperature or product size changes then the calculation has to be made a fresh to access the suitability of hydraulic system. In general a press capacity of 1.25–1.3 times that of required force for molding a product is applied to ensure complete mold filling and better product development.

For simple mold design or simple products the approximate mold force can be calculated by calculating the projected area which is based on overall maximum width and length dimensions of the cavity and equal to the surface area. The mold force (kN) is equal to the product of molding pressure (MPa) and projected area (cm^2)/10. However, for a complex structure projected area exceed the actual surface area which increased the calculated mold force and creates problem during molding [7–9].

6.4 Types of Mold

The molds are made of tool steel or stainless steel to withstand the high pressure during compression molding. Molds are generally chrome plated to give a mirror finish to facilitate product removal during ejection step. The molds are manufactured preferably by hobbing process and rarely by EDM. Mold design is a very important part of plastic processing and itself is a topic of great interest and research. The design of any of these molds must allow venting to provide for escape of steam, gas, or air produced during the operation. After the initial application of pressure the usual practice is to open the mold slightly to release the gases. This procedure is known as breathing. There are three basic types of molds are used for compression molding known as open flash, fully positive, and semi positive (Fig. 6.4).

In an open flash mold a slight excess of molding powder is loaded into the mold cavity. On closing the top and bottom platens the excess material is forced out of the mold as flash. The flash blocks the plastic remaining in the cavity as a result pressure exerted on the molded materials due to continuous movement of plunger. The closing of the mold must be done carefully so that the volatile gases can be expelled from the mold and a product without defects will be produced. Resins having high viscosity are preferred for this process. Since most rubbers have high melt viscosities, the flash mold is widely used for producing gaskets and grommets, tub and flash stoppers, shoe heels, door mats, and many other items. Because of lower pressure exerted on the plastic in the flash molds, the molded products are usually less dense. Moreover, because of the excess material loading needed, the process is somewhat wasteful as far as raw materials are concerned. However, the process has the advantage that the molds are cheap, and very slight labour costs are necessary in weighing out the powder.

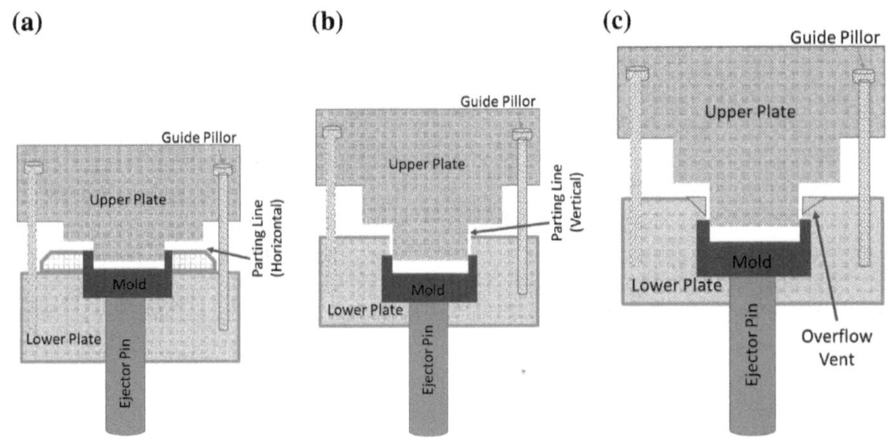

Fig. 6.4 Types of mold **a** flash type mold, **b** positive type of mold and **c** semi positive type mold

In the fully positive molds excess powder is not loaded in the cavity. While loading of excess powder will prevent mold closing; an insufficient charge will result in reduced thickness of the molded article. Hence, exact amount of charge has to be loaded to the mold which is a disadvantage of the positive mold. In this type of mold the volatile gases produced during the curing reaction are trapped inside and may show as blisters on the molded surface. Excessive wear on the sliding fit surface on the top and bottom forces and the difficulty of ejecting the molding are other reasons for discarding this type of mold. The mold is used on a small scale for molding thermosets, laminated plastics, and certain rubber components.

The semi positive mold combines certain features of the open flash and fully positive molds and makes allowance for excess powder and flash. It is also possible to get both horizontal and vertical flash. Semi positive molds are more expensive to manufacture and maintain than the other types, but they are much better from an applications point of view. Moreover, the mold is given a taper on each side. This allows the flash to flow on and the entrapped gases to escape along with it, thereby producing a clean, blemish-free mold component.

6.5 Mold Design

Molds can be classified based on their type and number of cavities present in them. They can be classified as single cavity molds, dedicated multi-cavity molds and family of multiple cavity molds as shown in Fig. 6.5.

Among these three types of mold single-cavity mold is the simplest one but the production rate is very low because for each cycle of molding only one product is formed. The multiple-cavity molds can be of two types. A dedicated multiple-cavity mold possessed multiple mold cavities with same geometry and hence produce the same part which increase the productivity depending on number of mold cavities. This type of mold is very popular because it is easy to balance the plastic flow and establish a controlled process. In a family multiple-cavity mold, each cavity may produce a different part. This type of mold are not preferred because of non-uniform mold filling. Since each part possessed different projected area, the time required for each part to fill each mold cavities are different. The mold with smaller area filled up faster than the big ones and create a no uniform flow throughout the mold. However, recent advances in mold making and gating technology make family molds appealing. This very important in case of multiple product assemble where each part has different geometry. These types of mold keeps all the parts ready in one cycle so that the final product can be assembled.

Mold design is a separate and important part of plastic processing. The mold designer must consider several factors that may affect the fabricated part, such as the plastics material, shrinkage, and process equipment. Apart from the mold design the mold must have provision to easy assembly so that it can be easily changed with another mold with the demand. Previously the molds are basically manufacture by manual lathe or later on CNC machines. But now the whole process has been

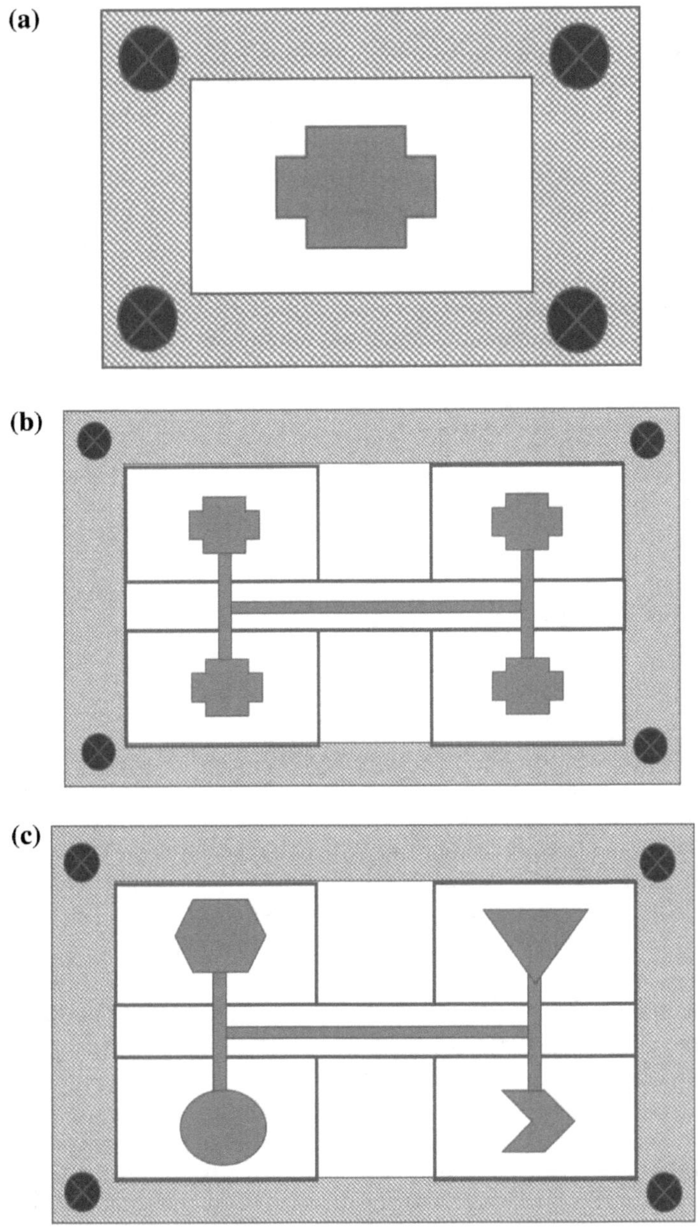

Fig. 6.5 **a** Single cavity molds, **b** dedicated multi-cavity molds and **c** family of multiple cavity mold

machined on computer-controlled mills, lathes, and electric discharge machines. The mold is first made with the available method and a prototype product has to be made to check the accuracy of m olds. But now several alternatives are available to prototyping, e.g., CAD, finite-element analysis (FEA), and rapid prototyping. While CAD allows a tool designer to work with a three-dimensional computer model of the mold being designed and to analyze the design, FEA allows the tool to be evaluated (on a computer) for production worthiness. The mold is then fabricated from the computer model, a process called computer-aided manufacturing (CAM). Rapid prototyping is a relatively new method of producing a plastics part by using a three-dimensional computer drawing. In this process a sophisticated prototyping apparatus interprets the drawing and guides an articulating laser beam across a specific medium such as a photopolymer plastic or laminated paper, the result being a physical representation of the computer-based drawing. Prototyped parts can be produced in less than 24 h, and part designs can be scaled to fit the size of the prototyping equipment. Another trend is the introduction of molds that accept interchangeable modules. Modules take less time to manufacturing, and in turn, cut down on the delivery time and costs.

6.6 Control and Operation

The sequence of compression molding includes following six steps:

Step 1: Opening and cleaning of mold and application of mold releasing agent.

This step is required to free the mold from any contamination adsorbed inside of the mold cavity to achieve better surface finished product. The mold releasing agents are used to facilitate the kick out of the product during ejection cycle. The releasing agent creates a layer on the mold surface and prevents the polymer to come in contact with the mold surface and hence avoids any adhesion between mold surface and molded part which facilitate easy removal of finished product. Selection of mold releasing agent or a lubricant has to be made in such a way that the lubricant should have little to no effect on the mold article aside from freeing it from the interior of the mold, otherwise the lubricant might chemically affect the material being used. The lubricant can also be selected to give an additional ability to change the surface makeup of a mold article such as to make it shiny or matted in appearance. However, the selection must be done carefully so that the lubricant does not degrade or chemically react during the complete compression molding cycle. Certain materials are much more likely to adhere to the interior of a mold, so increasing the slip capabilities of a given mold release agent can be preferable. In the case of combination materials (such as cross-linked polymers that have complex chemical structures) some areas of a mold may be much more difficult to achieve release than others, so parting agents can be combined to induce different slip levels in different areas. It's also necessary to choose a mold release agent that will not cause unsightly build up on either the mold or the mold article. Some agents require thicker coatings than others, while other agents necessitate a longer drying period to be effective. These variables must be considered while selecting a lubricating agent. Apart from this during the selection of a mold releasing agent the workplace and environment safety must be considered so that the person who is exposed to it could be protected from any health hazard. The lubricants must be odourless and hypoallergenic, so products exposed to them will not cause reactions to consumers or workers. Generally the mold releasing agents can be of water based, solvent based or Aerosol type and mostly based on fluoro or silicon polymers which possessed excellent anti-sticking property and high thermal stability. Recent research is going on to develop green lubricating agent to reduce the environmental hazard. Among various types of mold releasing agents, aerosol based mold release agents are more popular than other type of mold releasing agents. An aerosol can filled with pressurized mold release agent can be used by either a worker or by a CNC controlled machine on an assembly line, though application methods vary. Specialized machines also focus on applying mold release agents to mold interiors without releasing possibly dangerous aerosol plumes into the air.

Step 2: Loading the preform into the mold cavity

The next step is to load the preform in the mold cavity for molding. Plastics are generally heat–insulating in nature which creates a problem of uneven heat distribution in the preform. The section of preform in direct contact with mold heated first and then the heat is transferred to the other section. Due to the heat-insulating nature, the curing reaction will starts at the surface of the preform, in direct contact with mold, prematurely whereas the lack of temperature in the bulk material restricts the curing. This effect is very detrimental for product development. To avoid this problem the preforms are preheated. The preheating also reduces the curing time of the material and hence increases molding frequency. It also helps in rapid heating when a large pile of polymer powder or pellet are loaded to fabricate a big molding part. The preheating also removes the moisture or other volatile components prior to molding and reduces mold shrinkage. Preheating facilitate the flow of material in the mold and hence shorten the mold filling time and hence shorten the compression cycle. While preheating is beneficial in many ways, excessive preheating can cure the material before it reaches the mold cavities. Hence an optimum time and temperature must be selected for preheating.

There are various processes available by which preheat can be executed and depends mainly on type and form of material to be molded. Electronic or high frequency or dielectric preheating is the most rapid and efficient method for preheating. In this preheating process the molding compounds are placed in between two electrodes and an AC current is passed through the electrodes. Due to this material undergo dielectric polarization along the direction of current flow and due to AC current the direction of polarization changes rapidly which trigger the molecular friction and the frictional force converted into heat. During this process the material heated uniformly both from outside and inside. However, outer surface lose heat to the surrounding rapidly as compared to the bulk material. But during the molding process outer surface brought to the contact of hot mold surface first to compensate this loss. The rate of heating can be controlled by changing the AC frequency and voltage. In order to attain a uniform heating, preform rotators are used with round preforms. But the use of rotators decrease the heating rate slightly. The heating rate of the material increase with temperature and hence a variation in preheating time, even in seconds, significantly affect material temperature. Sometimes the preform are not heated unevenly during this preheating process which can be caused by uneven preform height, unparalleled electrodes, uneven preform density or distribution of moisture content in the preform. These factors must be judicially considered to decide the preheating condition for this preheating process.

The second preheating process is oven heating. This process can be used to preheat both thermoplastic and thermosetting polymers. In this process materials are placed with a depth not exceeding one inch inside a hot air oven fitted with air circulating fans and heaters. The fans circulates the temperature uniformly inside the oven to facilitate uniform preheating. This process require slower and often require agitation to achieve uniform heating. In infrared preheating method the materials are exposed to the IR radiation for heating. The materials can be put on a stationary tray and put under a IR lamp or they can be loaded on a conveyor belt and passed below the IR radiation. While a stationary tray method heat up the material up to 2 in. depth, conveyor only can achieve preheating up to 0.25 in. due to shorter exposure time. In a hot plate preheating process preforms are heated o the electric or steam heated plates or simply placed on the press plate.

Step 3: Closing of Mold

After loading of the preform the mold has to close by using the hydraulic system. Initially the mold is closed rapidly till both halves of mold touches each other. Then the closing speed is decreased to facilitate flow.

Step 4: Breathing cycle

One of the important factor which needs attention during closing of the mold is the evolution of volatile gases during crosslinking reaction. These gases must be expelled out from the closed mold to avoid any defects in molded products. Entrapment of these gases in the product reduced the cross linking density and can cause blisters or ruptures. To avoid this a ***breathing cycle*** has to be given during mold closing to allow easy escape of vapour, air, and other gases from the molding compound as the plastic experiences heating from the mold and from the exothermic reaction of the resin and catalyst during cure. Hence, the breathing time is the time interval between the opening and closing of mold for escape of volatile gases. The timing of starting a breathing cycle can be decided based on the material property where the material achieved maximum point of gas generation. The mold should be opened for a shorter time to allow all the gases to escape from the mold. During the breathing cycle the mold is opened a little and held for 2–4 s (called "breathedwell") and then closed with required pressure to complete the curing. The mold should not be opened for longer time otherwise the crosslinking reaction will make the preform rigid and cannot be pressed in the mold to desired density.

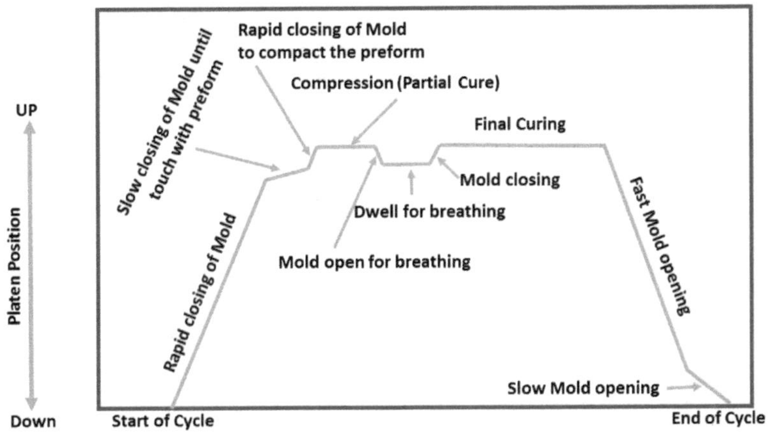

Fig. 6.6 Schematic diagram of platen position during compression molding

A schematic diagram of the movement of platen during compression molding has been represented in Fig. 6.6. The breathing cycle is repeated for 2–3 times for molding relatively larger parts. Apart from volatile gases generated during cross linking reaction the material absorbs moisture which converted to steam during molding and require to be removed during breathing cycle. Previously some 2–3% of moisture was added to the material which converted to steam at molding temperature and act as a heat distribution media during molding. However, modern preheating treatment methods eliminated this process.

Step 5: Curing

After the completion of breathing cycle the material is held under desired temperature and pressure for certain time to complete the curing process which is called dwell time for curing. The curing time can be decided based on the materials used and their composition or a DSC run of exact composition can be run prior to compression molding, to evaluate the curing time required for the material.

Step 6: Opening of mold and ejection of finished product

After the completion of dwell time of curing the mold opens up and the finished product is ejected out from the mold either by automatic ejection system or manually. In some cases the materials can be removed from the mold before the completion of curing reaction if the product achieved desired stiffness for the ejection. The curing can continue for the next several minutes as its temperature gradually returns to ambient temperature. After the completion of this cycle the mold is again filled with a fresh set of preform and the molding process repeated.

6.7 Flow Property of Preform

Flow properties of a preform is a very important parameter which require special attention. During compression molding, the preform flow inside the mold under the influence of pressure and temperature. With increase in temperature the viscosity of the polymer decreased with increase in fluidity which resulted in better mold filling within a short time period. However, increase in temperature also speed up the curing reaction which sets the preform and hence reduces the available flow time. Increase in curing rate also can induce lot of stress in the finished product. Similarly, increase in pressure also increase the flow of preform but excessive pressure can resulted in incomplete mold filling and the material will be flashed out. Hence, there are three parameters that must be considered to set the molding condition. First one is fluidity of the material which will decide the mold filling capability. Secondly, viscosity response of the material with respect to temperature and pressure which will decide the molding temperature and pressure and the third is the available time before the material completely sets in the mold. The total flow that will occurs before the material set is the product of flow time and fluidity. The preform or the molding powder can be classified as soft flow (easy flow) or stiff flow (hard flow) depending on their flowability. For products with delicate parts or complicated structures the small mold cavities must be filled effectively which can be achieved with soft flow materials. However, soft flow material can some time flashed out prematurely without filling the small mold cavities which resulted in product defects. In contrast, stiff flow material can give a better finished product where ever the mold design is less complex.

The flowability of a material depends mainly on molecular weight and curing rate of the thermosets. With increase in molecular weight, viscosity of a polymer increased if other variables remain unchanged. Hence, while a low molecular weight polymer can be classified as soft flow material, its high molecular weight form belongs to the stiff flow category. Apart from molecular weight, curing rate also affect the flowability of the thermosets. With increasing cure rate, the thermosets gets little time to fill the mold cavities. Molds which require little flow, a fast curing thermoset grade preform can be used but where extensive flow is required a slow curing grade is recommended. Similarly the ration of resin to curing agent also can affect the curing rate of thermosets. With an increase in catalyst ration the curing rate increased and hence the flow time reduces. The physical form of preform also plays an important role in compression molding. Thermosets in fine powder form can give a good finished product as compared to the coarse form of the same thermosets under similar molding condition. But finer form of thermosets can gives to dust problem which can be avoided with coarse form.

6.8 Advantages and Disadvantages of Compression Molding

There are both advantages and disadvantages associated with compression molding [4–12]. Among the advantages, the compression molding is a fabrication process which involves very less wastage. It requires low initial setup costs because the machine is very simple and the tooling costs are also low which makes this process very economical and can be setup easily within a short time. The process can be both automated and manual which gives a lot of independency for the selection of compression molding machine depending on one's requirement and economics. Heavy plastic parts can be molded by this process with good surface finish. The molding process is cheaper as compared to injection molding.

Among the disadvantages, the compression molding is not a preferred process for molding complicated parts. Product repeatability depends on worker efficiency for manual compression molding. It is a batch process hence the productivity is slow.

6.9 Transfer Molding

Transfer molding is a method of molding thermoset materials in a closed mold. The transfer molding has been proposed as an improvement that corrects the weaknesses in the compression molding method. This process combines the principle of compression (hydraulic pressing and the same molding materials—thermosets) and Injection Molding (ram-plunger and filling the mold through a sprue). The method is used primarily for molding thermosetting resins (thermosets), but some thermoplastic parts may also be produced by transfer molding. In this process, the thermoset resin is placed into a transfer chamber, immediately ahead of the mold cavity, where the resin is heated until liquid. A plunger then forces the liquid resin, from the transfer chamber, into the mold cavities. A schematic diagram of the whole process is shown in Fig. 6.7.

During this transfer process the mold remain closed to allow the part to cure and solidify. After an appropriate time, the mold is opened and the part is ejected. The part connected to the runner system and the transfer chamber is trimmed off and discarded to get the finished product. A small excess of charge is placed into the transfer chamber to ensure that the cavities are completely filled. Transfer molding is capable of molding part shapes that are more intricate than compression molding but not as intricate as injection molding. Transfer molding also lends itself to molding with inserts, in which a metal or ceramic insert is placed into the cavity prior to injection, and the heated plastic bonds to the insert during molding. For transfer molding, generally pressures of three times the magnitude of those required for compression molding are required. Apart from this, transfer molds can be more complicated than compression molds and are, therefore, costly as compared to compression molding. However, curing time is much less in case of transfer

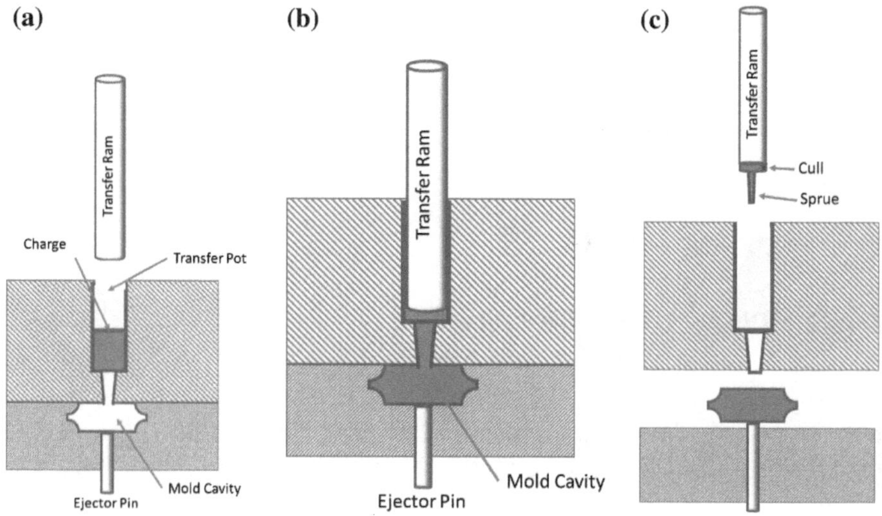

Fig. 6.7 Transfer molding process. **a** Assembly of transfer molding, **b** plunger force on the charge to mold cavity through sprue and **c** removal of finished product by ejector pin

molding as compared to compression molding due to effective heat dissipation during the flow through delivery system.

6.10 Types of Transfer Molding

There are basically two variants of the transfer molding: (a) pot transfer molding, in which the charge is injected from a "pot" through a vertical sprue channel into the cavity; and (b) plunger transfer molding, in which the charge is injected by means of a plunger from a heated well through lateral channels into the mold cavity [10–12]. In both cases, scrap is produced each cycle in the form of the leftover material in the base of the well and lateral channels, called the cull. In addition, the sprue in pot transfer is scrap material. Because the polymers are thermosetting, the scrap cannot be recovered.

6.10.1 Pot Type Transfer Molding (True Transfer Molding)

In true transfer or pot-type transfer molding the mold is closed and placed in an open press. The molding compound is placed into an open pot at the top of the mold and the press is closed. The plunger is placed into the pot before closing the mold. As the press closes, the plunger exert pressure on the molding compound which force it through a vertical sprue, runners, and gates into the cavities. The pot size is generally higher than the total mold area including runners to ensure complete mold filling and the excess material will exert pressure during the curing reaction. After the completion of curing reaction, the plunger is withdrawn, the mold opened, and the parts are ejected. A lot of material is wasted in the pot and hence it is not economical. Due to this short coming, plunger transfer molding is more popular than pot type transfer molding.

6.10.2 Plunger Transfer Molding

In pot type transfer molding the plunger is a part of mold assembly whereas, in plunger molding, the plunger is a part of the press itself which is operated by a hydraulic circuit and a cylinder attached to the head of the press which reduces the projected area as compared to pot type transfer molding. The clamping action of the press keeps the mold closed independently of the plunger movement or force. Even though the molding types are different, in both the cases the behaviour of the molding compound is identical and hence the molding process can be selected based on the availability of equipment, type of die desired, economics of material wastage, and press and mold costs.

The transfer molding can be semi-automatic or fully automatic. In semi-automatic transfer molding the operator actuates the press each cycle to close the mold and manually introduce the molding material in the transfer pot. Then the operator actuates the plunger with appropriate controls to force the material to the mold cavities. The press controls take over to fix the curing time and then open the mold and eject the parts from the cavities using ejector pins. Then the operator lifts the parts from the ejector pin and initiate another cycle. Whereas in a fully automatic transfer molding process the press recycles automatically thereby feeding moulding compounds into the transfer pot and ejecting the finished product after each cycle using the ejector pins. These types of systems are generally configured with a horizontal configuration so that the ejected products can be collected in the collector basket automatically.

Another variation of transfer molding, which belongs to the fully automatic category has been used recently and known as screw transfer molding (Fig. 6.8). In this process the molding material is preheated and plasticized in a screw chamber and pushed into the pot of an inverted plunger mold.

The preheated molding material is then transferred into the mold cavity by the action of the transfer ram which pushed the material into the mold cavity. The same process also can be carried out with a vertical plunger, with the mold placed at the bottom and the plunger pushed the material downward to the mold cavity. Since material can be loaded continuously into the screw chamber with the help of a hopper the complete process is a continuous and automated which increased the productivity of the whole process.

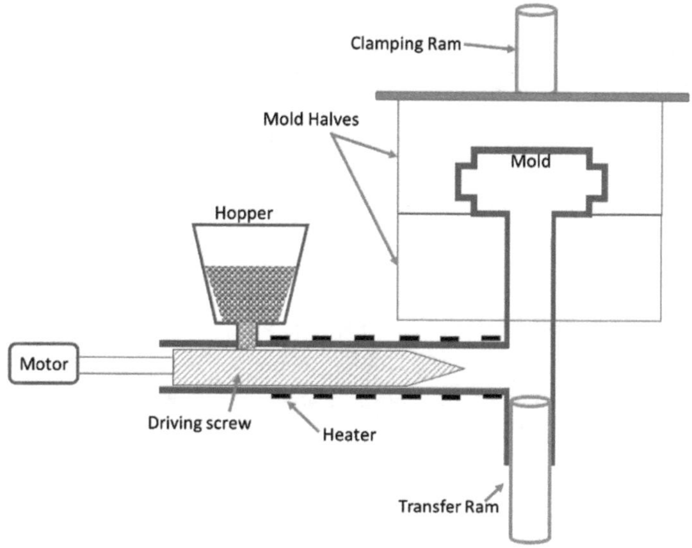

Fig. 6.8 Screw type transfer molding machine

6.11 Process Characteristics

Transfer molding process involves the following steps. A pre-weighed amount of a polymer mixed with additives and fillers (charge) is placed into the transfer pot. The charge may be in form of powders, pellets, putty-like masses or pre-formed blanks. The charge is heated in the pot where the polymer softens. The plunger, mounted on the top plate, moves downwards, pressing on the polymer charge and forcing it to fill the mold cavity through the sprue. The mold, equipped with a heating system, provides curing (cross-linking) of the polymer (if thermoset is processed).The mold is opened and the part is removed from it by means of the ejector pin. If thermosetting resin is molded, the mold may be open in hot state because cured thermosets maintain their shape and dimensions even in hot state. If thermoplastic is molded, the mold and the molded part are cooled down before opening. The scrap left on the pot bottom (cull), in the sprue and in the channels is removed. Transfer molding cycle time is shorter than compression Molding cycle but longer than Injection molding cycle. The method is capable to produce more complicated shapes than compression molding but not as complicated as Injection molding. Transfer molding is suitable for molding with ceramic or metallic inserts which are placed in the mold cavity. When the heated polymer fills the mold it forms bonding with the insert surface. Transfer molding of thermosets is used for molding parts encapsulating metal inserts, wear plates, pins, studs, electronic components with molded terminals. Transfer molding is also used for manufacturing radio and television cabinets and car body shells.

Throughout the transfer molding process the tool and auxiliary parts must be heated to the required temperature, depending on the powder being molded, and the temperature must be maintained during the complete molding cycle. Similar to the compression molding, the molds can be heated by steam, hot waters, and induction heaters. Preheating also used in case of transfer molding to reduce the curing time and to increase the flowability of material. The preheating methods discussed for compression molding can also be used for transfer molding process.

6.12 Test Methods Used Before Molding

Before going for molding, the materials can be tested with certain standards to access the conditions suitable for compression molding.

Spiral Flow test (ASTM D3123): This test method covers a procedure for measuring the spiral flow of thermosetting molding compounds (soft or very soft) designed for molding pressures under 6.9 MPa [1000 psi]. It is especially suited for those compounds that may be used for encapsulation or other low pressure molding techniques. It involves the use of a standard spiral flow mold in a transfer molding press under specified conditions of applied temperature and pressure with a controlled charge mass. This test will give the measure of fusion under pressure, melt viscosity and gelation rate under specific condition. This can be used as a quality control method. In this method the mold and plunger of a transfer molding press is heated to 150 °C. The plunger is equipped with a sealing groove and must be free from any contamination. The transfer pressure must be adjusted to the recommended value. Then the recommended amount of material is transferred into the transfer pot and transfer molded with recommended time, temperature and pressure for complete curing. The cured material then removed from the mold and spiral flow length is directly read from the product. The process required to be repeated 3 times to get an accurate reading.

Cup Flow Test (ASTM D731): This test method covers the measurement of the molding index of thermosetting plastics ranging in flow from soft to stiff by selection of appropriate molding pressures within the range from 600 to 6580 psi. This test method will provide a guide for evaluating the mold ability of thermosetting molding powders. The sensitivity of this test method decreased with the decrease in molding pressure below 950 psi. Hence, pressure below 950 psi is not recommended.

Torque Rheometer test (ASTM D2396 and D2538): This test is used to determine the melt flow values, stability and degradation time at varying shear rates. This can also give the effect of fillers and additives on flow pattern of polymers.

6.13 Advantages and Disadvantages of Transfer Molding

Similar to the compression molding, transfer molding also possessed certain advantages and disadvantages. Among the advantages it resulted in less mold erosion and hence the same mold can be used for longer time without deteriorating the shape of final product and require less tool maintenance. Complex part can be molded by transfer molding because the plunger force used to push the material into mold cavity can be fill the small diameter holes of a complex mold. The flash production is also lesser as compared to compression molding which makes this process more economical. Higher density can be achieved with transfer molding due to additional pushing force during molding. Multiple parts can be molded with this process due to better mold filling ability. The molding and material loading time also lesser which can increase the productivity.

With these advantages, there are several disadvantages are also associated with transfer molding. The wastage can increase because of runners and sprues. The machine cost is higher than compression molding and mold design also complicated as compared to compression molding. Since the process is carried in a closed mold, proper venting must be done to remove volatile gases otherwise a lot of defects will appear in the finished products. Longer travelling time of material can degrade the material and hence affects strength of finished product.

6.14 Comparison of Transfer and Compression Molding

Comparison of compression and transfer molding along with some trouble shooting methods has been summarized in Tables 6.1 and 6.2, respectively.

Table 6.1 Comparison of transfer and compression molding [1–7]

Property	Compression molding	Transfer molding
Material used	Basically developed for Thermosetting but some thermoplastics can also be molded	Basically developed for Thermosetting but some thermoplastics can also be molded
Loading of material	Powder/preforms	Powder/preforms
Preheating of material	Yes	Yes
Mold placement	Open	Closed
Molded product	Simple products can be molded	Complex part can be molded
Product size	Larger size product can be molded	Limited size product can be molded
Resin viscosity	Moderate to high	Low to moderate

(continued)

Table 6.1 (continued)

Property	Compression molding	Transfer molding
Mold filling of small holes	Difficult	Easy
Shrinkage or product	Least	Greater than compression Shrinkage is low across the line of flow as compared to with the line of flow
Productivity	Low	High
Cure time	High	Low
Breathing	Frequently used	Not used because of proper venting arrangements during molding
Molding pressure	Low	High

Table 6.2 Trouble shooting of common problems in compression and transfer molding [3–9]

Fault	Cause	Remedy
Thin flash	Insufficient powder	Increase charge
Thick flash	I. Premature cure before complete mold filling II. High temperature of mold which increases flowability III. Incomplete mold filling IV. Incorrect mold setting	Check curing agent ratio, Molding temperature and pressure Check mold setting
Thick flash with good molding	Excessive material charge	Reduce charge
External blisters	I. Mold temperature too high II. Moisture III. Component undercured IV. Poor degassing or venting V. Insufficient pressure	I. Increase curing time II. Reduce mold temperature III. Increase preheat temperature and check breathing cycle IV. Increase pressure
Poor finish product	Incomplete curing	Use fine powder with proper preheating
Marks from ejector pins	I. Insufficient cure II. Part not cold enough III. Excessive ejection pressure	I. Increase curing time II. Increase dwell time of cooling III. Reduce ejection pressure
Internal voids	I. Insufficient material load II. Improper location of material III. Mold too hot IV. Incorrect degassing V. Insufficient cure time VI. Insufficient pressure	I. Increase material charge II. Decrease mold temperature III. Increase pressure IV. Check breathing cycle V. Increase preheating time and temperature VI. Increase curing time

(continued)

Table 6.2 (continued)

Fault	Cause	Remedy
Poor gloss	I. Lightly insufficient charge II. Excessive or incorrect release agent III. Poor mold finish IV. Old Material V. Long Degassing time	I. Increase charge II. Check releasing agent III. Check mold finish IV. Change material V. Decrease degassing time
Orange peel surface	I. Insufficient pressure II. Press closing too slowly III. Mold temperature too high	I. Increase pressure II. Increase mold pressure III. Decrease mold temperature
Sticking	I. Insufficient or incorrect release agent II. Non-uniform or low temperature III. Lack of mold polish or scratched mold IV. Ejection pins not working V. Insufficient cure time	I. Change releasing agent or apply more II. Check temperature of molding III. Check mold surface and change it IV. Check ejection system V. Increase cure time
Flow marks on surface	I. Material distributed incorrectly giving asymmetric flow II. Insufficient preheating III. Flash line uneven encouraging flow in one direction IV. Material too easy flowing	I. Adjust distribution II. Increase preheat time III. Modify flash line IV. Select stiffer flow powder
Brittle parts	I. Press opening or dosing too slowly II. Mold temperatures too high III. Cure too long IV. Old material	I. Check mold closing and opening cycle II. Decrease mold temperature III. Decrease cure time IV. Change material
Broken product	I. Under cure or over cure II. Sticking to the mold III. High ejection pressure	I. Change cure time II. Apply mold releasing agent III. Decrease ejection force
Porosity	I. Material charge insufficient II. Lack of venting on blind ribs III. Mould temperature too high IV. Flash line too tight V. Material flow too easy	I. Increase weight load to tool II. Add vents or vented ejector pins III. Reduce temperature IV. Ease flash line V. Select stiffer flowing materials

6.15 Molding Temperatures of Common Thermosetting Polymers

Molding temperatures of thermosets vary with their nature and property. Following table will give an idea of common thermosets which are used globally. The molding temperatures listed are only an approximation and will vary with the addition of filler, percentage of curing agent and desired product density (Table 6.3).

Table 6.3 Molding temperatures of thermosets

Polymer	Molding temperature (°C)
Epoxy	140–200
Polyester	80–150
Urea-Formaldehyde	135–160
Phenol-Formaldehyde	140–160
Melamine-Formaldehyde	135–160
Silicone	150–190

6.16 Molding of Composites

Composite material is a heterogeneous material system consisting of two or more physically distinct materials. In a composite material system, the individual materials exhibit their unique properties and the composite as a whole shows properties that are different from its constituents. In addition to the constituent's unique properties, the properties of composites are also dependent on the form and structural arrangements of the constituents and the interaction between the constituents. Broadly speaking, composites consist of two components, a binder or matrix and are inforcement. Polymer matrix composites (PMCs) are comprised of a variety of fillers like short or continuous fibres and bound together by an organic polymer matrix. The matrix functions as the body constituent, serving to bind the reinforcement together and giving the composite its bulk form. The reinforcements are the structural constituents, providing high strength to the internal structure of the composite. Advanced polymer composites are now being applied extensively for all types of applications in the industrial and automotive markets. Molding of these PMC also can be done by both compression and transfer molding. Depending on the filler, the resin can be mixed with the reinforcement and can be compression or transfer molded. In another way the reinforcement in the form of mat, cloth or any other continuous form can be placed in the mold and the resin can be force to fill the mold. This will produce a continuous fiber reinforce composite. The operational parameters vary with the percentage of fillers and resin.

References

1. Processing and Finishing of polymeric materials. Wiley
2. Richardson TL, Lokensgard E (2004) Industrial plastics: theory and applications. Cengage learning
3. Isayev AI (1987) Injection and compression molding fundamentals. CRC Press
4. Lokensgard E (2008) Industrial plastics: theory and applications, engage learning
5. Peters ST (1998) Handbook of composites. Chapman & Hall
6. Kutz M (2011) Applied plastics engineering handbook: processing and materials. PDL handbook series
7. Baird DG, Collias DI (2014) Polymer processing: principles and design. Wiley
8. Rosato DV, Rosato DV (1997) Plastics processing data hand book. Springer, Netherlands

9. Harper CA (2006) Handbook of plastic processes. Wiley
10. http://www.bip.co.uk/
11. http://www.longmold.com/
12. http://www.daikin.com/

Chapter 7
Paints and Coating of Multicomponent Product

P. Sathish Kumar, R. Rajasekar, Samir Kumar Pal, Ganesh Chandra Nayak and S.M.R. Syed Ismail

Abbreviations

MFT	Minimum film forming temperature
Tg	Glass transition temperature
TGIC	Triglycidyl isocyanurete
TMA	Trimellitic anhydride
IPDI	Isophorone diisocyanate
IR	Infrared
UV	Ultra-violet
PRC	Pulsed radiation curing
wft	Wet film thickness
dft	Dry film thickness
ICATS	Industrial coating applicator training scheme
VOC	Volatile organic compounds
PTFE	Polytetra fluoro ethylene
SEM	Scanning electron microscope
PHBV	Poly(hydroxyabutyrate-co-hydroxyvalerate)
CED	Cathodic electro-deposition

P. Sathish Kumar · S.K. Pal
Department of Mining Engineering, Indian Institute of Technology Kharagpur, Kharagpur 721 302, West Bengal, India

G.C. Nayak
Department of Applied Chemistry, Indian School of Mines—Dhanbad, Dhanbad 826 004, Jharkhand, India

S.M.R. Syed Ismail
Rubber Technology Centre, Indian Institute of Technology—Kharagpur, Kharagpur 721 302, West Bengal, India

R. Rajasekar (✉)
Department of Mechanical Engineering, Kongu Engineering College, Erode 638 052, Tamil Nadu, India
e-mail: rajasekar.cr@gmail.com

J.K. Kim et al. (eds.), *Multicomponent Polymeric Materials*,
Springer Series in Materials Science 223, DOI 10.1007/978-94-017-7324-9_7

SAM	Self-assembled monolayer
LBL	Layer-by-layer
AFM	Atomic force microscopy
XPS	X-ray photoelectron spectroscopy
ACF	Activated carbon fibers
PET	Polyethylene terephthalate
HA	Hyaluronic acid
CG	Cationized gelatin
FTIR	Fourier transform infrared spectra
ATR	Attenuated total reflectance
EDX	Energy dispersive X-ray spectroscopy
MWCNT	Multi-walled carbon nanotubes
TEM	Transmission electron microscopy
PSC	Polymer solar cells
PVDF	Poly(vinylidene fluoride)

7.1 Introduction

Paint is a liquid phase, whereas coating and thin film deposition consists of various processes that apply a layer of material onto a surface. Paint is generally considered as a type of coating in which the paint coating have some attractive features such as ease of processing, low cost, ease of field maintenance and commercial availability [1]. Many good research and review outcomes based on paints and coatings for solar thermal applications [2, 3] and military applications for radar detection were published [4, 5]. Products made of metal are almost coated by using electroplating technique (e.g., chrome plating), painting or other process. The major reasons for coating a metal are to (1) provide corrosion protection, (2) enhance product appearance i.e., color and texture, (3) increase wear resistance and to reduce friction of the surface, (4) increase electrical conductivity, (5) increase electrical resistance, (6) prepare a metallic surface for subsequent processing and (7) rebuild surfaces worn or eroded during service. Nonmetallic materials are also sometimes coated. The significant examples include (1) plastic parts coated to give them a metallic appearance, (2) antireflection coatings on optical glass lenses and (3) certain coating and deposition processes used in the fabrication of semiconductor chips and printed circuit boards. In all the cases, the substrate surface must be very clean to achieve good adhesion between coating and substrate. Particle size [6, 7], volume concentration [8], dispersion [7], thickness of the layer [9], application technique [10] influenced the performance of the coating surfaces.

Organic coatings are generally polymers and resins that produced either naturally or synthetically to apply as liquids that dry or harden as thin surface films on substrate materials. The unique advantages include low cost, variety of colors and textures, easy to apply and capacity to protect the substrate surface. However, most organic coatings are applied in the form of liquid and powder. Organic coatings are formulated with (1) binders to enhance the coating properties, (2) dyes or pigments to obtain color to the coating, (3) solvents to dissolve the polymers/resins and add proper fluidity to the liquid and (4) additives [11].

Binders in organic coatings are polymers and resins that determine the solid-state properties of the coating such as strength, physical properties and adhesion to the substrate surface. The binder holds the pigments and other ingredients in the coating during and after application to the surface. The most common binders in organic coatings are natural oils and resins of polyesters, polyurethanes, epoxies, acrylics and cellulose. *Dyes and pigments* provide color to the coating. Dyes are soluble chemicals that color the coating liquid but do not conceal the surface beneath. Generally, dye-colored coatings are transparent or translucent. Pigments are solid particles of uniform, microscopic size that are dispersed in the coating liquid but insoluble in it. Since pigments don't color the coating and are particulate in matter, they also tend to strengthen the coating. *Solvents* are used to dissolve the binder and certain other ingredients in the liquid coating composition. Common solvents used in organic coatings are aliphatic and aromatic hydrocarbons, alcohols, esters, ketones and chlorinated solvents. Different solvents are required for different binders. *Additives* in organic coatings include surfactants, biocides and fungicides, thickeners, freeze/thaw stabilizers, heat and light stabilizers, coalescing agents, plasticizers, defoamers and catalysts to promote cross-linking. These ingredients are formulated to obtain a wide variety of coatings such as paints, lacquers and varnishes [11].

Inorganic coatings [12, 13] and some polymer coatings [14] are suitable for semiconductor applications. In addition, the various factors such as composition of the coating liquid, required thickness of the coating, production rate and cost considerations, part size and environmental requirements influenced the method of applying organic coating onto the surface. Surface preparation involves cleaning and possible treatments of the surface such as phosphate coating are most important for all the applications. For most common applications, the produced film has a thickness range of 0.5–500 μm. In some cases, metallic surfaces are plated prior to organic coating for maximum corrosion protection.

7.1.1 Film Formation

A film is a thin layer of coherent solid matter or liquid bounded to a substrate. Polymer flow viscosity and uniform distribution onto the substrate plays a vital role in developing the good film. The flow viscosity (range of 0.05–1.0 Pa s) mainly depends on the coating technique and the substrate. The main factors affecting film formation are minimum film forming temperature, effect of solvents, stabilizers, substrate, polymer composition, substrate composition, physical characteristics such as particle size and quality of dispersion and the environment such as temperature and relative humidity [15].

The complex film formation from an aqueous polymeric dispersion was examined by several researchers [16–18]. In the wet state, discrete polymer particles come together in close contact, deform, coalesce and ultimately fuse together to form a discrete film. During processing, the substrate surface will be wetted using diluted dispersion of liquid or solid matter. However, removal of water brings the polymer particles close together to form a group because of capillary action. Minimum film-forming temperature (MFT) is the minimum temperature to form a film under individual processing conditions. It also depends on glass transition temperature (Tg) of the polymer. In this temperature, the hard glassy form of a largely amorphous polymer changes to softer, more rubbery and in consistency state. Lehmann [19] reported the concept of MFT includes the plasticizing effect of water on the film-forming process. He also suggested that for aqueous dispersions, coating temperature of 10–20 °C or above the MFT must be kept to ensure the optimal conditions for film formation. MFTs of Eudragit aqueous dispersions are described by Lehmann [20]. Upon application by spraying, brushing or various industrial processes, surface coatings undergo film formation. In most film-formation processes, a liquid coating of relatively low viscosity is applied to a solid substrate and cured. High molecular weight polymer based adherent films possess the desired properties. Coatings before the 1960s were often liquids of low solids content, from which considerable organic solvent was emitted into the atmosphere during film formation. Environmental and economic pressures have forced a reduction of solvent levels in coatings and coating designers redesign to improve the film-formation processes. Finally, there are three major types of film processes: evaporation of solvent or carrier liquid, cross-linking of low-molecular-weight, low-viscosity polymer precursors and coalescence of small particles. For a specific coating, the mixture of these processes may be adopted for overall film-formation.

7.1.1.1 Crosslinking Film Formation

Three-dimensionally cross-linked network were build using highest-performance coatings films based on the chemical reaction of polymer precursors. This is one of the very old techniques for film formation and still using now-a-days. During the middle ages, drying oils were used without solvent to formulate a paint that formed films totally by oxidative cross-linking. Drying oils are natural products such as linseed oil or tung oil that contain at least 50% unsaturated fatty acid triglycerides. They react with oxygen in the air to form crosslinked polymer networks to obtain decorative and protective properties. Drying oils modified with soluble natural resins such as tree gum, rosin and naturally derived solvents such as turpentine are known as varnishes. When cast and allowed to dry on various substrates, varnishes form films by evaporation of the solvent and by the cross-linking reactions of the unsaturated fatty acids in the oils. The cross-linking reactions are quite complex and by adding atmospheric oxygen leads to the formation of hydroperoxide derivatives of the fatty acids. These hydroperoxides decompose especially in the presence of driers such as white lead or cobalt naphthenate to form free radicals for further crosslinking of unsaturated fatty acid [21]. New cross-linking technologies are based on two-component 100% solids reactive systems that are mixed just prior to or during application and form the final polymer coating by rapid cross-linking. In many cases, solvents are used to control the flow viscosity, in which rapid polymerization enhanced to the higher extent. Furthermore, a catalyst is required to complete the reaction on time and temperature required for the specific application.

7.1.1.2 Evaporation Based Film Formation

Evaporation-based film formation is based on low solids content and large amounts of organic solvents. It is one of the fastest and simplest methods of film formation and was the basis of the nitrocellulose lacquers used in automotive production lines from the 1920s to the 1950s. It is still the mode of film formation of many spray paints by releases large quantities of solvent into the atmosphere. For this reason, the use of lacquers has been banned by environmental legislation. In this technique, the molecular weight and the properties of the coating polymers were thoroughly studied before starting the process. In addition, pigments and additives are then incorporated to develop the fully formulated coating. The liquid coating is applied to a substrate and the film forms solely by solvent evaporation method [21].

7.1.1.3 Coalescence Based Film Formation

Polymer particles with a size of 0.05–1.0 μm, dispersed in water or organic solvent, rubbery and higher Tg forms a clear polymer film onto the substrate. The polymer particles suspended in the water flow together or coalesce to form a film because of surface-mediated forces. If the polymer is rigid, glassy and has lower Tg, a small amount of coalescent is added to the system to assist film formation. This coalescent later evaporates and leaves the solid polymer film. Coalescence-based film formation takes place mainly with latex polymers and some organic solvent dispersed polymer particles. However, limitations on the use of organic solvents have made water the predominant carrier solvent. Another mode of film formation closely related to water-based coalescence is the melting and fusing of solid paint particles known as powder coating. It is a process in which an object is coated by a spray or fluidized bed of pigmented polymer particles and the particles are fused by heating to form a continuous film. Some other reactions may also be visualized during the melting and fusing processes. However, the predominant film-formation reaction is the fusing or coalescence of the dispersed particles [21].

7.1.2 Thermoplastic Polymers

Thermoplastics melt and flow with the application of heat but retain the same chemical composition when cooled. Common thermoplastics are polyvinyl chloride, nylon, polypropylene, and thermoplastic polyester. They are used as encapsulation materials for a variety of parts [22]. There are a number of ways to produce effective, coherent films. The easiest to visualize is that of dissolving a polymer in a suitable volatile solvent, applying the coating and then allowing the solvent to evaporate. Solvent selection is vital in developing film integrity from solution polymers. Ideally a theta solvent that is a solvent that completely dissolves the polymer or a combination of solvents to produce complete dissolution should be employed. The polymer then exists as discrete polymer molecules in solution which are forced together as the solvent evaporates and the polymer concentration correspondingly increases [15].

Film formation requires that the polymer molecules remain mobile and stable, at least in the early stages of solvent evaporation and/or solvent absorption into the substrate. The result is a layer of polymer which is built up on the substrate. The solvent, or solvent mix, needs to solvate the polymer and resist precipitation of the polymer. If the polymer precipitates before film formation occurs, then the final coating may appear cloudy or powdery.

In order to maintain good application properties, solution polymers are often formulated with low molecular weight species (less than 50,000) at low concentrations. In this respect, initial solvent release is not affected by the presence of the polymer. In solvent mixtures a highly volatile solvent may be employed to develop an early touch-dry coating. The remaining solvents need to be efficient in order to allow good film integrity. Initial solvent release is therefore a function of vapor pressure, temperature, surface area of the coating and air flow over the surface. However, as the relative polymer concentration increases, the solvent release is no longer controlled by evaporation but by diffusion through the polymer film. Total solvent release may not occur for hours, days or even years. Films may be initially plasticized by residual solvent but this could lead to brittle films over a period of time. Residual solvent may also be responsible for the development of cloudy or hazy films. Thermoplastic polymers are capable of flow under temperature. Heating can there fore improve film formation by further entanglement of polymer chains and by greater removal of residual solvent [15]. Films formed from thermoplastic solution polymers are, by their nature of formation, susceptible to solvent attack. Excellent cohesion between coats can be obtained with solution polymers as a result of solvent penetration into the previous film.

7.1.3 Thermoset Polymers

In the discussion on thermoplastic solution polymers it was noted that relatively low molecular weight polymers had to be used in order to get the desired application viscosities. Such coatings can have relatively poor thermal or solvent resistance. More-resistant films can be obtained if the molecular weight of the coating increases upon application. Alkyd resins, two-pack epoxies and moisture-cured polyurethanes are examples of such coatings and are discussed elsewhere in this book. Thermoset polymers have high molecular weight and do not flow upon heating. Ideally the entire film is one polymer molecule with a high degree of crosslinking, although in practice the molecular weight depends on the functionality of the reactant species. The concepts of film formation are similar to those discussed in the previous section where the thermoset polymers are initially applied from solution. Solvent evaporation still plays an important role in film integrity. However, the reactivity of the chemicals involved will also have an effect on the film properties. The viscosity of the film will increase due to both solvent evaporation and increasing polymer molecular weight. Formulations therefore need to take into consideration good flow properties before the coating increases in viscosity. Residual solvent can be more of a problem in thermoset polymer films due to entrapment by the growing polymer. Similarly, some thermosetting polymers continue to react for long periods eventually producing a very hard, yet brittle, film. In 100% active systems (such as some two-pack epoxies) it is necessary only to have the system at a viscosity suitable for application, and flow, before any significant reaction and increase in viscosity occur.

The most common some of the important thermoset polymers and their characteristics are listed below [23]:

Epoxies—Epoxies are used in powder coating industry with excellent mechanical surface properties, pencil hardness, corrosion resistance, chemical resistance and impact resistance. These resins also can be bent around 1/4 in. mandrel with no loss of adhesion. Epoxies have poor weathering resistance and consequently best suited to indoor applications [24].

Hybrids—The combinations of epoxy and polyester resins designed for a good mix with poor weather resistance. The presence of polyester resins reduces yellowing of the film produced by overbaking.

Urethanes—Due to its superior exterior durability, it is best suited for outdoor use. It has very good surface properties (hardness, flexibility, corrosion protection, etc.) better than epoxies. It can be applied in thin coats, whereas heavier coats form thicker films.

Acrylics—It is more common in U.S. with very good to fair surface properties. It is suited for exterior use because of its good weather resistance.

TGIC-TGIC polyesters coat the thickness of about 75–125 μ by incorporating the cross-linking agent i.e., triglycidyl isocyanurete. It produces films with excellent surface properties such as hardness, flexibility, exterior durability, corrosion and overbaking protection [23].

It causes allergic reactions when expose to trimellitic anhydride (TMA), a monomer used in polyester resins. Hybrid resins, TGIC resins with TMA based powders can help to reduce these reactions [25]. In response, several chemical companies have launched alternative hardeners, including caprolactam-blocked isophorone diisocyanate (IPDI) adducts [26].

7.1.4 Curing Methods

The coating performance depends mainly on the appropriate curing conditions. If the curing is adequate, mechanical and other related properties associated with a given thermally crosslinking coating can be assured [27–35]. Bonding to a subsequent layer may be affected in under-cured coatings. However, the maximum curing temperature and curing time may results in good technical properties. The type of resin determines the type of chemical reaction that takes place in curing [11]. Curing process may be either physical or chemical depends on the suitable combination of binders.

7.1.4.1 Physical Curing

Polymers dissolved in organic solvents gradually adhere to form a solid film and then a network. Cohesion occurs solely as a result of solvent evaporation without chemical cross-linking. Such polymer films are generally reversible, i.e., they dissolve in the original solvent [28]. Physically drying binders (e.g., nitrocellulose and its esters, vinyl resins, polystyrene, acrylate esters, chlorinated rubber, bitumen) are mostly chain-like or thread-like molecules with short side chains.

7.1.4.2 Chemical Curing

Chemically cured film occurs as a result of formation of chemical bonds between the binder molecules. As crosslinking starts, the binders become insoluble and form thermoset films [28]. In solvent-containing systems, physical curing also occurs simultaneously.

The three common curing methods [11] are as follows:

1. Curing with a heat carrier (air)
2. Curing with infrared (IR), ultra-violet (UV), electron beams, laser beams, plasma arc radiation
3. Curing by means of electrical processes (inductive curing, resistance, high-frequency and microwave curing).

7.1.4.3 Curing with Heat Carriers

In this method, circulating hot air stream evaporates the solvent vapors to from coated film [27]. Heat transfer and film formation occur from the exterior to the interior. The parts having a complicated shape may also be cured using this process. However, long curing times and large ovens are required for this curing technique. Heat consumption is relatively high because the workpiece, coating film, transporting device, parts of the conveying system and the fresh air all have to be heated simultaneously.

7.1.4.4 Curing with Radiation

Radiation curing methods have become increasingly important in the last few years. Paint curing proceeds more rapidly than in circulating air curing since the whole workpiece does not have to be heated. However, only large flat parts can be satisfactorily treated [27]. Microwaves (frequencies 3–600 GHz, wavelength 0.5 mm–10 cm) are generated to exhibit wave effects such as interference which lead to localized concentrations of energy (wave peaks). Interaction between the polar material in the coated film and the electromagnetic alternating field is manifested macroscopically as a heating effect. Since microwaves are reflected by electrically conducting surfaces, this method can be used only for non-conducting substrates (plastics, wood, or paper).

IR radiation (wavelength of 1 mm) is absorbed, reflected or transmitted by an object. In the coated film, absorbed radiation is converted into heat. The coated film cures and does not harden initially. The wavelength and intensity of the IR radiation must be matched with the hardened coating. Absorption behavior is determined by the pigment, pigment volume concentration and binder. Long wave IR radiation is absorbed by the pigments at the surface, while shortwave radiation can penetrate onto the coating layer. Normal thermal outputs are 5–25 kW/m^2 and utilizing suitable regulating machineries may increased up to 100 kW/m^2. *UV radiation* with a wavelength range of 0.0003–0.0004 nm initiates photochemical reactions which lead to cross-linking. An added photo initiator (sensitizer) decomposes the coating into free radicals that initiate polymerization of the binder. Curing with UV radiation is of practical importance for hardening colorless polyester putties or primers, offset and printing inks. The hardening times are of the order of a few seconds. *In pulsed radiation curing* (*PRC*), exothermically reacting coating and printing inks are cured by UV pulses at a wavelength of ca. 197 nm. The pulses break the carbon double bonds of the binder because this wavelength range corresponds to their resonance frequency. A chain reaction starts throughout the whole layer and curing takes place within a few seconds. *Electron beams* are generated by applying an accelerating voltage (150 kV) to a thermionic cathode. An electron beam (ca. 6 mm diameter) is spread out into a curtain beam by a beam splitter. The electrons leave the beam distribution housing through a very thin metal sheet. When these electron beams strike binder monomers, they initiate polymerization in the coated film. Polymerization occurs in a fraction of a second and must be performed in a vacuum or in an inert gas atmosphere. The equipment must be screened to protect the operators [27]. Sharply localized curing can be achieved only with the help of *laser beams.* The carbon dioxide laser, which has a total beam output of 100 W/cmZ at a wavelength of 0.01 nm can be used for this purpose. *A plasma arc* may also initiate cross-linking in the coated film. The high temperatures produced in the interior of the plasma arc are transmitted only to a small extent.

7.1.4.5 Curing by Electrical Methods

In this method, electric current is directly converted into heat in the workpiece or in the coated film. In inductive curing, an induction coil is located close to a metallic workpiece and generates eddy currents in it. Therefore, the workpiece becomes hot and the coating is heated from beneath the film to evaporate the solvent or curing to occur. In high frequency curing, the workpieces are arranged between two capacitor plates in a radiofrequency field (108–109 Hz). Dipoles align themselves and are polarized in the alternating electric field [27].

7.2 Paint Coating

Paint is the most commonly used material to comply with industrial environmental legislation and to overcome the demands with improved durability performance. The advanced formulation for coat application with individual film thickness replaced the conventional coating systems. The most important illustrations are epoxy and polyester glass flake coatings for high build thickness in one or two coat applications and single coat high build elastomeric urethane coatings up to 1000 μm thick [36–38].

Modern specifications consist of sequential coating application of paints or alternatively paints applied over metal coatings to form a duplex coating system. The protective paint systems usually comprise of primer, undercoats and finish coats. Each coating layer in any protective system has a specific function and the different types of primer are applied followed by intermediate/build coats in the shop and finally the finish or top coat either in the shop or on site.

7.2.1 Composition of Paints and Film Formation

Paints are made by mixing and blending three main components [36–38]:

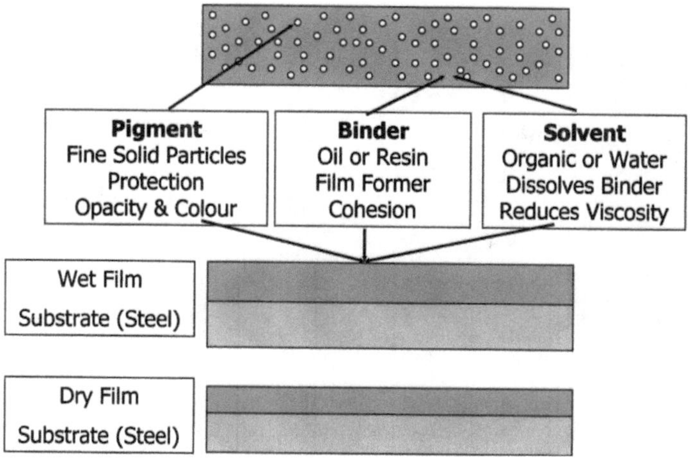

Fig. 7.1 Paint constituents and their function. Reused from [36]

- **Pigments**: Finely ground inorganic or organic powders which provide color, opacity, film cohesion and sometimes corrosion inhibition.
- **Binders**: The binder is the film forming component in the paint. For example, organic compounds such as resins or oil.
- **Solvents**: Organic liquids or water used to dissolve the binder and to facilitate application at the paint.

The major paint constituents and their functions are displayed in Fig. 7.1.

Paints are applied to the steel surfaces by means of many methods to produce a wet film. The thickness can be measured using a comb-gauge before the solvent evaporates. As the solvent evaporates, film formation occurs tends to leave the binder and pigments on the surface to form dry film. The thickness of the dry film can be measured using electro-magnetic induction gauge. The corrosion protection afforded by a paint film is directly proportional to its dry film thickness [36–38].

The relationship between the applied wet film thickness (wft) and the final dry film thicknesses (dft) is determined by the percentage volume solids of the paint.

$$Dry\,film\,thickness\,(dft) = wet\,film\,thickness\,(wft) \times\,vol.\%\,of\,solids$$

7.2.2 *Classification of Paints*

In detail, paint comprise of a pigment, dispersed in a binder, dissolved in a solvent limits the number of generic types of paint. The most common methods of classifying paints are either by their pigmentation or through binder type [36–38]. Primers for steel are usually classified according to the main corrosion inhibitive pigments and range of binder resins used in their formulation, e.g. zinc phosphate

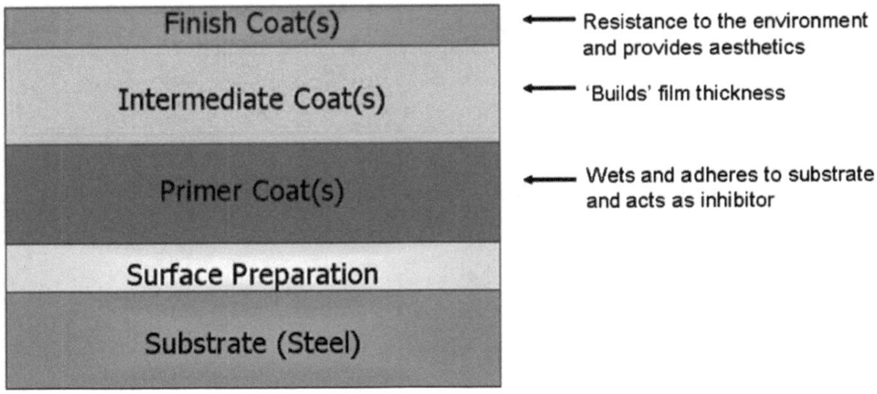

Fig. 7.2 Schematic diagram of a paint system. Reused from [36]

primers and metallic zinc primers, zinc phosphate alkyd primers, zinc phosphate epoxy primers, etc. Intermediate and finish coats are usually classified based to their binders, e.g. epoxies, vinyl, urethanes, etc.

7.2.3 Paint Coatings

Paints are usually applied one coat on top of another and each coat has a specific function purpose (Fig. 7.2). These are described as follows [36–38].

7.2.3.1 Primers

The primer is applied directly onto the cleaned steel surface or in the sealed metal coating for duplex systems. Its purpose is to wet the surface and to provide good bonding for subsequently applied coats. It is required to provide corrosion inhibition onto the surfaces. There are two basic types of primer namely (a) Primers pigmented with metallic elements anodic to steel. Zinc-rich primers are the most commonly used of this type and (b) Primers relying on the high adhesion and chemical-resistance properties of the binder.

Two-pack epoxy primers are typical of this type. This contains inhibitive pigments to interfere with the corrosion process. Zinc phosphate, commonly used inhibitive pigment in modern primer formulations.

7.2.3.2 Intermediate Coats

Intermediate coats are applied to develop the total film thickness of the system. Generally, the thicker coating has longer life. Intermediate coats enhanced the overall protection and lowered the permeability to oxygen and water. The incorporation of laminar pigments reduced the moisture penetration in humid atmospheres and shows improved tensile strength. Along with the conventional laminar pigments, glass flakes were added to improve the technical properties. Undercoats must remain compatible with the finishing coats.

7.2.3.3 Finish Coats

The finish coats furnish the required appearance and surface resistance of the system. Depending on the conditions of exposure, it must also render the first line of defense against weather and sunlight, open exposure and condensation.

7.2.3.4 Stripe Coats

Stripe coats are additional coats of paint that are applied locally to welds, fasteners and external corners. The main purpose is to build a satisfactory coating thickness at edges and corners where paint has a tendency to contract and thin upon drying. The type, duration and number of stripe coats required decided the performance characteristics of this coat.

7.2.3.5 The Paint System

The various superimposed coats of the same generic type or different within a painting system have to be compatible with one another. For example, chemical resistant types such as a re-coated polyurethane finish coat may be applied onto epoxy primer and intermediate coats [36–38]. An important factor in the coating system is the definition and measurement of the dft. Dry film thicknesses are generally checked for the complete paint system and individual films. For nominal dry film thicknesses, individual values less than 80% of the nominal thickness are not acceptable. Values between 80 and 100% are acceptable provided that the overall average is equal to or greater than the nominal. Specifications for minimum dry film thicknesses require careful paint application to avoid excessive film thickness. This may also results in the formation of high stresses and may cause premature failure of the system. The calculation of wet film thickness for the coating helps to verify the dry film thickness.

7.2.4 Main Generic Types of Paint and Their Properties

Air drying paints: Oxidation process helps to form dry thin films by means of absorbing oxygen from the atmosphere. The developed film usually had limited solvent resistance and poor chemical resistance [36–38].

One pack chemical resistant paints: Solvent evaporation technique is used to form thick films by settling of solvent in the film. No oxidation process is involved. The produced film remains relatively soft with poor solvent resistance and good chemical resistance. Bituminous paint solutions with asphaltic bitumen or coal-tar pitch in organic solvents may dry through solvent evaporation method.

Two pack chemical resistant paints: Two separate components such as base and the curing agent were mixed prior to the process to initiate the chemical reaction. These materials have minimum pot life before the coating. The polymerization reaction and solvent evaporation technique produced a densely cross linked hard film with good solvent and chemical resistance. Liquid resins of low viscosity can be used in the formulation thereby avoiding the need for a solvent. Such coatings are referred to as solvent less or solvent free coatings for very thick films. Some of the paint binders and their technical properties were listed in Table 7.1.

7.2.5 Prefabrication Primers

It is also known as blast primers, shop primers, temporary primers, holding primers, etc. [36–38]. The application on structural steelwork after blast cleaning tends to maintain the reactive blast cleaned surface in a rust free condition. They are mainly applied to steel plates and sections before fabrication. The main requirements of a prefabrication primer are as follows:

- The primer should be capable for airless spray application to produce a very thin even coating. Dry film thickness is in the range of 15–25 μm. Below 15 μm, the peaks of the blast profile are not protected and rust rashing may occur during weathering. Above 25 μm, the primer affects the quality of the weld and produces excessive weld fume.
- The primer must dry very quickly. Priming is often done in-line with automatic blast cleaning plant followed by handling. The interval between priming and handling process is of 1–10 min and hence the primer film must dry within this time.
- Normal fabrication procedures (e.g. welding, gas cutting) must not be significantly impeded by the coating and the primer should not cause excessive weld porosity.

Table 7.1 Summary of the generic types of paint and their properties

Binder	System cost	Tolerance of poor surface	Chemical resistance	Solvent resistance	Water resistance
Black coatings	Low	Good	Moderate	Poor	Good
Alkyds	Low–medium	Moderate	Poor	Poor–moderate	Moderate
Acrylated rubbers	Medium–high	Poor	Good	Poor	Good
Epoxy	Medium–high	Very poor–good	Good	Good	Good
Polyurethane	High	Very poor	Very good	Good	Very good

Reused form [36]

- Weld fume omitted by the primer must not exceed the appropriate Occupational Exposure Limits. Proprietary primers are tested and certified by the Newcastle Occupational Health Agency. A health and safety certificate should be available from the paint manufacturer.
- The primer coating should provide adequate protection. In aggressive conditions, durability can often be measured in weeks rather than months. Zinc rich and zinc silicate primers provide the highest order of protection of all prefabrication primers.
- The primed surface after weathering should require the minimum of re-preparation for subsequent painting and must be compatible with the intended paint system. Many proprietary prefabrication primers are available but they can be classified under the following main generic types.

7.2.5.1 Etch Primers

Etch primers based on polyvinyl butyral resin reinforced with a phenolic resin render enhanced water resistance. It can be supplied in a single pack or two pack form to provide better durability.

7.2.5.2 Epoxy Primers

Epoxy primers are two pack materials utilizing epoxy resins and have either polyamide or polyamine curing agents. They are pigmented with a variety of inhibitive and non-inhibitive pigments. Zinc phosphate epoxy is chosen as a best primer with better durability.

7.2.5.3 Zinc Epoxy Primers

Zinc epoxy primers can be either zinc rich or reduced zinc types. Zinc rich primers produce films with 85 wt% of metallic zinc powder and the reduced zinc type as low as 55% by weight. In marine or highly industrial environments, zinc epoxy primers form insoluble white zinc corrosion products which must be removed from the surface before over-coating.

7.2.5.4 Zinc Silicate Primers

Zinc silicate primers develop a level of protection which is in accordance with the zinc rich epoxy types and render similar drawbacks, e.g. formation of zinc salts and production of zinc oxide fume during welding. Based upon the binder and the zinc content, it can be classified into different categories. Low zinc primers were developed and display improved weldability, reduced durability and minimized weld porosity. Finally, organic silicate primers are the most important prefabrication primers.

7.2.6 Application of Paint Coatings

The quality and durability of the coating mainly depends on the method of application and coating conditions [36–38]. Standard methods used to apply paints include application by brush, roller, conventional air spray and airless spray. Airless spraying has become the most commonly used method of applying paint coatings to structural steelwork under controlled shop conditions. Brush and roller application are more commonly used for site application, though spraying methods are also used.

7.2.6.1 Brushing

Brushing is the simplest, slowest and most expensive method. Compared to other methods, it has certain advantages like better wetting of the surface, used in restricted spaces, useful for small areas, less wastage and contamination of surroundings.

7.2.6.2 Roller

Roller process is bit quicker than brushing that needs suitable rheological properties of the paint and found application in large flat areas.

7.2.6.3 Air Spray

Spray paint coatings were normally employed for structural steelwork materials. The paint is divided into fine droplets, projected onto the surface and joined together to form a continuous film. The atomization can be accomplished in a number of ways.

In air spraying, the paint is atomized and applied using conventional spray gun. The paint can be either sucked into the air stream or fed to the spray gun under pressure from a pressure pot. For ideal application, careful adjustments of the spray nozzle and air pressures must be made based on the composition of paint product and film thickness. The application rates for air spray are quicker for brushing or rolling results in high paint wastage.

7.2.6.4 Airless Spray

For airless spraying, the paint is hydraulically compressed, released through airless spray gun, atomized and projected onto the surface. Depends on the orifice size, shape and hydraulic pressure, atomization develop thin to thick coating with a wide range of deposition. The equipment required is much more expensive, highly skilled labor and render high hydraulic pressure compared to conventional air spraying. However, the application rates are higher with reduced overspray wastage. This find application for solvent-free materials such as two-pack products, in which mixing occurs at the moment of application.

7.2.7 Conditions of Application

Temperature and humidity are the two main principal conditions that affect the application of paint coatings. These can be more easily controlled under shop conditions than on site [36–38].

- **Temperature**: Air temperature and steel temperature affect solvent evaporation, brushing and spraying properties, drying and curing times and the pot life of two-pack materials, etc.
- **Humidity**: Paints applied affect the application or drying of the coating due to relative humidity of the atmosphere. The optimum minimum steel temperature is maintained at least 3 °C above the dew point.

However, moisture cured paints are available. These paints are specifically formulated for application in damp and humid conditions.

7.2.8 Coating Applicator Training and Certification

To achieve the desired performance, industry formulated training and certification scheme (ICATS—Industrial Coating Applicator Training Scheme) for paint applicators, highway agency and network rail bridges [36–38].

The scheme has the following 6 modules:

- Health and Safety
- Site Access
- Plant and Equipment
- Surface preparation
- Paint Types and Application
- Quality Control

There are also 4 optional specialist modules:

- Abrasive Blast Cleaning Operator
- Paint Sprayer
- Thermal (Metal) Sprayer
- Water Jetting Operator

ICATS enables contractors to provide best value service to maximize coating performance over structure life and to reduce whole life costs.

7.3 Coating of Fabrics and Textile or Leather

Top finishes began to be used already in the 18th century when fabrics were coated with linseed oil to produce oilcloth. This was the first procedure of coating several agents to the textile substrate and can be considered as the predecessor of multi-layered materials [39]. Multilayered materials can be produced in different ways:

- by laminating a polymer layer to the textile surface material
- by direct applying a polymer to the textile material
- by indirect applying a polymer to the textile material

The chemical compositions of polymer coatings are constantly developed and new types of polymer additives are increasingly introduced. The influence of industry globalization, the requirements of suppliers and consumers as well as new technologies change the market and expand the use of polymer coated textile materials.

The key factor for the success of polymer coating is its versatility and long durability [39]. Coating of fabrics with rubber is an important process in the rubber industry. These composite materials are used in automobile tires, conveyor belts, inflatable rafts and waterproof cloth for tarpaulins, tents and rain coats. The coating of rubber onto the substrate fabrics includes a variety of processes. The most common fabric coating method is calendaring in which the fabric is fed into the calendaring rolls to obtain a reinforced rubber sheet [39–41]. Polymer layers can be polyurethane, polyvinylchloride or polyacrylate layers. To improve their properties, appropriate additives are added: softeners, porosity-generating agents, filling materials, binders, fungicides etc. Coated polymers are applied to the textile material directly, and indirectly using paper or coagulation procedure [39]. Abrasion resistance and strength are by far higher in polyurethane coating compared to other polymers. Polyurethane has the property of good adhesion which can be strengthened by addition of cross-linking agents. Hardness or softness can be achieved by variation of polymer structures without using a plasticizer. It is also possible to reduce fragility by the impact of light [42, 43]. Textile composite materials are composed of two or more different materials with at least one textile layer (woven fabric, knitted fabric or nonwoven material). All components composing the final product affect the properties of multilayered composites. This kind of composites have multiple advantages over the classic fabric since they are more durable and stronger, their body protection against meteorological effects (rain, wind, UV radiation), they did not lose their comfort (they are airy and have good sweat permeability), they are more resistant to abrasion and load and they have less anisotropic properties in contrast to the classic fabrics [39]. The mass losses of the coated textile materials after and before abrasion were depicted in Fig. 7.3.

Alternatives to calendaring process include skimming, dipping and spraying. In the skimming process, a thick solution of rubber compound in an organic solvent is applied to the fabric as it is unreeled from a supply spool. The coated fabric passes under a doctor blade that skims the solvent to the proper thickness and then moves

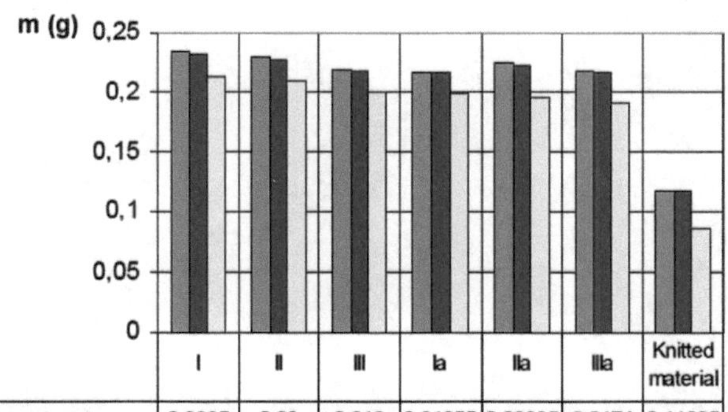

	I	II	III	Ia	IIa	IIIa	Knitted material
Mass before abrasion	0,2335	0,23	0,219	0,21655	0,22385	0,2174	0,11825
Mass after 5,000 cycles	0,23325	0,22845	0,2186	0,216	0,22315	0,2164	0,1171
Mass after 10,000 cycles	0,213	0,21	0,201	0,1983	0,1955	0,191	0,0855

Fig. 7.3 Mass loss of the coated textile materials. Reused from [39]

into a steam chamber to drive off the solvent by heat. Dipping involves temporary immersion of the fabric into a highly fluid solution of rubber, followed by drying. In spraying, a spray gun is used to apply the rubber solution onto the fabric substrate.

Polymer coated textile materials have a wide range of application, from the textile industry to technical textiles. The coating procedure can be direct or using siliconized paper. When polyurethane is coated directly, the polymer is coated using special coating blades indirectly to the textile material. When coating is indirect, the polyurethane polymer is coated first to the paper, and then is laminated with the substrate and the textile material respectively. When it is first coated to the paper, it can be in several layers. After each coating, the polymer is dried and cooled down. Upon completion of the coating procedure, the paper is separated from the finished material. The paper returns to the machine entry and can be used for further coatings, approximately from 8 to 10 times [39]. By coating polyurethane paste to textile materials, materials known as artificial leather is obtained. They occupy an important place on the market. Artificial leather is unthinkable without the textile substrate. The scanning electron microscope (SEM) images of acrylate based carbon nanotubes coated onto the textile materials were shown in Fig. 7.4. Coated textile products either as artificial leather or as laminates assume properties of the materials they are made of. Since they are partially made of the textile that is in its properties mostly anisotropic, coated material as a whole is also anisotropic, meaning that the coated material behaves differently in different direction when stressed. A change in polymer coatings and their properties related to textile materials affect final properties of multilayered materials. Likewise, adding a target polyurethane coating and after treatment, even the selection of color can provide a target product with appropriate properties [39].

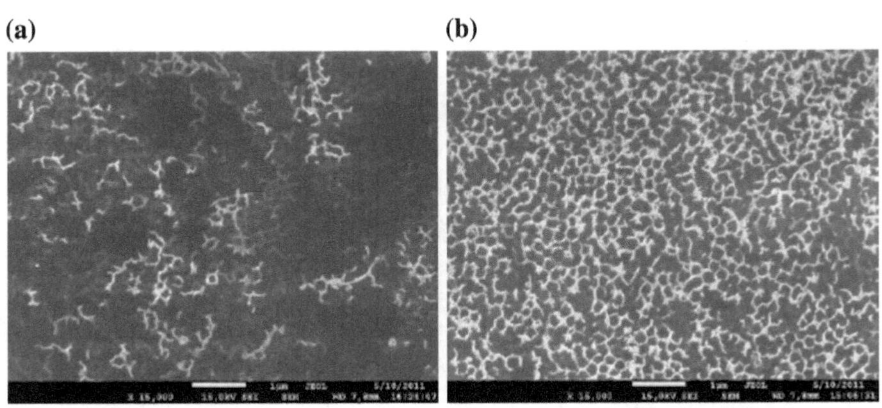

Fig. 7.4 SEM images of acrylate coating with 6 wt% CNT (**a**) and 13 wt% CNT (**b**). Reused from [41]

7.4 Spray Coating

7.4.1 Overview

Spray coating is the commonly adopted technique for applying organic coatings. The main process involves forcing the coating liquid to atomize into a fine mist immediately prior to deposition onto the part surface. When the droplet reaches the surface, they spread and form a uniform coating over the entire work surface. The process can be performed manually in spray painting booths or through many automated process. One of the most important applications of spray coating is in automotive industry for applying external paint coats to car bodies. The typical sequence of coating sheet-metal automobile car bodies in mass production are (1) phosphate and primer coat applied by dipping, (2) color paint coat using spray coating and (3) clear coat (for high gloss and added protection) through spraying techniques. It is also used for coating appliances and other consumer products [27].

Immersion technique utilizes large amounts of liquid coating to the work surface and allows the excess to drain off and recycled. The two common methods of immersion technique are dip coating and flow coating. The easiest method is dip coating, in which a work part is immersed in an open tank of liquid coating material. Suddenly, the part is removed and the excess liquid drains back into the tank. A change of dip coating technique is electrocoating, in which the part is electrically charged and then dipped into a paint bath to produce opposite charge. This enhances the adhesion and allows use of water-based paints to reduce fire and pollution hazards. In flow coating process, the work parts are moved through an enclosed paint booth which contains a series of nozzles shower to coat the liquid. Excess liquid drains back into a sump and reused.

Once the coating is applied, the conversion of organic coating from liquid to solid state may happen. Many organic coatings commonly dry by evaporating their solvents. However, in order to form a durable film on the substrate surface, a further conversion called curing is necessary. Curing involves a chemical change in the organic resin in which polymerization or cross-linking occurs to harden the coating. The type of resin used determines the type of chemical reaction that occurs during curing [27]. The principal methods for effective curing in organic coatings are (1) ambient temperature curing—evaporation of the solvent and oxidation of the resin (e.g., lacquers), (2) elevated temperature curing—elevated temperatures are used to accelerate solvent evaporation as well as polymerization and cross-linking of the resin, (3) catalytic curing—starting resins require reactive agents mixed immediately prior to application to bring about polymerization and cross-linking (e.g., epoxy and polyurethane paints) and (4) radiation curing—various forms of radiation such as microwaves, ultraviolet light and electron beams are required to cure the resin [27]. Wu et al. [44] reviewed the preparation of super amphiphobic polymer based coating through single step and two step coating methods with and without pretreatments. Various methods employed for the fabrication of super amphiphobic polymer-based surfaces were explained in Fig. 7.5.

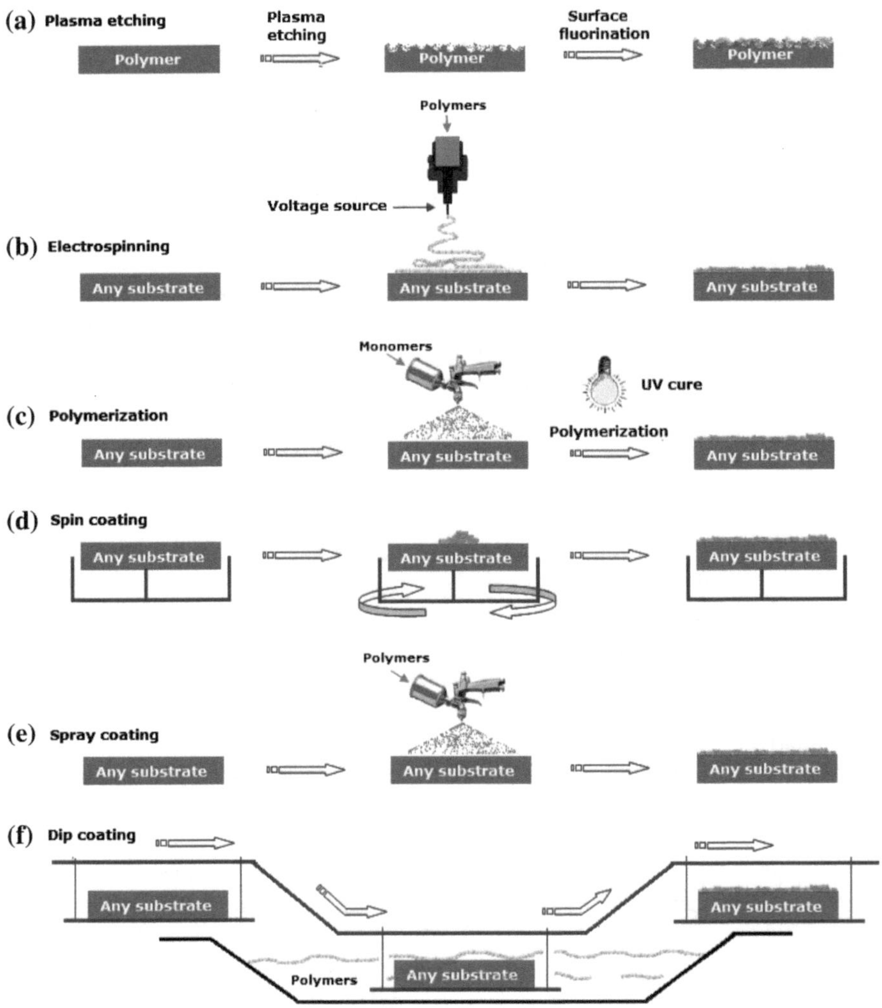

Fig. 7.5 Various methods employed for the fabrication of superamphiphobic polymer-based surfaces, including **a** plasma etching, **b** electrospinning, **c** polymerization, **d** spin-coating, **e** spray-coating and **f** dip-coating. Reused from [44]

In comparison, spray- and dip-coating strategies using polymers as components provide the advantages of great simplicity, scalability, speed, low cost (thus facilitatinglarge-scale fabrication), and compatibility with almost any type of substrate [45, 46]. Some attempts have been made to improve the water- and oil-repellency of polyacrylates [47, 48], polyurethane [49–51], and acrylate-modified polyurethane [52, 53] by introducing fluorinated units into these traditional polymers, which have traditionally been applied as coating materials onto textiles, furniture, stone and various other substrates via spray- and dip-coating techniques. Rungwasantisuk and Raibhu [54] explained the process variables and property analysis of UV curable polyurethane acrylate spray coating coating on polycarbonate substrate. Zabihi et al. [55], Aziz and Ismail [56], Gilissen et al. [57], Wu et al. [44], Wang et al. [58] and Barletta et al. [59] prepared corrosive resistant, robust, wear durable polymer based thin film coating through spray coating technique for light emitting diode and solar cell applications.

7.4.2 *Airless Atomization*

Airless spraying atomization is the method of spraying that does not directly use compressed air to atomize the coating material. It was developed during early 1960s. It atomized the fluid by pumping hydraulic air at high pressure (approx. 500–4500 psi) through a small orifice spray nozzle tip located at the front of the airless gun (Fig. 7.6). The fluid is thrown into small droplets to produce finely atomized spray. The spray pattern size or fan angle (3–21 in.) and orifice are usually predetermined. It produces uniform deposition of paint from different angles on various locations. To improve its efficiency, largest fan angle and the smallest

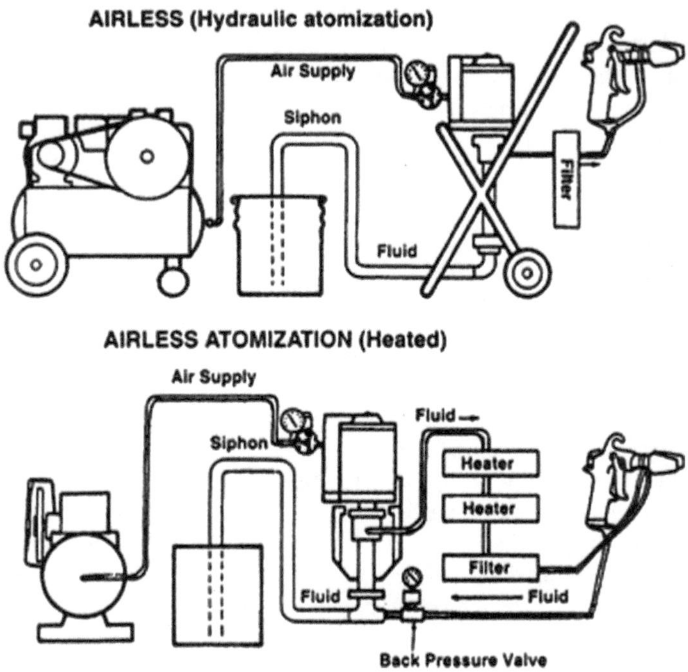

Fig. 7.6 Airless atomization. Reused from [60]

orifice were required. Its unique advantages found applications for maintenance painters, ship builders and highway stripers [60].

7.4.2.1 Advantages

- Speed of application
- Fastest method of spray atomization
- Deliver twice the amount of material
- Improved transfer efficiency
- Reduced fog and overspray
- Reduced spray booth maintenance
- Reduce wastage and easy to reuse.

7.4.2.2 Disadvantages

- Less control of spray pattern
- Airless gun is either on or off
- No feathering capability
- Need high equipment maintenance
- Wear of nozzle tips
- Exposure of human body will cause serious injury
- Foreign particles will block the orifice
- Cause runs or sags.

7.4.3 Flame Spray Coating

Various researchers [61–139] have explained that flame spray coating is a process in which melted materials are sprayed onto a surface [140]. The feedstock or coating precursor is heated by electrical (plasma or arc) or chemical means (combustion flame).

During the coating process, there is no distortion of the part being coated. The part temperature generally is below 250 °F (121 °C) during the spray operation. The substrate of the part that is coated does not metallurgically get altered. Virtually any material can be applied to any metal substrate and some plastics including ceramics, carbides, pure metals, alloy metals and much more. Coating buildup can exceed 2.54 mm in thickness, with some materials able to be applied over 5.08 mm thick. The application time is very fast, unlike plating applications that take hours to accomplish a minimal buildup. This can reduce downtime during repairs. Flame sprayed coatings will increase component life, helping to extend the life cycle of

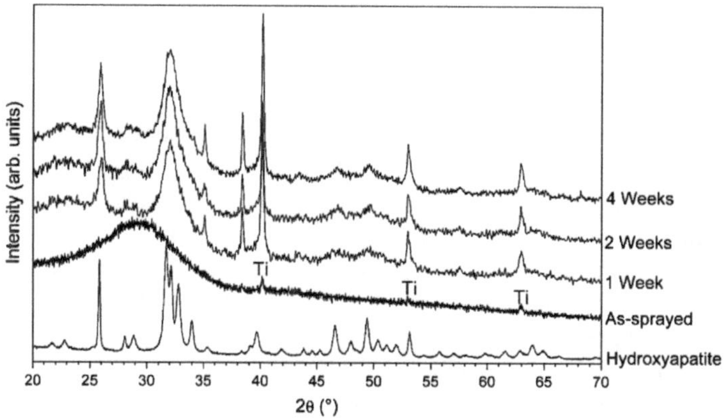

Fig. 7.7 XRD patterns of the bio-active glass coating in as-deposited condition and after soaking for 1, 2 and 4 weeks, compared to the reference pattern of pure hydroxyapatite. Reused from [61]

equipment, therefore reducing maintenance costs. When necessary, the coating can be removed with virtually no degradation to the part base material and reapplied to restore and place the component back in service. Finished parts are protected by masking, so that only the surface requiring coating buildup gets coated, with the rest of the part remaining free of any coating attachment. X-ray diffraction (XRD) of bioactive glass coating before and after soaking was shown in Fig. 7.7. In addition, research overview on plasma sprayed coating materials [141–150] and their techniques [151–182] have been reviewed. The technical properties related to that coating in comparison with the flame spray coating have been analyzed.

7.4.4 Spray Transfer Efficiency

Transfer efficiency is a critical measure for all coating methods [11]. It is the proportion of coat supplied to the process that is actually deposited onto the work surface. The conventional methods of applying liquid organic coatings are brushing and rolling, spray coating, immersion and flow coating. In some cases, several successive coatings are applied to the substrate surface to achieve the desired result. The transfer efficiency is relatively high (approaching 100%) for brushing and rolling coating methods [11]. Spray transfer efficiency is relatively as low as 30% (i.e., 70% of the paint is wasted and cannot be recovered) for manual adopting methods. Manual brushing and rolling methods are suited to low production, not for mass production. While brushing is quite versatile and rolling is limited to flat surfaces. Efficiency can be improved by electrostatic spraying technique, in which the work surface is grounded electrically and the atomized droplets are electrostatically charged onto it. However, the droplets are drawn onto the part surfaces and develop uniform distribution to enhance the transfer efficiencies up to 90%.

7.5 Powder Coating

Powder coating technique was first developed in Australia during 1967 and has many advantages over traditional liquid painting. It is better for the environment and so far the youngest surface finishing techniques commonly used today [23]. Dry, solid, finely pulverized, free-flowing powders such as thermoplastic or thermoset polymers are coated onto the substrate by means of fluidized bed and electrostatic spraying process without any solvent. It is usually applied in a single coat with thickness greater than solvent based coatings. The powder melts, cured under heat and cooled to form a continuous uniform film and hard coating. Due to the absence of liquid carrier, it doesn't emit any volatile organic compounds (VOC). It produces thicker coatings with smooth, texture free surfaces and easier to apply compared to smooth thin films. During the process, coloring agent is also introduced onto the substrate. Generally, the powder coating has high transfer efficiency compared to all other coating techniques.

Hester and Nicholson [183] showed reduction in the potential VOC with powder coatings. A large conventional coating facility covers 12 million ft/year of substrates with 1.2 m and produce thick coats. The plant uses a VOC treatment system with capture efficiency of 70%. In general, a powder coating facility using electrostatic application of polyester-urethane resins will emit only 0.6 tons/year of VOCs and avoids the need for emission control equipment.

Thermoplastic powders include polyvinylchloride, nylon, polyester, polyethylene and polypropylene re-melt upon heating with thick coating of 0.08–0.30 mm, while thermosets may not [11, 23]. It doesn't undergo any change in molecular structure. Thermoplastic resins are used mainly in functional coatings such as thick, protective coatings on dishwasher trays [24] not as a replacement of thin film coatings from solvent-borne paints. The chemical crosslinking with other ingredients occurred in fine thermoset powders like epoxy, hybrids, urethanes, polyester and acrylic at high temperatures, polymerization render desirable technical properties with thickness range of 0.025–0.075 mm. It can't be re-melt upon heating. They are used for decorative, protective, coatings in architecture, on appliances, furniture and elsewhere. The recent technological advancement produce a surface coating that is comparable to liquid coatings on these resins [24]. The Tg of the powder influenced the film thickness, particle size and texture.

The physical and chemical properties of the powder have to be carefully controlled. The effectiveness of powder coating depends on obtaining a smooth and nonporous film. Formation of a good coating with free of voids, pinholes and orange peel distortions depends on controlling the particle size distribution, glass transition temperature, melting point, melt viscosity and electrostatic properties. Well-controlled size distribution is important in achieving good powder-packing on the surface [23].

It usually has a particle size of 30–50 μm, Tg of 80 °C, Tm around 150 °C and curing at about 200 °C. The presence of orange peel texture hides the metal defects and less prone in showing fingerprints. In addition, powder slurry process (powder coating and liquid coating) produced very smooth, low film thickness coating by dispersing fine powders with particle size of 1–5 μm into water. Powder coating is used to develop high specification coating which is relatively hard, abrasion resistant and tough. Powder coatings can be applied over a wide range of thickness. Thermoset powder coatings recommend 25 μ minimum for mild interior applications and up to 60 μ minimum for exterior applications. Color matching is quite acceptable from batch to batch. The powder coating, although not metallurgically bonded to the metal will not develop crack, chip or peel as like conventional paint films. The increased environmental standards pave a major driving force in the growth of powder coatings [184].

7.5.1 Powder Production and Part Preparation

The most commonly used polymers for powder preparation are polyester, polyurethane, epoxy, acrylics and polyester-epoxy [23]. The granules were mixed, heated in an extruder, rolled, cooled, milled and sieved to make a fine powder. The three main processes involved in powder coating are as follows: part preparation or pre-treatment, powder application and curing. The variety of chemical and mechanical methods were used to remove the impurities such as oil, soil, dirt, lubrication greases, metal oxides, welding scale etc., prior to coating. The selection of methods mainly depends on size, part material, type of impurities and required efficiency of the final product. The common part preparation treatments for interior and exterior applications include rinsing, etching, de-rusting, chromate or phosphate and demineralization. The powder coating technique mainly utilized for coating of household appliances, electronic devices, aluminium extrusions, drum hardware, automobile, bicycle parts, medium-density fiberboard, outdoor furniture manufacturing, architectural, building industries and telecommunication.

7.5.2 Application Techniques

Powder coating systems are designed to reclaim powder that has not formed part of the coating. The application techniques offer much higher transfer efficiency than wet spray liquid coating methods for the following reasons [23]:

- An electrostatic charge causes an attractive force between the powder coating material and the substrate.
- Wasted powder can be readily reused.

7.5.2.1 Electrostatic Spray

The most commonly used powder coating method is electrostatic spraying. The dry powder is fluidized and electrostatic spray gun is used to coat the powder onto the part substrate to separate the individual grains of powder and to improve the electrostatic charge which in turn the powder can flows more easily to the gun and produce thicker coating [11, 185, 186]. The dry powder particles are electrostatically charged and propelled by compressed air through the nozzle. A transformer supplies high voltage (typically 100 kV) low-amperage current to an electrode in the spray gun nozzle (Fig. 7.8). The current ionizes the surrounding air, transferring a negative charge to the powder particles as they pass through the corona of ionized air. The substrate to be coated is grounded, allowing powder particles to follow electric field lines and air currents from the gun to the substrate [24, 26, 185, 186]. The powder passed through air off take system in which the powder sticks to the back of the part. The powder passed though the job must be collected, filtered and reused to increase the efficiency of the process. Increase in electrostatic charge enhanced the removal of powder from the part surface. Manually or robots operated spray gun designs are available to impart the charge to the powders. The powder coated samples were placed in an oven at a temperatures range of 160–210 °C to obtain the final solid, tough, abrasion resistant coating. Due to heat treatment of thermosetting powder, it undergoes four stages such as melt, flow, gel and cure to complete the process. Depending on powder design, the final coating is continuous and will vary from high gloss to flat matter. Compared to conventional air spraying of wet coatings, electrostatic spraying achieves greater coverage of the substrate because the powder tends to wrap around corners and coat surfaces that are not

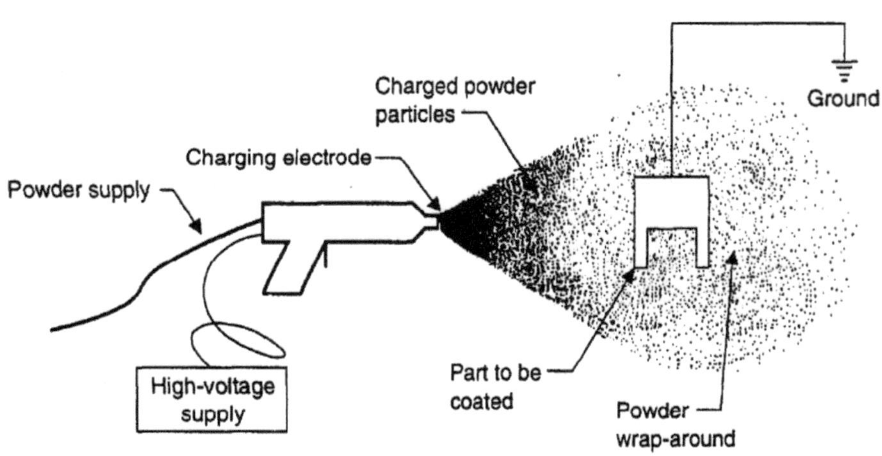

Fig. 7.8 Electrostatic spray gun for powder coating. Reused from [23]

in-sight with the spray gun. These results in less overspray and higher transfer efficiency of the coating.

The following shortcomings of electrostatic powder coating led to the development of tribocharging:

1. Reduced coating thickness.
2. Entrapped air ions develop surface imperfections like orange peel-like distortions and cratering [26].

Powder Coating Guns

- Corona charging guns used electric power to generate the electrostatic charge either internally or externally [23, 185, 186].
- Tribo charging guns developed friction between the powder and the gun barrel to generate the charge.
- Bell charging guns cast away the powder from the perimeter of the bell for charging.

In addition, electrostatic tunnels were also used to charge the powder.

7.5.2.2 Fluidized Bed

It provides another way of powder coating, alternative to electrostatic spraying process [11]. In this process, the work part to be coated is preheated and passed through powder suspended fluidized bed (Fig. 7.9). Low pressure dry air is

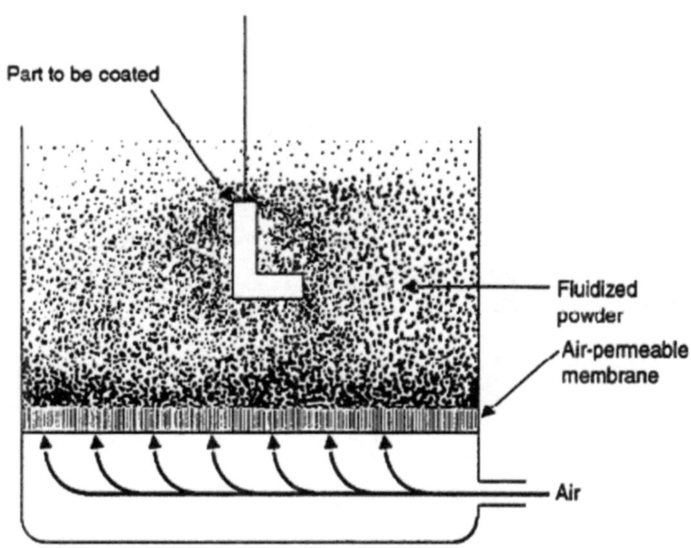

Fig. 7.9 Fluidized bed for powder coating. Reused from [23]

circulated through the bed, causing the rest powder to attain a lofted and fluid-like state. The powder attracted towards the part surface to develop fairly complete, uniform coating of complex shaped parts. In some cases, electrostatically charged powders were used to enhance the adhesion between the powders and part surfaces [23].

Fluidized bed systems are primarily used to apply coatings of thermoplastic powder in desired range of thicknesses. The substrate is heated to a higher temperature than the melting point of the resin to strike the hot surface, they melt and coalesce to form a thick, continuous film on the substrate. During fluidized bed coating, powder is added to replace material that has formed a coating on the substrate. Because very little powder is lost or degraded during coating, powder utilization is near 100%. The fluidized bed method is the original method used for applying powder coatings and is still the method of choice for heavy functional coatings [24].

A modification of this system is the electrostatic fluidized bed. In this, the powder receives a charge from air which flows through a high voltage charging system. However, the grounded object is lowered or suspended over the tank. Variations on this principle allow wire mesh or other endless type products to be coated. Electrostatic fluidized beds (Fig. 7.10) are limited to an effective depth of about 2–3 in. so that they are best suited to coating two-dimensional parts [187].

Barletta and Tagliaferri [188] briefly explained the influence of process parameters of various polymer film coating produced through electrostatic fluidized bed coating process.

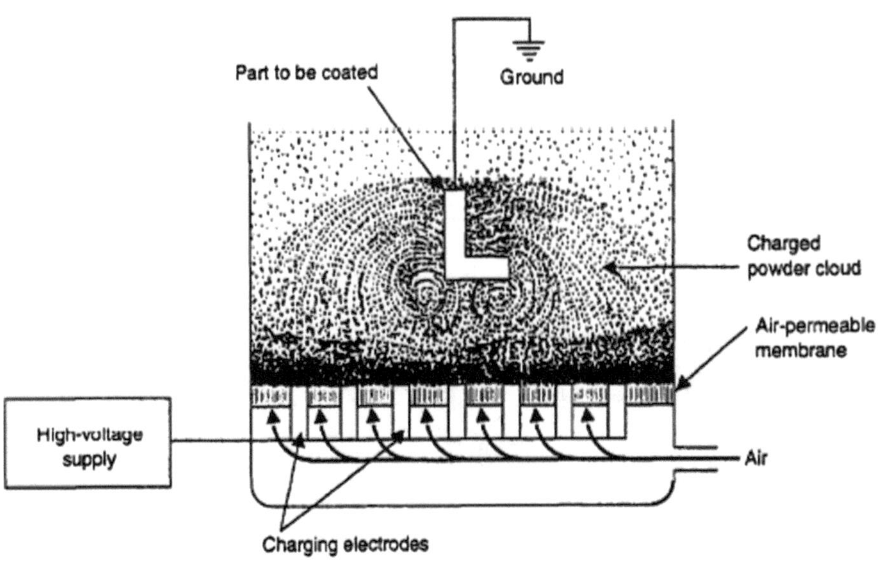

Fig. 7.10 Electrostatic fluidized bed for powder coating. Reused from [23]

7.5.2.3 Tribocharge Spraying

The basic principle of tribocharging relies on friction between the powder and the spray gun. The action of the powder flowing through the barrel of the gun generates a frictional charge on the powder. The charged powder is carried by air stream to the substrate where it deposits and sticks due to electrostatic attraction. There is no high-voltage system generating a field between the spray gun and the substrate, the electric field is substantially smaller and the powder tends to follow air currents rather than field lines. The smaller electric field results in a much reduced Faraday cage effect. Consequently, tribo guns produce smoother finishes, allow deposition of thicker films and provide better coverage of intricately-shaped objects [23]. The frictional charge is generated because the powder and the gun have different dielectric constants. The tribocharging annular chamber is made up of polytetrafluoroethylene (PTFE). It has a low dielectric constant, a positive charge will be imparted to most powders. Some powders have low dielectric constants (e.g., mixtures of polyester and TGIC) and do not pick up electrical charges readily. Therefore, attempts have been made to modify the powder composition. Small quantities of additives such as amines or quaternary ammonium phosphate salts increase the ability of the powder to accept positive charges. Additives, however, can modify the reactivity of the powder and also create a non-uniform composition. Uneven composition can lead to powder segregation and create problems with recycling. Steps are underway to produce a polyesteramide that has increased ability to accept frictional charge, reducing the need for additives [26, 189].

Tribocharging is less complex than traditional electrostatic powder coating systems because it does not use high-voltage transformers for applying the charge on the powder. Tribocharging guns do wear out faster than regular guns because of the abrasion of the powder on the PTFE surfaces. Because of the absence of electric field lines and a reduced dependence on leakage to ground of free ions, tribo guns are more suitable for painting nonconductive surfaces.

7.5.3 Powder Coating Equipment

In this process, powder delivery system used spray gun to supply the powder and powder storage container or feed hopper with a pumping device to transport a stream of powder to the gun through hoses or feed tubes. Compressed air is pumped to separate the powder particles for easier transport. Spray guns can be classified as either manual or automatic with variety of styles, sizes and shapes. The selection of gun type mainly depends on the performance characteristics of the products being coated. Improvements have been made to enhance the coating transfer efficiency. Many of these changes include spray patterns variations and nozzles to resist clogging have been introduced for using one gun on multiple parts of different configurations [23].

The powder delivery system can supply powder to one or several guns from the powder supply. Delivery systems are available in different sizes depending on the application, the number of guns to be supplied and the volume of powder to be sprayed during a given time period [183, 190]. Recent advances in powder delivery systems, coupled with improvements in powder chemistries that can reduce clumping to deliver a consistent flow of particles to the spray gun. Improvements in design of spray booths, enabling color changes to be made with a minimal downtime and recovery of a high percentage of the overspray were also described.

Agitating or fluidizing the powder in the feed hopper also helps to prevent clogging or clumping in the transport lines. Innovations in powder delivery systems allow the powder supply reservoir to be switched easily to another color powder when necessary. If the overspray collection system is also not changed, the collected powder containing all the colors applied between filter replacements or booth cleaning.

For collected over-sprayed powder should be free of cross-contamination between colors. Numerous systems are now available for segregating colors and that allow several colors to be applied in the same booth. In this system, a moveable dry filter panel or cartridge filter is used to dedicate for one color and that can be removed. Color changes are accomplished by:

1. Disconnecting the powder delivery system and purging the lines
2. Cleaning the booth with compressed air or a rubber squeegee
3. Exchanging the filter used with the filter for the next color
4. Connecting the powder delivery system for the new color

Unlike most solvent based coatings, electrostatic and tribo-charged spraying both results in collecting and reusing the overspray powder. Figure 7.11 shows a

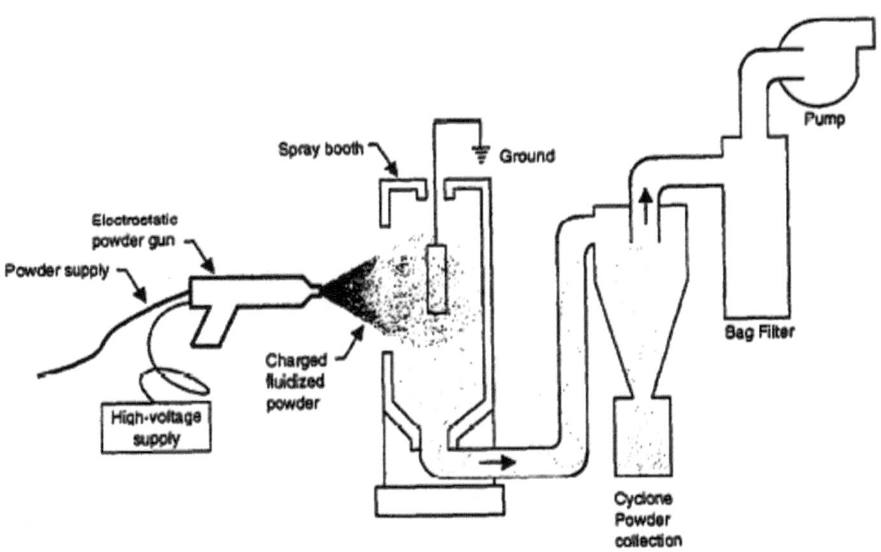

Fig. 7.11 Powder coating recycling system. Reused from [23]

schematic of a recycling system for powder coating. Recycling systems coupled with high transfer efficiency results in reduced waste paint disposal, lowering both costs and environmental impacts.

Powder coating equipment and techniques are easier compared to conventional dip coating or spray painting. The operator needs less skill than a liquid spray painter. However, the operator needs experience in powder properties to obtain good film before the coating enters the oven [23, 189, 190].

7.5.4 Advantages and Disadvantages

There are several advantages and limitations of powder coating over conventional liquid coatings and other coating process:

- Doesn't emit VOC
- No need of solvents
- Compressed air is used
- Don't require mixing or stirring
- Produce thicker coatings
- Harder and more resilient coating
- Overspray can be recycled
- Increased edge protection
- More textured finish
- Factory applied process
- Less hazardous waste
- Available in an anti graffiti specification for external applications
- Less operating costs and disposal costs
- Higher capital cost and material cost
- Does not require control equipment
- Less differences between horizontally and vertically coated surfaces
- A wide range of special effects is available
- Barrier protection system
- High transfer efficiency and material utilization rates
- Require less energy to cure

Thermoplastics are used for many items like park benches that require both plasticity and durability [23, 189, 190]. It has many unique advantages such as high lubricity, improved impact resistance, recyclability and ability to remold or reshaped. The main drawbacks includes very high expensive and melt upon heat. Thermosets are often used to coat appliances because of the high heat withstanding capability. The main advantages are cheap, resistant to extreme temperatures, beautiful look with enhanced strength and stability. It can't be reshaped, recycled and difficult to finish. The most common limitations of powder coating process as follows:

- Handling of heated parts, a cool down zone is required.
- Color matching from batch to batch is difficult.
- Shading or tinting cannot be done by the end user.
- Uniform coating of the complex shaped parts require special process.
- The parts must be electrically conductive or covered with primer.
- Time consuming and produce cross-contamination of color.
- Difficult to incorporate metal flake pigments.

7.5.5 Curing of Powder Coatings

Powder coatings must be heat-cured or melted on to the substrate. For thermoplastic resins, the substrate can be heated prior to coating so that the resin melts directly on application. Thermoset powders melt, flows out and then chemically reacts to form a higher molecular weight polymer in a network-like structure at high temperatures. The crosslinking requires a particular time and temperature for complete curing and to achieve desired film properties. The curing schedule may vary based to the application and specification of the powder. Convection cure ovens or infrared cure ovens are used for energy related applications [23]. The substrate must be able to withstand temperatures of 260 °F or higher. Delicate substrates like certain thermoplastics or wood cannot be cured in ovens. The substrate must also be of a size and shape to allow immersion coating or heating in a curing oven. This prevents general indoor or outdoor application where heating options are not available. As an alternative to blocked isocyanate cross-linking agents, melamine resins mounted on polymer support materials can be used for curing solid polyester resins. Advantages of these resins include lower curing temperatures, lower hardener content and lower volatile emissions [23].

7.5.6 Developments in Powder Coating

Certain thermoplastic powder coatings can be applied by a flame-spraying method. This technology, developed by Plastic Flamecoat Systems Inc. in Texas, uses a propane and compressed air flame in the spray gun to melt the powder as it is propelled towards the substrate. The molten powder hits the substrate and flows into a smooth pinhole-free coating [184]. Thermoplastic powder coatings are used in military applications such air force weapon systems and aircraft. The ability to field repair these coatings using spray guns is valued in these applications [191]. Gas-fired ovens have considerable economic advantages over other energy sources, however, they can produce nitrous oxides which come in contact with the coating as it cures. Some grades of powder are more susceptible to yellowing under these conditions than others. Grilesta has developed several powders that can be used in gas-fired ovens without yellowing [25]. If only partial coating of a part is required, the part needs to be masked to prevent powder from adhering to the entire surface. Large numbers of manufacturers need to mask parts and this can be a major problem, depending on the part and the degree of masking required. Powder coatings traditionally had poor weathering properties by the use of epoxies. Certain acrylic powders give better ageing and weathering resistance. One of the better systems for high wear resistance uses carboxylated polyesters cured with TGIC. In an accelerated weathering experiment, this system showed wear resistance similar to an acrylic resin cured with TGIC hardener. After 4 years exposure in IR radiation, the gloss retention value as high as 60–70% was obtained. Florida curing [26] by IR radiation induces very rapid development of the crosslinked film, enhancing line speed, increased viscosity of the film, hindering transport of water during the curing process. Water evolved as a byproduct from the crosslinking reaction can be trapped in the film leading to problems of severe pin holing and gloss reduction [26].

Based on epoxy/polyester hybrids, thin-layer coatings are available with good hiding power. Currently these are only suitable for indoor applications because the epoxies degrade on exposure to outdoor weathering [184]. Thin coats may be desirable where protection from corrosion and other environmental factors is not so important. The reduced thickness directly reduces the amount of powder needed and consequently the cost of the coating. For thin coat application, the powder particle size is approximately 25 μ and must have a narrow size distribution [26]. When thin coats are applied, impurities in the powder can give rise to visible surface defects such as cratering. For this reason, the resin is carefully filtered to remove traces of gel, unreacted monomers and other non-soluble materials. Thin coatings show surface imperfections more readily than thick coatings. Small amounts of thermoplastic resin can be added to the formulation and this will act as a permanent plasticizer and reduce the melt viscosity, giving better flow to the finished coating [26].

Growing concerns over solvent and solid waste emissions from solvent-borne coil coating operations have led to use of powder for coil coating. Powder formulation technology has advanced and current powders are now highly formable and stain resistant, allowing their use in coating coil stock manufactured for domestic appliances such as refrigerators, washing machines and microwave ovens [26]. Blank coating by powder offers similar benefits to coil coating. Flat metal blanks are cut, cleaned, formed into the part, coated and cured in high speed operation. Other benefits include uniform film thickness, high transfer efficiency and a compact finishing operation [184]. Low-gloss coatings are now available with good mechanical surface properties and appearance. Gloss values range from 1% or less with epoxies to approximately 5% for weather-resistant polyesters [184]. Textured powder finishes range from fine textures with low gloss to rough textures suitable for hiding an uneven surface on the substrate. Textured powder coatings have shown large improvements in mechanical and processing abilities compared to those of several years ago [184]. Metallic powder coatings incorporate metal flakes which are blended with the powder before being sprayed onto the substrate. Aluminum extrusions are commonly coated and efforts are concentrated on matching anodized parts. A clear topcoat over the metallic base improves exterior durability of the coating [184]. In-mold powder coatings allow manufactures of certain molded plastic products to coat with powder during the molding operation. Powder is sprayed into the mold before the molding compound is added, the powder then melts and cures, chemically bonding to the molding compound and producing a finish with excellent surface properties such as chip and impact resistance. Suitable substrates are sheet molding compounds and bulk molding compounds, used to produce automotive body panels and other items [184]. Polyester and acrylic powder coats have been developed for outstanding weather resistance. Powder coatings can meet all the requirements of AAMA 603 and 605 specifications, except the Florida five-year exposure test which is still underway [184]. Shell Development Company in Houston is working on low-temperature cure epoxies that have adequate physical and chemical properties to allow exterior use. Adding a selective catalyst such as Crylcoat 164, to conventional polyester allows curing at reduced temperatures of approximately 260–280 °F [184]. Two major drawbacks are a risk of premature reaction during the extrusion process and poor surface appearance because of partial crosslinking before complete fusion of the powder [26]. American Cyanamid Co. produces a solid amino cross-linking resin for powder coatings. This resin can be used with either hydroxyl function polyesters or acrylics to produce highly durable, light-stable coatings with good mechanical properties. The cross-linking resin has low toxicity, low environmental impact, performance and economic advantages. Internally catalyzed polyester designed for use with 1174 is commercially available. The polyester has wide cure responses along with good film appearance and other properties [192]. Kunaver et al. [193] studied particle distribution of powder coating using selective plasma etching technique under oxygen conditions. The density of positive ions, plasma parameters and density of neutral oxygen atoms were measured under controlled neutral gas and electron temperature. Plasma was pretreated and applied to remove the polymer matrix in-order to determine the particle distribution using SEM analysis.

7.5.7 Applications

Powder coatings are applied in small to medium sized metal parts including lighting fixtures, equipment cabinets, outdoor furniture, heat exchangers, microwave antennas, shelving, hand carts and wagons [23]. Radiation Systems, Inc., based in

Virginia, develop a coat using polyester powder to resist physical damage, solar radiation and environmental effects [194]. Industrial applications of powder coatings are metal furniture manufacture, lawn and garden equipment, store shelving, exercise equipment and aluminum extrusions. Job-shop electroplaters have adopted powder coatings to meet customer specifications and environmental regulations [195]. Clear coatings, topcoats are increasing in importance. The automotive industry use clear coats on wheels, trim and other fittings. Bicycle manufacturers also use powder clear coat when performance and quality count more than small cost increases [25]. Automotive components coated by powder include vacuum booster housings, door handles, steering columns, oil filters, aluminum wheels, shock absorbers and antennae. Toyota, General Motors and Chrysler use powder coatings for door frames and in some instances for lower body anti-chip coatings. Powder coating is used as an effective primer surfactant and for blackout finish. Current activity is focused on applications for light trucks and sport-utility vans. General Motors applies primer surfactant and blackout finishing to Chevrolet S-10s and Blazers [196]. Industrial and Architectural Finishing Powder coatings are used to protect many parts of buildings such as exterior and interior, residential and commercial. Polyester powders are applied in controlled factory settings to aluminum and galvanized steel profiles and sheet products for use in windows, doors, curtain walls and exterior cladding. Fusion-bonded epoxy powder coatings are used for protection of steel reinforcing bar and mesh in concrete structures. Polyurethane or polyester powders finish many ancillary components in the building industry, such as downpipes, lampposts, fencing, railing, street furniture and other metal parts. Epoxy and epoxy/polyester hybrid powders are not suitable for exterior use and therefore coat interior components including air-conditioners, light fittings, partitioning and radiators [197, 198]. Polyester powder coats can be applied to galvanized steel, producing strong, corrosion-resistant and visually pleasing architectural components. Pin holing is unique to galvanized steel and although unsightly, it does not impair corrosion resistance. Pretreatment of the metal surface is necessary to reduce pin holing. A vigorous cleaning regime involving a soak clean, an etching cleaner, a chromate solution and various cold and de-mineralized water rinses is recommended [199]. Industrial powder coatings can provide a corrosion-resistant finish without pretreatment such as priming or even sandblasting. Powder coatings applied to substrates that had not been pretreated found to be resistant to conditions encountered in the North Sea and eastern Asia [200, 201]. Powder coating without any pretreatment should be considered quite risky and proper testing needs to be applied [23, 201].

7.6 Electrostatic Coating

Electrostatic coating or Electrostatic spraying was introduced in 1989 by Warrington based Electrostatics Ltd as a methods to re-coat plain and polyester coated substrate. It uses an electrical charge to enhance the strong bonding between the substrate and the coating. It doesn't require the object to be placed in an oven for curing after coating. It was used in the car industry for many years until the introduction of robots [201–207].

Many industries adopted this technique for various substrates in their commercial products with specific advantages as follows: It provides an amazing durable finish superior than traditional coating methods, less air pollution, quick drying, uniform coating, reduced overspray, prevents corrosion, improved adhesion and long service life. In order to remove the weak spots in powder coating, electrostatic coating may be used.

7.6.1 Working

Positive electrostatic charged coating is fed through a spinning ceramic disc at the end of a paint applicator. It comes into contact with the spinning disk to atomize the paint and broken up into small droplets. The coating substrate is then negatively charged and positively charged paint becomes adhered to the surface [201–207]. Once this positively charged liquid has been attracted to the negatively charged object, the magnetism is dispelled. The improved interaction shows that the working substrate completely wrapped by the paint. However, the remaining negatively charged paint will expels out to provide uniform coating compared to conventional spraying.

7.6.2 Benefits and Applications

Electrostatic coating process has some unique advantages [201–207] such as:

- No Overspray
- Fast Drying
- Strength and durability
- Cost Effective
- Durable
- Factory Finish
- Go Green
- Little maintenance
- Allergy Friendly

The most common applications of electrostatic coatings are lockers, window mullions, elevators, machinery, office equipment, hand rails and bathroom partitions [201–207].

7.7 Electrodeposition Coating

The surface for spray painting and dip coating were prepared carefully to develop uniform thin layer throughout the substrate. In contrast to this, electro-deposition process forces the paint solids onto the workpiece. Except the paint bath increments, the tank is full of aqueous paint solution. These volumes of bath cannot be rinsed back into the tank by the use of water to avoid overflow. An ultra-filter is used to retain the paint solids and drain out the solution through filtration or permeation [208–210]. The freshly coated workpiece rinsed by this ultra filtrate to get back the lifted paint. For this reason, high concentration paints are used to replace the coated solids.

7.7.1 Process Parameters

7.7.1.1 Throwing Power

Throwing power is the depth of deposited paint and the measure of ability to extend durable paint films. In this process, slits are used to obtain the higher efficiency of the paint compared to round holes. The higher applied voltage, higher bath conductivity, lower coulomb per gram paint requirement, larger openings for entrance into a cavity and a larger perimeter of the opening results enhanced throwing power [209].

7.7.1.2 Maintaining a Steady State

This process is similar to electroplating in balancing the incoming and outgoing materials. Paint solids, evaporated water and other incidentally lost materials must be replaced to meet all required solidities [208, 210]. For improved efficiency of the feed coating, a certain combination in different proportions of resin, pigments, solvents, cross-linkers and other materials were used. Generally, a tank fill of 70 wt % resin and 30 wt% pigment may produce a deposited coating of 68 wt% resin and 32 wt% pigment. This in turn replaced the conventional materials leaving the tank.

7.7.1.3 Rupture Voltage

The increase in voltage leads to enhanced throwing power. However, rupture voltage seems to be a dielectric breakthrough of the forming film and it depends on the surface of the metal being coated. Blemished and reduce salts spray resistant films were produced due to higher voltages compared to applied rupture voltage.

7.7.2 *Equipment*

7.7.2.1 Conveyors

Overhead conveyors at 30° angle were used for high volume production. Similarly for bulk coating free and power conveyors were used at 90° vertical to save the energy and space.

7.7.2.2 Metal Preparation

Seven- to nine-stage zinc phosphate or iron phosphate are widely used. Some special pretreatments are required for nonferrous metals.

7.7.2.3 Tank Enclosures

Safety is interlocked with the power source and conveyor to de-energize the system.

7.7.2.4 Dip Tanks

The workpiece is immersed and tanks are operated on a continuous basis using free or power conveyors.

Figure 7.12 explains a typical electrodeposition coating process and equipments [208–218].

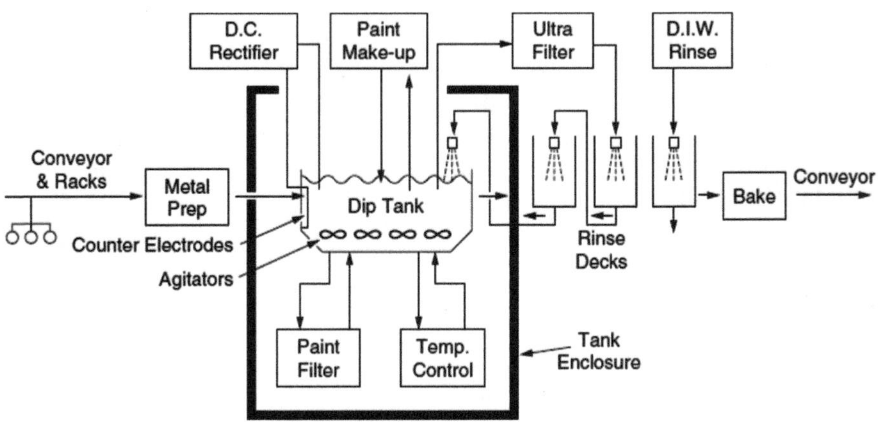

Fig. 7.12 Schematic diagram of electrodeposition process. Reused from [210]

7.7.2.5 Rectifiers

Direct current power sources of less than 5% ripple factor are used.

7.7.2.6 Counter Electrodes

Most electrocoating tanks are coated with an epoxy in which the counter electrodes have opposite polarity, whereas the tank, the conveyor, the merchandise and so on are all grounded [211–218]. Electrodialysis process is carried out to collect the counter-ions by separating the counter electrodes from the tank.

7.7.2.7 Agitation

Ejector nozzles and other agitating devices are used to prevent the settling of solids. The law of probability explains that pumping stability is required in which the original fill material still dwells in the tank. The dwell time in the bath is reported as turnover rate.

7.7.2.8 Temperature Control

The operating temperature is around 75–95 °F. Chillers must be used to remove the heat generated through electrical and agitational energy [211–218].

7.7.2.9 Ultrafilter

The ultrafiltrate is used for removing dissolved impurities and paint savings.

7.7.2.10 Paint Filters

Filter bags, wound filters and indexing filters with pore size in the range of 5–50 μm were used.

7.7.2.11 Paint Makeup

It is actually the last step in manufacturing water-borne paints in which all the needed paint components were mixed with tank contents.

7.7.2.12 Deionized Water

During final rinse, evaporated or lost water is replaced.

7.7.2.13 Bake or Cure

Electrocoating materials usually require curing. In which, ambient temperature curing materials are also available.

Wang and Tan [219] explained the electrodeposition of polyaniline coating using wire bean electrode method for corrosion applications. Hamid and Ghayad [220] studied the technical properties and morphology of nickel-polyethylene coating composites. Cheng et al. [221] understand the characterization of semicrystalline chitosan hydrogel using cathodic electrodeposition technique. Pierleoni et al. [222] used various techniques for the preparation of graphene coated polymer films for gas barrier applications. Herrasti et al. [223] understand the preparation of nanotube coated polypyrrole-titanate composite for corrosion resistance. Lin et al. [224] described the preparation of solar cells using cobalt sulfide electrodes through cathodic electrodeposition technique. Rapecki et al. [225] coated silver onto the polypyrrole films for application as conducting polymers on metals. Moral Vico et al. [226], Chen et al. [227] adopted the electrodeposition process for the preparation of corrosion resistant and conducting bilayers. Similarly, several researchers [228–253] explained the two-coat, single-coat and cathodic electrodeposition procedure, morphological characterization and various technical property relationship on several polymers like polyethylene, polyimide, carbon fibers, alginate, chitosan, gelatin, nanosilver, poly(hydroxyabutyrate-co-hydroxyvalerate) (PHBV), epoxy-amine, polypyrrole, polyacrylonitrile, PTFE etc.

7.7.3 Cathodic Electro Deposition

Cathodic Electro Deposition (CED) is the organic coating process in which a conductive surface connected to a circuit act as the cathode and positively charged paint particles suspended in aqueous medium under direct current. It is widely used for coating automobile components. CED coating is used as the base coat for top coat of wet paint, polyurethane paint and powder. It is semi glossy black in color. Advance technology of 6th generation CED paint has more edge corrosion resistance.

Bichler et al. [254] reported that adhesion and permeation barrier mainly depends on the chemical composition between the substrate and coated barrier layer. To improve the functionality of polypropylene films in packaging applications, vacuum roll to roll coating technique is adopted. The surfaces of polypropylene films were modified with different oxygen plasma pretreatments. Untreated and treated films were subjected to atomic force microscopy (AFM),

X-ray photoelectron spectroscopy (XPS) and contact angle studies. Hao et al. [255] adopted vacuum coating, thermal decomposition and electrochemical sedimentation methods to modify the glass substrates using platinum. The surface resistance and photoelectrical efficiency of the solar cells were measured. Compared to other two methods, electrochemical sedimentation method depicts higher photoelectrical efficiency with low surface line resistance. The surface resistance, open circuit voltage, short current and fill factor is 12.4 Ω cm^{-1}, 6.2 mA/cm^2, 0.662 V and 52.3% respectively.

7.7.4 Special Features

- Good corrosion resistance
- Uniform coating thickness without porosity
- Parts of complex shapes and cavities can be fully covered
- Efficient utilization of coating materials
- Covers a wide range of materials
- Relatively high speed coating
- Aqueous process, hence free of fire risk
- Environmentally friendly.

7.8 Floc Coating

Principle: Glue is first applied by direct coat-carpets, some cords and velvetsing on the textile surface, flocs (short textile filaments) are then shaken out and electrostatically directed on the glued surface [256, 257]. Examples: Carpets, some cords and velvets.

7.8.1 Carbon Fiber Based Floc Coatings

Activated carbon fibers (ACF) were difficult for uniform coat using electrostatically floc coating method, because of its electrical conductivity. Normally, ACF fibre has 0.8–1 mm in length and 10 μm in thickness can be coated to a density of 20,000–50,000 pieces of floc/cm. This method has been used to prepare the coating tank to avoid the build-up of static charges. The metallic electrode is coated with an insulator, residual static charges in the tank are dispersed and the inner surface of the tank is coated with an antistatic agent. It adsorbs odors and solvents faster than activated charcoal and has 20 billion fine pores/cm^2. Further, the fine pores can be filled with insecticides or anti-mildew agents. It found applications in building materials and clothing because of its fire-retardant and deodorization properties [256].

7.9 Dip Coating

Dip coating technique is the most commonly used technique in both academia and industry for many chemical and nanomaterial engineering research projects. The process normally refers to the immersing of a substrate into a tank containing coating material, withdrawn the piece from the tank under controlled speed, temperature and atmospheric conditions and finally allowing it to drain. The coated piece can be dried either by force-drying or baking. In accordance with the spin coating technique, this also creates the thin film coated materials. It is an industrial coating process used to manufacture coated fabrics [258], condoms, candles and 3D objects. It serves as an alternative to spin coating technique. This technique is employed to develop thin films from sol-gel precursors for applying films onto flat or cylindrical surfaces [259]. It is excellent for producing high-quality, uniform coatings, complex shapes which requires precise control and a clean environment. The applied coating may remain wet for several minutes until the solvent evaporates by heating. In addition, the applied coating may be cured using conventional thermal, UV or IR techniques depending on the coating solution formulation. Once a layer is cured, another layer may be applied on top of it with another dip-coating/curing process. In this way, a multi-layer stack is constructed.

The final product may incorporate the substrate and the coating or the coating may be peeled off to form an object after complete drying. In the dip-coating method, the substrate is removed from the sol with a constant velocity to obtain a uniform coating. Coating thickness influenced with faster withdrawal speed. The thickness of coating is usually affected by fluid viscosity, fluid density, surface tension, functionality of the initial substrate surface, submersion time, withdrawal speed, number of dipping cycles, solution composition, concentration and temperature, number of solutions in each dipping sequence and environment humidity [260, 261]. The greater film thickness should be achieved with higher metal temperature and the longer immersion time. The automated computerised control systems were used for dipping operations. The whole experimental set-up is isolated on a vibration damping table to ensure that the liquid surface remains completely undisturbed to obtain uniform thickness at each deposition. The overview and different stages of dip coating process are shown in Figs. 7.13 and 7.14.

7.9.1 *Process*

The dip coating process can be classified into five stages [262], whereas initially the parts to be coated should be chemically cleaned to enhance primer adhesion,

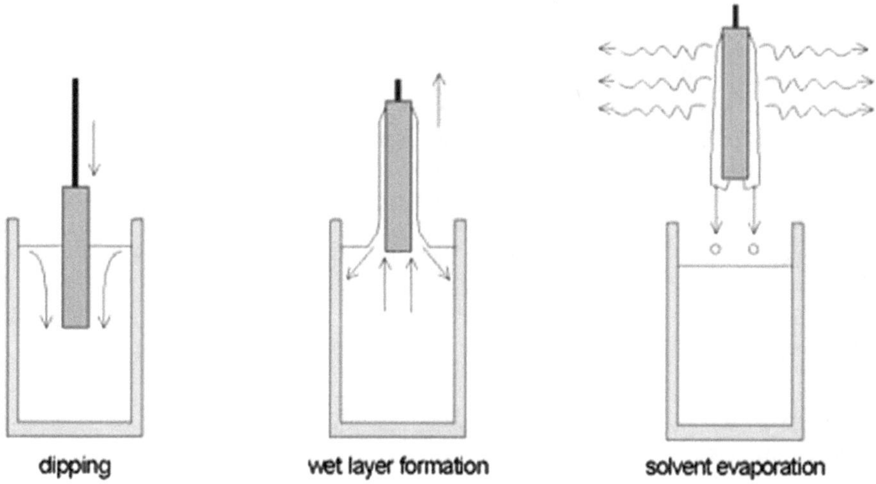

Fig. 7.13 Overview of the dip coating process: dipping of the substrate into the coating solution, wet layer formation by withdrawing the substrate and gelation of the layer by solvent evaporation. Reused from [260, 261]

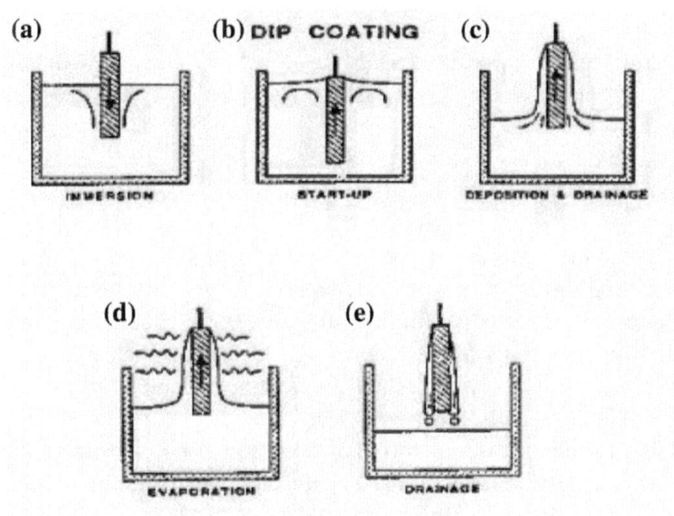

Fig. 7.14 Stages of the dip-coating process. Reused from [260, 261]

corrosion resistance and to remove contaminants. The part is flash dried and baked for the process to begin.

- **Immersion**: The substrate is immersed in the solution of the coating material at a constant speed.

- **Start-up**: The substrate has remained inside the solution for a while and is starting to be pulled up.
- **Deposition**: The thin layer deposits itself on the substrate while it is pulled up at a constant speed to determine the thickness of the coating [263].
- **Drainage**: Excess liquid will drain out from the surface.
- **Evaporation**: The solvent evaporates to form the thin layer. For volatile solvents (alcohols), evaporation starts during the deposition and drainage stages.

Finally, the finished product is conveyed through conveyor belt for packing and shipment. In the continuous process, these steps are carried out directly one after each other.

The waveguide preparation through dip-coating technique has four stages:

- Preparation or choice of substrate
- Thin layers deposition
- Film formation
- Densification throughout thermal treatment.

7.9.2 Dip Coating Techniques

The dip coating technique is used to make thin films through self-assembly, the sol-gel technique and layer-by-layer assembly.

7.9.2.1 Self-assembly

Self-assembly is the process to produce more complex objects with minimum energy using solution or gas phase. It generally produces film thickness of around one monolayer. It is also a powerful bottom-up assembly method beyond nanoscale level. Their structural property determines the location of self-assembled components. Self-assembly is most important in biomolecular nanotechnology applications to assemble atomically precise devices.

Surfactant molecules form monolayer onto the solid substrate by means of physisorption, covalent binding or electrostatic interactions. The developed self-assembled monolayer (SAM) is a two-dimensional film, one molecule thick, covalently organized or assembled at an interface. Some of the examples of SAM are (1) reaction of alkanethiols with a gold surface and (2) reaction of silanes with glass, quartz or SiO_2 surfaces.

7.9.2.2 Sol-Gel Technique

sol-gel technique is the deposition method used to create protective coatings, optical coatings and ceramics. Normally, hydrolysis of sol followed by polycondensation process produce a gel that contains both a liquid phase and a solid phase. However, the proportion of liquid is reduced gradually, remaining liquid can be removed by means of drying and finally coupled with a thermal treatment to obtain the suitable technical properties. In this technique, deposition speed and solution viscosity mainly controlled the increased film thickness.

7.9.2.3 Layer-by-Layer Assembly

The layer-by-layer assembly (LBL) technique is very simple, cheap and used to deposit alternate layers of materials in opposite charges. This technique for thin film coating render controlled thickness over the substrate [260, 261, 264, 265].

Li et al. [265] modified polyethylene terephthalate (PET) by using hyaluronic acid (HA) and cationized gelatin (CG) through layer by layer self assembled technique. The surface properties and cell compatibility were analyzed through SEM, Fourier transform infrared spectra (FTIR) in attenuated total reflectance (ATR) mode, energy-dispersive X-ray spectroscopy (EDX), contact angle and biomechanical analysis. In vitro and in vivo studies of the coating render enhanced cell adhesion, cell growth and reduced gene expression compared to pure. The animal experiments using rabbit and porcine ligament promotes tissue formation and cell infiltration. The formation of type I collagen leads to ligament reconstruction.

In this the coating thickness is based on the withdrawal speed, solid content and viscosity of the liquid. The coating thickness can be calculated using Landau–Levich equation [258].

James and Strawbridge [259] demonstrated that the experimental film thickness value for an acid catalyzed silicate sol is in line with the calculated values. The selection of appropriate viscosity determines the coating thickness with high optical quality. The suitable atmospheric conditions evaporate the solvent quickly and destabilize the sols (Fig. 7.15). It leads to formation of a transparent film due to small particle size in the sols [262].

In general, sol particles are stabilized by surface charges, and the stabilization condition follows the Stern's potential consideration [263]. The author demonstrated that during densification reaction, the alkaline diffusion occurs onto the coated layer such as conventional glasses like soda lime glasses. In this method, the adhesion between the layers was perfect and has some effect on the refractive index. Based on developments of Schroder [266] and Dislich [267], plate glass for solar energy control systems and anti-reflective coatings on windows were developed through this process. Some optical applications of dip coating techniques are bulbs, optical filters or dielectric mirrors and fabricating multilayer systems with very high precision.

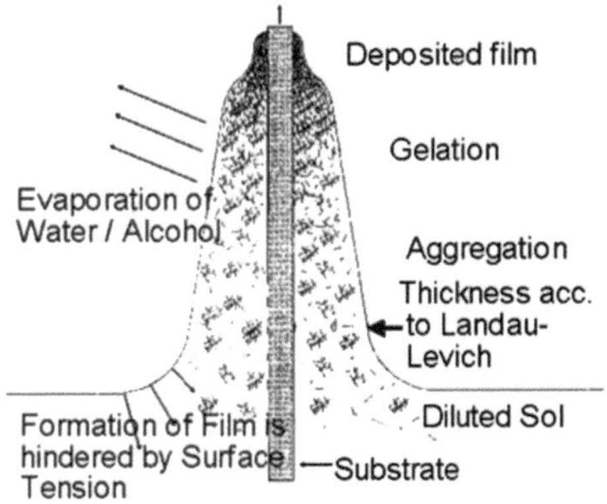

Fig. 7.15 Gelation process during dip coating process, obtained by evaporation of the solvent and subsequent destabilization of the sol. Reused from [262]

7.9.3 Advance Developments

Dip coating combined with brush coating and wire wound rod coating is one of the oldest coating methods. The two common basic dip coating processes are as follows.

In the wet coating process, the substrate is partially submerged into a coating pan to obtain wet film. In the discrete method, the irregularly shaped discrete objects like nails, circuit boards, architectural steel members to optical components were coated.

The continuous wet coating process is still widely used because of the following reasons:

- It can apply a range of coating thickness with reasonable coating quality
- The coating equipment is inexpensive and relatively simple to operate
- High-speed capability of the process should lead to low costs and high productivity
- Scale-up from laboratory coaters is much easier than for precision coaters such as slot-die and curtain coating
- Excess coating material can be removed by doctoring devices such as mayer rod, air knife, blades and squeegee rolls to allow wide range of coating weights
- Understanding the limits of dip coating is important to understanding the behavior of many other coating techniques.

The wet thickness for the unassisted dip coater is determined by fluid properties, coating speed and withdrawal angle. The dip coating process have viscosity in the range of 20–2000 cP or mPa-s, wet thickness between 10 and 200 μ (0.4–8 mil), Coverage uniformity is ±10% and line speed of 0.5–7.5 m/s (100–1500 fpm). The enhanced thickness of wet coating should be controlled by means of viscosity and coating speed. In addition, doctoring device is used to produce thin coatings.

da Conceicao et al. [268] adopted dip coating method to coat polyether imide onto the pretreated magnesium alloy substrate to investigate the optimal performance of hydrofluoric acid and the corrosion properties. The samples exposed to 3.5 wt% sodium chloride solution for 3300 h and showed impedance in the order of 10^7 Ω cm^2. X-ray photoelectron spectroscopy (XPS) analysis showed higher performance at the interface due to acid-base interaction. Fang et al. [269] paved easier and economical way of dip coating process to coat polylactic acid onto the metal surface for various applications (Fig. 7.16). They formulated a mechanism to observe the special behavior in dip coating process. The film thickness and other related properties mainly depend on the withdrawal velocity, drag force, gravity, concentration and viscosity of the solution. They optimized the process on medical implant surface to enhance the functionality. The dip coating process found applications in thin films in sol-gel technology, photoresists films, lubricant layers for magnetic hard disks, coat thin films onto plates, slabs and almost any irregular shaped objects. The process, space required and equipment used can be simple.

Li et al. [270] observed low shell formation time and temperature with enhanced shell thickness (10–25 nm), resistivity (0.12–430 Ω cm) and multi-walled carbon nanotubes (MWCNT) dispersion through dip coating process. The experimental investigation on the process variables such as spin speed, retraction speed as well as solution parameters influencing thickness and uniformity of thin LEP films prepared using dip and spin coating were examined by Yimsiri and Mackley [271]. The modifications that include solvent evaporation and viscosity for both the techniques is in line with the final film thickness. Benkreira and Khan [272] examined the effect of reduced air pressure on dynamic wetting based on the roughness and coat viscosity at high speeds. To reduce the experimental error and analyze the wetting performance, substrate with varying roughness were used. The variation in air entrainment speed depends on the viscosity of the coating solution. Cengiz and Erbil [273] described the contact angle measurements and SEM of styrene perfluoromethacrylate random copolymer coated glass substrate. The sizes of coating, degree of phase separation, total surface free energy were controlled through variation in copolymer composition and solvent/non-solvent ratio. Contreras et al. [274] characterized the surface properties of one-step dip coated superhydrophobic injection molded polypropylene films through profilometer measurements, SEM, FTIR-ATR, and XPS analyses. The static and dynamic contact angle measurements were done to determine the wetting property. The control of roughness and polar group content produce more efficient superhydrophobic dynamic component. Rioboo et al. [275] prepared superhydrophobic polypropylene surfaces by dip-coating various molecular weight and tacticity of polypropylene. The hydrophobic surfaces were developed depends on the initial film thickness and initial solution concentration. The desired film thickness, transition between hydrophobic and superhydrophobic surface is induced by increasing the polymer concentration. Gao and He [276] developed superhydrophobic and antireflective coating through different sol-gel dip coating process with 94.6% transmittance at 614 nm wavelength and exhibit excellent broad band antireflection with low substrate transmittance. The prepared samples were subjected to transmission electron microscopy (TEM), SEM, AFM, UV-vis IR and contact angle measurement to observe the morphology, optical property, wettability and surface

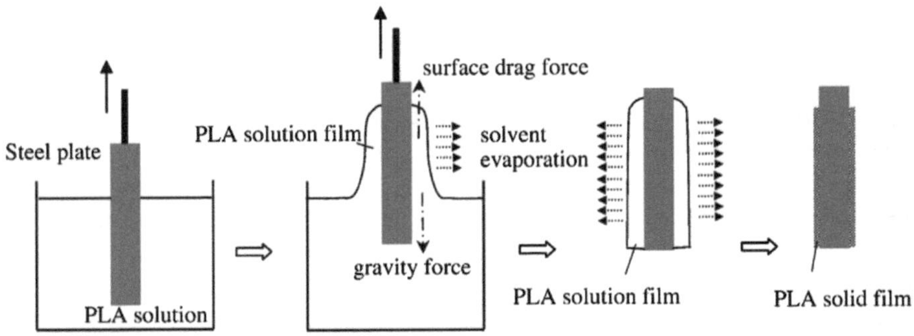

Fig. 7.16 Schematic of dip coating assisted polylactic acid deposition on steel plate and the factors affecting the solid film formation in a dip coating process. Reused from [269]

roughness of the coatings. Xu et al. [277] obtained transparent superhydrophobic coatings with the transmittance of 88% at the wavelength of 550 nm through facile dip-coating of an organic–inorganic composite. The antireflective (maximum transmittance: 94% at 550 nm) superhydrophobic coatings were discovered through layer-by-layer assembly [278]. The authors [279] followed simple dip-coating of hollow silica nanoparticles on glass substrates to fabricate superhydrophobic coatings with the maximum transmittance of 90.2%. Recently, Camargo et al. [280] combined dip-coating and physical vapor deposition to fabricate superhydrophobic and antireflective coating in the range of 1000–2000 nm. In addition, hydrophobically modified hollow silica nanoparticles and poly (methylmethacrylate) displayed superhydrophobic and visible/near-IR antireflective coatings through simple dip coating process for large size fabrication [281]. Catauro et al. [282] examined the mechanical, biological properties such as coating bioactivity and biocompatibility of the dip coated titanium di-oxide/poly-ε-caprolactone hybrid materials. Increase in coating thickness or applying protective coating leads to improved substrate performances. SEM analysis confirmed the preparation of crack-free and porous coating due to the addition of poly-ε-caprolactone. Hao et al. [283] developed highly porous mesh like outer surface of the hollow fiber membrane by dip coating with poly(vinylidene fluoride) (PVDF) solution.

7.9.4 Advantages

- Provide safe ergonomics and beautiful finishes
- Automated and available in all shapes and sizes
- Control film thickness using viscosity
- Improved handling and appearance
- Protective shield that resists corrosion
- Insulates against heat, cold, stress and electrical currents
- Adaptable to high volume orders requiring fast delivery

- Durable and UV resistant
- Alternative colors and finishes can be created easily and economically
- A wide range of thicknesses, textures and hardness are available
- Sound damping and eliminates the need to debur base materials.

7.9.5 Disadvantages

- Light parts tend to float and fall from the conveyor
- Film thickness can vary from top to bottom
- Fatty edges develop on the bottom of parts as excess coating drains
- Refluxing by the solvent vapors above the tank removes some of the coating.

7.10 Spin Coating Process

Spin coating technique is used to develop thin, uniform coating for solution-based anti-reflection applications. In detail, the substrate is mounted horizontally on a rotating platform and spins at high speed until the coating solution is uniformly dispensed as shown in Fig. 7.17. The desired thickness of the coating is based on the substrate rotation.

7.10.1 The Key Stages in Spin Coating

Based on overview of all the researchers, there are four distinct stages in the spin coating process. Stage 3 (flow controlled) and Stage 4 (evaporation controlled) are the two stages that have the most impact on final coating thickness.

Stage One: *Deposition of the coating fluid onto the wafer or substrate.*

It can be done using a nozzle that poured or sprayed the coating solution onto the surface. Sub micron size filter be used to eliminate the flaw particles in the solution. This dispense stage utilize more amount of coating solution and it is important to wet the surface completely.

Stage Two: *The substrate is accelerated up to its final, desired, rotation speed.*

The aggressive fluid on the top surface ejected out from the wafer surface in higher twisting motion. The wafer reaches its desired speed and the fluid is thin enough to balance the viscous shear drag and the rotational accelerations.

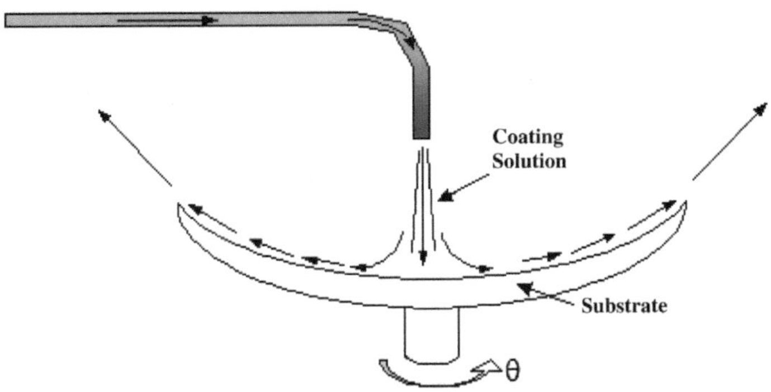

Fig. 7.17 Diagram of the spin-coating process. Reused from [11]

Stage Three: *The substrate is spinning at a constant rate and fluid viscous forces dominate fluid thinning behavior.*

The substrate in slow spin used gradual fluid thinning to obtain uniform and reduced coating thickness. Edge effects and small variation in coating thickness are the most common defects found due to the surface tension, viscosity, rotation rate, etc. The presence of Newtonian viscosity leads to a uniform coating throughout the process.

Stage Four: *The substrate is spinning at a constant rate and solvent evaporation dominates the coating thinning behavior.*

In this stage, the coating gels and solvent evaporation leads to increase in viscosity of the solution. Some of the researchers quantified the coating thickness based on the spin speed, viscosity and its relationship to the evaporation rate.

Many of the applications required heat treatment or firing to undergo coating after spinning process. Depends on the desired applications or usage, photoresists undergo many other process. Viscous flow and evaporation process must occur simultaneously to control the effect on coating thickness.

7.10.2 Common Defects and Developments

A number of defects/features that are characteristic of spin coated films are as follows,

- Comets
- Striations
- Chuck Marks
- Environmental Sensitivity
- Wafer Edge Effects

Kohoutek et al. [284] prepared amorphous chalcogenide films using organic solution by means of spin coating technique. Composition and physico-chemical properties of the prepared films were investigated under vacuum conditions at elevation temperatures. The enhanced holographic diffraction gratings, optical and thermal properties found application in optical recording media. Flack et al. [285] investigated that numerous process variables, radial flow and solvent evaporation have effect on the design, improvements, film formation, film thickness and wafers resistance. The author formulated mathematical equations to dominate the conventional film formation techniques and to clearly understand the film thickness using spin coating process. The non-newtonian character, shear rate, change in resist viscosity and solvent diffusivity determines the derived model based on the polymer concentration. The experimental results based on film thickness, film profile, spin speed and spinner acceleration is in line with the predicted results. Dong et al. [286] investigated the effect of poly (allylamine hydrochloride) and dextran on the surface wettability of the film, deposition of aqueous solution, adhesion, bonding strength and conductive nature. The semi transparent inverted polymer solar cells (PSCs) with improved properties were prepared through spin coating technique under vacuum conditions. It provides way to fabricate solution processed PSCs by roll-to-roll processing. The enhanced photovoltaic performance and PCE were observed. It is mainly used as a subcell of solar cells, power generating thin film coating and in other architectural surfaces. Tsai et al. [287] used single component polymeric materials to fabricate white LED through spin casting process for electronic displays, large and flexible panel applications. The aggregates occurred in solid films and wavelength emissions were investigated. Hole or electron transporter was introduced to analyze the efficiency of these developed devices. The desired properties of the devices were obtained through thermal treatment. Jiang and Mcfarland [288] developed colloidal crystal, macroporous polymer, polymeric nanocomposites with wafer-scale sizes and high crystalline qualities through fast spin coating process. The shear and polymerization process leads to formation of 3D crystals. The spin speed and time controlled the desired film thickness and uniform distribution. Removal of polymer matrix and silica leads to formation of macroporous polymers. The potential device applications paves researchers a way to study shear induced crystallization, melting and relaxation. Barletta et al. [289] explained the formation of thin continuous films by depositing two thermoplastic (acid modified polyolefins) and thermoset organic powders onto the aluminum substrate through hot dipped fluidized bed and electrostatic spray coating methods. The scratch performance depends on coating material, deposition technique and curing process. The prepared samples were subjected to progressive mode scratch tests and contact conditions analysis. The difference in morphology, crystallinity, thickness and adhesive strength were identified. Electrostatically sprayed thermoset coatings render enhanced properties and minor deformation compared to thermoplastic coatings. Rungwasantisuk and Raibhu [54] applied spray coating with design of experiments, residual examination, variance analysis and t-test on UV-curable, polyurethane modified, acrylate based film to control the coating defects and thickness. The most important process variables that gave defect free coatings are as follows: re-spraying, spraying path, overlapping distance, drying and curing times, viscosity of formulation, distance between nozzle and substrate, atomizing pressure, liquid feeding pressure and travel speed. The coated samples results enhanced total transmittance, adhesion, roughness, hardness, scratch hardness, abrasion resistance, durability and haze compared to uncoated samples and commercial PC coated lenses. Gong et al. [290] fabricated the uniform and thin organosilica coating of polysulfones film through spin coating technique. Dimitriou et al. [291] understand the effect of this technique for the preparation of amphiphilic block copolymer surface composition. Soz et al. [292] and Dario et al. [293] understand the nanostructured film formation and their properties based on surface roughness, molecular weight and solvent composition. Iyore et al. [294] compared the spin coating neat and photo crosslinked PVDF-co-hexafluoropropylene films based on dielectric applications. Herrera et al. [295] studied the gas permeability of cellulose nanocrystals films. Syed et al. [296] observed the enhancement in corrosion resistance of polyacrylonitrile/polyether imide multilayer films produced through spin coating process. Similarly, morphology, conductivity and wetting behavior of thin films were studied by Zabihi et al. [55], Dong et al. [286] and Kim et al. [297]. However, in accordance to the above studies, Miao et al. [298], Wu et al. [299], Jasinski et al. [300], Chang et al. [301], Xu and Yamamoto [302], Jayamurugan et al. [303], Seah et al. [304], Sabitha and Joe [305], Kobayashi and Oda [306] adopted spin coating process in various polymers to understand the thickness, modifications and optical properties for thin film transistor and solar cell fabrication.

7.10.3 Advantages and Disadvantages

- Fast process time
- High uniformity over the surface of curved parts
- Uniform coating
- Minimal thickness variation
- Low equipment cost

It can only be performed on one part at a time compared to dip coating in which many parts may be processed simultaneously [307]. Spin-coating is used in low volume operations whereas, dip-coating in high volume operations.

7.11 Pinhole Free Thin Coating

Bottom-up film growth mechanism concluded that most ALD thin coatings are naturally pinhole free. The unique characteristics found application in barrier and passivation applications. ALD also enables high-quality, high-yield insulators over large areas. The TFEL displays are high electrical field devices and ALD-based insulators have been a key factor in the high reliability of the products and for the high production yield. In addition, ALD technology is widely investigated as an option for oxygen and humidity barriers for organic and flexible electronics devices. Major improvements in the barrier properties are already achieved in the 10 nm film thickness range.

7.12 Coating Based on the Application

Based on the application of a polymer coating onto a substrate material [11], it can be classified as (1) wire and cable coating, (2) planar coating, (3) contour coating and (4) roll coating.

7.12.1 Wire and Cable Coating

The most important polymer extrusion processes involved in insulation application is wire and cable coating. As shown in Fig. 7.18 for wire coating, the melt polymer is applied to the bare wire which is pulled at high speed through a die. To enhance adhesion of the coating, light vacuum is drawn between the wire and the polymer [11]. The taught wire provides rigidity during cooling, which is usually aided by

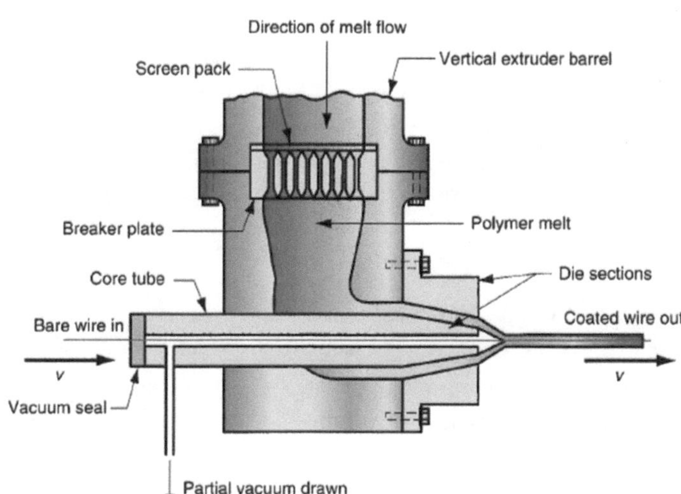

Fig. 7.18 Side view cross section of die for coating of electrical wire by extrusion. Reused from [11]

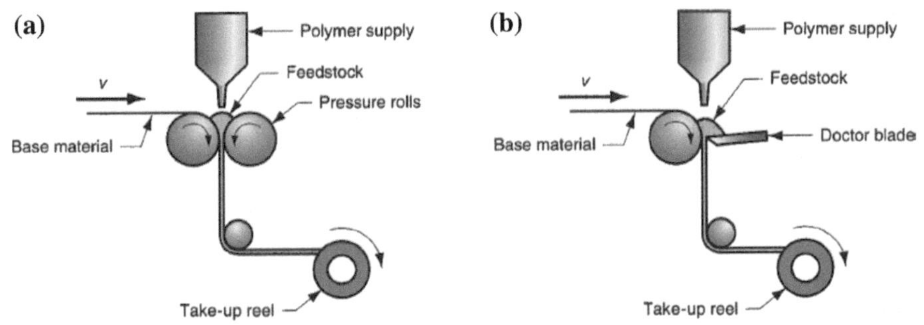

Fig. 7.19 Planar coating processes: **a** roll method and **b** doctor-blade method. Reused from [11]

passing the coated wire through a water trough. Finally, the product is wound onto the large spools at a speed of 50 m/s.

Youssef et al. [308] adopted wire coating technique to understand the electron beam modifications of polyvinyl chloride films. Matallah et al. [309] and Shah et al. [310] simulated and mathematically derived the viscoelastic behaviour of wire coating process.

7.12.2 Planar Coating

Planar coating is used to coat thin films and major product of some plastics such as fabrics, paper, cardboard and metal foil. The polymers involves in planar coating include polyethylene and polypropylene, with lesser applications for nylon, PVC and polyester. The two major planar coating techniques such as roll method (Fig. 7.19a) and doctor blade method (Fig. 7.19b) have coating thickness of around 0.01–0.05 mm. The coating material can be supplied by means of slit-die extrusion or calendaring process. In the roll method, opposing rolls squeezed the coating material against the substrate. In the doctor blade method, the amount of polymer melt that is coated onto the substrate can be controlled using sharp knife edges [11].

7.12.3 Contour Coating

The contour coating of three-dimensional objects can be carried out through dipping or spraying [11]. Dipping involves immersion of the object into a suitable bath of polymer melt or solution, followed by cooling or drying. Spraying (such as spray painting) is an alternative method for applying a polymer coat onto a solid object. White [311] explained the spun-on contour coating behaviour for advanced optical lithography.

7.12.4 Roll Coating

Roll coating is a mechanized technique in which a rotating roller is partially immersed in a pan of liquid adhesive and picks up a coating of the adhesive. Finally, the coating is then transferred to the work surface [11]. The materials used for roll coating are commonly thin and flexible (e.g., paper, cloth, leather and plastic). The major applications involve coating adhesive onto wood, wood composite, cardboard and similar materials with large surface areas. Owens et al. [312] and Echendu et al. [313] observed the misting behaviour, wetting lines, air trapping of liquid solutions in roll coating process.

7.13 Conclusion

Paints and coatings for film formation posses a clear advantage regarding the cost, ease of processing and simplicity. The various techniques that could produce clear and effective film formation had been extensively discussed. Some vital improvements that could obtain improved performance of the coating compared to conventional coating techniques had been addressed. This chapter explains the beneficial advantages of the specific coatings to achieve the required level of performance. It also elaborates the important theoretical background and general criteria for coating film formation and described the challenges and possible methods for achieving the required tasks.

References

1. Wijewardane S, Goswami DY (2012) A review on surface control of thermal radiation by paints and coatings for new energy applications. Renew Sustain Energy Rev 16:1863–1873
2. Bogaerts WF, Lampert CM (1983) Review—materials for photo-thermal solar energy conversion. J Mater Sci 18:2847–2875
3. Agnihotri OP, Gupta BK (1981) Solar selective surfaces. Wiley, New York, NY
4. Pipoly R (1989) Polym Paint Colour J 179:357
5. Supcoe RF (1974) Coatings for infrared camouflage, Rep. C-3946. David Taylor Research Center, Annapolis, MD
6. Thiele ES, French RH (1998) Light-scattering properties of titania particles studied using a finite-element method. J Am Ceram Soc 81:469–479
7. Perera YD (2004) Effect of pigmentation on organic coating characteristics. Prog Org Coat 50:247–262
8. Gunde MK, Kozar Logar J, Crnjak Orel Z et al (1996) Optimum thickness determination to maximize the spectral selectivity of black pigment coatings for solar collectors. Thin Solid Films 277:185–191
9. Williams DA, Lappin TA, Duffie JA (1963) Selective radiation properties of particulate coatings. J Eng Power Trans ASME 85:213

10. Malhotra LK, Chidambaram K, Chopra KL (1981) Partly selective black paint coatings. Energy Res 5:393–396
11. Groover MP (2010) Fundamentals of modern manufacturing: materials, processes and systems, 4th edn. Wiley, New York, NY
12. Granqvist CG (1985) Spectrally selective coatings for energy efficiency and solar applications. Phys Scr 32:401–407
13. Randich E (1981) Thin Solid Films 83:393
14. Rabek JF (1988) Applications of polymers in solar energy utilization: review article. Prog Polym Sci 13:83–188
15. Beatty LW, Penboss I (1993) Chapter 21. Film formation. In: Oil and Colour Chemists Association (ed) Surface coatings: raw materials and their usage, vol 1. Chapman and Hall, London, U.K., p 325
16. Bindschaedler C, Gurney R, Doelker E (1983) Notions theoriques sur la formation des films obtenus a partir de micro-dispersions aqueuses et application a lenrobage. Labo-Pharma Probl Technol 31:389–394
17. Zhang G, Schwartz JB, Schnaare RL (1988) In: Proceedings of 15th international symposium controlled release of bioactive materials, Basel
18. Zhang G, Schwartz JB, Schnaare RL (1989) In: Proceedings of 16th international symposium controlled release of bioactive materials, Chicago
19. Lehmann K (1992) In: Donbrow M (ed) Microcapsules and nanoparticles in medicine and pharmacy. CRC Press, Boca Raton, FL, pp 74–96
20. Lehmann K (1989) In: Mc Ginity JW (ed) Aqueous polymeric coatings for pharmaceutical dosage forms. Marcel Dekker, New York, pp 153–247
21. Bierwagen GP (2015) Encyclopaedia Britannica. http://www.britannica.com/technology/surface-coating. Accessed 10 Oct 2015
22. Asthana R, Kumar A, Dahotre NB (2005) Materials science in manufacturing. Elsevier, Boston, MA
23. Williams D, Randall P (1994) Guide to cleaner technologies: organic coating replacements. Environmental Production Agency, Washington DC, U.S., pp 1–89
24. Lehr WD (1991) Powder coating systems. McGraw-Hill Inc, New York, NY
25. Reich A (1993) Market trends in powder coatings. Eur Polym Paint Colour J 297
26. Loutz JM, Maetens D, Baudour M et al (1993) New developments in powder coatings, Part 1 and Part 2. Eur Paint Polym J 584
27. Stoye D, Freitag W (1998) Paints, coatings and solvents, 2nd edn. Wiley, New York, NY
28. Goldschmidt A, Hantschke B, Knappe E et al (1984) Glasurit-Handbuch, 11th edn. Curt R. Vincentz Verlag, Hannover
29. Ondratschek D, Ortlieb K (1989) Taschenbuch fur Lackierbetriebe, 47th edn. Curt R. Vincentz Verlag, Hannover
30. Thomer KW, Ondratschek D (1987) Lackieren. In: Handbuch der Fertigungstechnik, vol 4/1. Carl Hanser Verlag, Munchen
31. Oppen D (1983) Die Vorbehandlung des metallischen Anstrichuntergrundes. Seminar: Lackierungen in der Metallindustrie, Technische Akademie Esslingen, Esslingen
32. Rausch W (1988) Die Phospliutierung von Metallen, 2nd edn. Eugen G. Leuze Verlag, Saulgau
33. Hoffmann U (1984) Grundsatzliche verfahrenstechnische Gesichtspunkte bei der Kunststofflackierung. Seminar: Lackieren von Kunststoffoberflachen. Wurttembergischer Ingenieurverein Stuttgart, Stuttgart
34. Wittel K (1986) Ind Luckierhetr 54:121–127
35. Hartmann R (1987) Physikalische Vorbehandlungsmethoden von Polymeren. In: Seminar: 4th Deutsche Kunststoff-Finish, Bad Nauheim, Praxis-Forum, Berlin
36. Paint coatings (2015) Encyclopedia of UK steel construction information. http://www.steelconstruction.info/Paint_coatings. Accessed 12 Oct 2015
37. Hendy CR, Iles DC (2010) Steel bridge group: guidance notes on best practice in steel bridge construction and high performance paint coatings, 5th issue, p 185

38. Steel buildings (2003) Chapter 12—Corrosion protection. The British Constructional Steelwork Association Ltd
39. Kovacevic S, Ujevic D, Brnada S (2010) Coated textile materials. In: Dubrovski PD (ed) Woven fabric engineering, pp 241–254. http://www.intechopen.com/books/woven-fabric-engineering/coated-textile-materials. doi:10.5772/10468
40. The Freedonia Group (2012) Coated fabrics. In: US industry study with forecasts for 2016 & 2021. www.freedoniagroup.com
41. Govaert F, Vanneste M (2014) Preparation and application of conductive textile coatings filled with honeycomb structured carbon nanotubes. J Nanomater. doi:10.1155/2014/651265
42. Soljacic I (1993) Textile coating. Tekstil 42:673–686
43. Skoko M (1998) Investigations of properties and multiaxial strength and deformations of coated textile fabrics. Tekstil 47:339–344
44. Wu X, Wyman I, Zhang G et al (2015) Preparation of superamphiphobic polymer based coatings via spray and dip coating strategies. Prog Org Coat. doi:10.1016/j.porgcoat.2015.08.008
45. Steele A, Bayer I, Loth E (2009) Inherently superoleophobic nanocomposites coatings by spray atomization. Nano Lett 9:501–505
46. Xiong D, Liu GJ, Hong LZ et al (2011) Superamphiphobic di-block copolymer coatings. Chem Mater 23:4357–4366
47. Yang W, Zhu LQ, Chen YC (2014) Synthesis and characterization of core–shell latex: effect of fluorinated acrylic monomer on properties of polyacrylates. J Fluor Chem 157:35–40
48. Castelvetro V, Francini G, Ciardelli G et al (2001) Evaluating fluorinated acrylic lattices as textile water and oil repellent finishes. Text Res J 71:399–406
49. Wu WL, Yuan GC, He AH et al (2009) Surface depletion of the fluorine content of electro-spun fibers of fluorinated polyurethane. Langmuir 25:3178–3183
50. Hwang HD, Kim HJ (2011) UV-curable low surface energy fluorinated polycarbonate-based polyurethane dispersion. J Colloid Interface Sci 362:274–284
51. Jiang M, Zhao XL, Ding XB et al (2005) A novel approach to fluorinated polyurethane by macro-monomer copolymerization. Eur Polym J 41:1798–1803
52. Li GH, Shen YD, Ren QH (2005) Effect of fluorinated acrylate on the surface properties of cationic fluorinated polyurethane-acrylate hybrid dispersions. J Appl Polym Sci 97:2192–2196
53. Wang C, Li XR, Du B et al (2014) Preparation and properties of a novel waterborne fluorinated polyurethane-acrylate hybrid emulsion. Colloid Polym Sci 292:579–587
54. Rungwasantisuk A, Raibhu S (2015) Spray coating process variable and property analysis of UV-curable polyethylene acrylate coating on polycarbonate substrate. Prog Org Coat 89:132–142
55. Zabihi F, Xie Y, Gao S et al (2015) Morphology, conductivity and wetting characteristics of PEDOT: PSS thin films deposited by spin and spray coating. Appl Surf Sci 338:163–177
56. Aziz F, Ismail AF (2015) Spray coating methods for polymer solar cells fabrication: a review. Mater Sci Semicond Process 39:416–425
57. Gilissen K, Stryckers J, Verstappen P et al (2015) Ultrasonic spray coating as deposition technique for the light emitting layer in polymer LEDs. Org Electron 20:31–35
58. Wang H, Gao D, Meng Y et al (2015) Corrosion-resistance, robust and wear-durable highly amphiphobic polymer based composite coating via a simple spraying approach. Prog Org Coat 82:74–80
59. Barletta M, Gisario A, Rubino G (2011) Scratch response of high-performance thermoset and thermoplastic powders deposited by the electrostatic spray and 'hot dipping' fluidized bed coating methods: the role of the contact condition. Surface Coat Technol 205:5186–5198
60. Airless Atomization. In: Finish systems, New Berlin, WI. http://www.finishsystems.com/airless.html
61. Bolelli G, Bellucci D, Cannillo V et al (2015) Comparison between suspension plasma sprayed and high velocity suspension flame sprayed bioactive coatings. Surf Coat Technol 280:232–249

62. Nistal A, Garcia E, Garcia-Diego C et al (2015) Flame spraying of adherent silicon coatings on SiC substrates. Surf Coat Technol 270:8–15
63. Puranen J, Laakso J, Honkanen M et al (2015) High temperature oxidation tests for the high velocity solution precursor flame sprayed manganese–cobalt oxide spinel protective coatings on SOFC interconnector steel. Int J Hydrogen Energy 40:6216–6227
64. Monsalve M, Lopez E, Ageorges H et al (2015) Bioactivity and mechanical properties of bioactive glass coatings fabricated by flame spraying. Surf Coat Technol 268:142–146
65. Belamri A, Ati A, Braccini M et al (2015) Hypereutectoid steel coatings obtained by thermal flame spraying-effect of annealing on microstructure, tribological properties and adhesion energy. Surf Coat Technol 263:86–99
66. Chen X, Yuan J, Huang J et al (2014) Large-scale fabrication of superhydrophobic polyurethane/nano-Al_2O_3 coatings by suspension flame spraying for anti-corrosion applications. Appl Surf Sci 311:864–869
67. Bolelli G, Cannillo V, Gadow R et al (2014) Properties of Al_2O_3 coatings by high velocity suspension flame spraying (HVSFS): effects of injection systems and torch design. Surf Coat Technol 270:175–189
68. Ye J, Bu C, Han Z et al (2015) Flame-spraying synthesis and infrared emission property of Ca^{2+}/Cr^{3+} doped $LaAlO_3$ microspheres. J Eur Ceram Soc 35:3111–3118
69. Chaliampalias D, Andronis S, Pliatsikas N et al (2014) Formation and oxidation resistance of Al/Ni coatings on low carbon steel by flame spray. Surf Coat Technol 255:62–68
70. Saremi M, Valefi Z (2013) The effects of spray parameters on the microstructure and thermal stability of thermal barrier coatings formed by solution precursor flame spray (spfs). Surf Coat Technol 220:44–51
71. Redjdal O, Zaid B, Tabti MS et al (2013) Characterization of thermal flame sprayed coatings prepared from FeCr mechanically milled powder. J Mater Process Technol 213:779–790
72. Bergant Z, Trdan U, Grum J (2014) Effect of high-temperature furnace treatment on the microstructure and corrosion behavior of NiCrBSi flame-sprayed coatings. Corros Sci 88:372–386
73. Torres B, Campo M, Lieblich M et al (2013) Oxy-acetylene flame thermal sprayed coatings of aluminium matrix composites reinforced with $MoSi_2$ intermetallic particles. Surf Coat Technol 236:274–283
74. Bao Y, Gawne DT, Zhang T (2012) Cure kinetics of flame-sprayed thermoset coatings. Surf Coat Technol 207:89–95
75. Yuan J, Zhan Q, Lei Q et al (2012) Fabrication and characterization of hybrid micro/nano-structured hydrophilic titania coatings deposited by suspension flame spraying. Appl Surf Sci 258:6672–6678
76. Bellucci D, Bolelli G, Cannillo V et al (2012) High velocity suspension flame sprayed (HVSFS) potassium-based bioactive glass coatings with and without TiO_2 bond coat. Surf Coat Technol 206:3857–3868
77. Ashrafizadeh H, Mertiny P, McDonald A (2014) Determination of temperature distribution within polyurethane substrates during deposition of flame-sprayed aluminum–12silicon coatings using Green's function modeling and experiments. Surf Coat Technol 259:625–636
78. Sharma S (2014) Parametric study of abrasive wear of Co–CrC based flame sprayed coatings by response surface methodology. Tribol Int 75:39–50
79. Gonzalez R, McDonald A, Mertiny P (2013) Effect of flame-sprayed Al–12Si coatings on the failure behaviour of pressurized fibre-reinforced composite tubes. Polym Testing 32:1522–1528
80. Kalacska G, Fazekas L, Keresztes R et al (2011) Cold flame-sprayed and oil-impregnated porous metallic coatings. Appl Surf Sci 257:9532–9538
81. Singrathai S, Rachpech V, Niyomwas S (2011) A thermal coating process using self-propagating high-temperature synthesis assisted flame spray coating process. Energy Procedia 9:398–403
82. Soveja A, Sallamand P, Liao H et al (2011) Improvement of flame spraying PEEK coating characteristics using lasers. J Mater Process Technol 211:12–23

83. Bolelli G, Cannillo V, Gadow R et al (2010) Microstructure and in-vitro behaviour of a novel high velocity suspension flame sprayed (HVSFS) bioactive glass coating. Surf Coat Technol 205:1145–1149
84. Vargas F, Ageorges H, Fournier P et al (2010) Mechanical and tribological performance of Al_2O_3–TiO_2 coatings elaborated by flame and plasma spraying. Surf Coat Technol 205:1132–1136
85. Maury-Ramirez A, Nikkanen JP, Honkanen M et al (2014) TiO_2 coatings synthesized by liquid flame spray and low temperature sol-gel technologies on autoclaved aerated concrete for air-purifying purposes. Mater Charact 87:74–85
86. Teisala H, Tuominen M, Aromaa M et al (2010) Development of superhydrophobic coating on paperboard surface using the liquid flame spray. Surf Coat Technol 205:436–445
87. Bolelli G, Bonferroni B, Cannillo V et al (2010) Wear behaviour of high velocity suspension flame sprayed (HVSFS) Al_2O_3 coatings produced using micron- and nano-sized powder suspensions. Surf Coat Technol 204:2657–2668
88. Bolelli G, Cannillo V, Gadow R et al (2009) Microstructural and in vitro characterisation of high-velocity suspension flame sprayed (HVSFS) bioactive glass coatings. J Eur Ceram Soc 29:2249–2257
89. Zhuang MA, Wei W, Zou JF et al (2011) Preparation and properties of flame-sprayed Mo–FeB–Fe cermet coatings. Trans Nonferrous Metals Soc China 21:1314–1321
90. Feng Z, Xu H, Yan F (2008) Preparation of flame sprayed poly(tetrafluoroethylene-co-hexafluoropropylene) coatings and their tribological properties under water lubrication. Appl Surf Sci 255:2408–2413
91. Bolelli G, Rauch J, Cannillo V et al (2008) Investigation of high velocity suspension flame sprayed (HVSFS) glass coatings. Mater Lett 62:2772–2775
92. Cano C, Garcia E, Fernandes AL et al (2008) Mullite/ZrO_2 coatings produced by flame spraying. J Eur Ceram Soc 28:2191–2197
93. Bradai MA, Bounar N, Benabbas A et al (2008) Study of microstructure, phases and microhardness of metallic coatings deposited by flame thermal spray. J Mater Process Technol 200:410–415
94. Sainz MA, Osendi MI, Miranzo P (2008) Protective Si–Al–O–Y glass coatings on stainless steel in situ prepared by combustion flame spraying. Surf Coat Technol 202:1712–1717
95. Akasaka H, Ohto M, Hasebe Y et al (2011) Yttria coating synthesized by reactive flame spray process using Y-EDTA complex. Surf Coat Technol 205:3877–3880
96. Uyulgan B, Dokumaci E, Celik E et al (2007) Wear behaviour of thermal flame sprayed FeCr coatings on plain carbon steel substrate. J Mater Process Technol 190:204–210
97. Liu HY, Huang JH, Yin CF et al (2007) Microstructure and properties of TiC–Fe cermet coatings by reactive flame spraying using asphalt as carbonaceous precursor. Ceram Int 33:827–835
98. Harsha S, Dwivedi DK, Agrawal A (2007) Influence of WC addition in Co–Cr–W–Ni–C flame sprayed coatings on microstructure, micro-hardness and wear behaviour. Surf Coat Technol 201:5766–5775
99. Li Y, Ma Y, Xie B et al (2007) Dry friction and wear behavior of flame-sprayed polyamide 1010/n-SiO_2 composite coatings. Wear 262:1232–1238
100. Cano C, Osendi MI, Belmonte M et al (2006) Effect of the type of flame on the microstructure of $CaZrO_3$ combustion flame sprayed coatings. Surf Coat Technol 201:3307–3313
101. Zhang G, Liao H, Yu H et al (2006) Deposition of PEEK coatings using a combined flame spraying—laser remelting process. Surf Coat Technol 201:243–249
102. Bemporad E, Bolelli G, Cannillo V et al (2010) Structural characterization of high velocity suspension flame sprayed (HVSFS) TiO_2 coatings. Surf Coat Technol 204:3902–3910
103. Habib KA, Saura JJ, Ferrer C et al (2006) Comparison of flame sprayed Al_2O_3/TiO_2 coatings: their microstructure, mechanical properties and tribology behavior. Surf Coat Technol 201:1436–1443

104. Liu HY, Huang JH (2006) Reactive flame spraying of TiC–Fe cermet coating using asphalt as a carbonaceous precursor. Surf Coat Technol 200:5328–5333
105. Lisjak D, Lintunen P, Hujanen A et al (2011) Hexaferrite/polyethylene composite coatings prepared with flame spraying. Mater Lett 65:534–536
106. Zhang C, Zhang G, Vincent JI et al (2009) Microstructure and mechanical properties of flame-sprayed PEEK coating remelted by laser process. Prog Org Coat 66:248–253
107. Bolelli G, Cannillo V, Gadow R et al (2010) Effect of the suspension composition on the microstructural properties of high velocity suspension flame sprayed (HVSFS) Al_2O_3 coatings. Surf Coat Technol 204:1163–1179
108. Sharma SP, Dwivedi DK, Jain PK (2009) Effect of La_2O_3 addition on the microstructure, hardness and abrasive wear behavior of flame sprayed Ni based coatings. Wear 267:853–859
109. Kratschmer T, Aneziris CG, Gruner P (2011) Mechanical properties of flame sprayed free-standing coatings. Ceram Int 37:2727–2735
110. Bolelli G, Cannillo V, Gadow R et al (2009) Properties of high velocity suspension flame sprayed (HVSFS) TiO_2 coatings. Surf Coat Technol 203:1722–1732
111. Rodriguez RMHP, Paredes RSC, Wido SH et al (2007) Comparison of aluminum coatings deposited by flame spray and by electric arc spray. Surf Coat Technol 202:172–179
112. Lia JF, Lib L, Stott FH (2004) Fractal characteristics of apparent pores present on polished cross sections of alumina coatings prepared by laser-assisted flame hybrid spraying. Thin Solid Films 453–454:229–233
113. Pimenoff JA, Hovinen AK, Rajala MJ (2009) Nanostructured coatings by liquid flame spraying. Thin Solid Films 517:3057–3060
114. Zhang G, Liao H, Cherigui M et al (2007) Effect of crystalline structure on the hardness and interfacial adherence of Xame sprayed poly(ether–ether–ketone) coatings. Eur Polym J 43:1077–1082
115. Arcondeguy A, Gasgnier G, Montavon G et al (2008) Effects of spraying parameters onto flame-sprayed glaze coating structures. Surf Coat Technol 202:4444–4448
116. Kahraman N, Gulenc B (2002) Abrasive wear behaviour of powder flame sprayed coatings on steel substrates. Mater Des 23:721–725
117. Krishna BV, Misra VN, Mukherjee PS et al (2002) Microstructure and properties of flame sprayed tungsten carbide coatings. Int J Refract Metal Hard Mater 20:355–374
118. Gonzalez R, Cadenas M, Fernandez R et al (2007) Wear behaviour of flame sprayed NiCrBSi coating remelted by flame or by laser. Wear 262:301–307
119. Linton JD, Jiang Q, Gatti CJ et al (2013) Erratum to "Simultaneous optimization of flame spraying process parameters for high quality molybdenum coatings using taguchi methods" [Surf Coat Technol 79(1–3) (1996) 276–288]. Surf Coat Technol 214:173–174
120. Kusoglu IM, Celik E, Cetinel H et al (2005) Wear behavior of flame-sprayed Al_2O_3–TiO_2 coatings on plain carbon steel substrates. Surf Coat Technol 200:1173–1177
121. Kim HJ, Hwang SY, Lee CH et al (2003) Assessment of wear performance of flame sprayed and fused Ni-based coatings. Surf Coat Technol 172:262–269
122. Lin L, Han K (1998) Optimization of surface properties by flame spray coating and boriding. Surf Coat Technol 106:100–105
123. Zhang Z, Wang Z, Liang B (2006) Tribological properties of flame sprayed Fe–Ni–RE alloy coatings under reciprocating sliding. Tribol Int 39:1462–1468
124. Guo DZ, Li FL, Wang JY et al (1995) Effects of post-coating processing on structure and erosive wear characteristics flame and plasma spray coatings. Surf Coat Technol 73:73–78
125. Hidalgo VH, Varela JB, Menendez AC et al (2001) High temperature erosion wear of flame and plasma-sprayed nickel–chromium coatings under simulated coal-fired boiler atmospheres. Wear 247:214–222
126. Rauch J, Bolelli G, Killinger A et al (2009) Advances in high velocity suspension flame spraying (HVSFS). Surf Coat Technol 203:2131–2138
127. Lian Y, Yu L, Xue Q (1995) The effect of cerium dioxide on the friction and wear properties of flame spraying nickel-based alloy coating. Wear 181–183:436–441

128. Navas C, Colaco R, de Damborenea J et al (2006) Abrasive wear behaviour of laser clad and flame sprayed-melted NiCrBSi coatings. Surf Coat Technol 200:6854–6862
129. Chaliampalias D, Vourlias G, Pavlidou E et al (2009) Comparative examination of the microstructure and high temperature oxidation performance of NiCrBSi flame sprayed and pack cementation coatings. Appl Surf Sci 255:3605–3612
130. Bhadang KA, Holding CA, Thissen H et al (2010) Biological responses of human osteoblasts and osteoclasts to flame-sprayed coatings of hydroxyapatite and fluorapatite blends. Acta Biomater 6:1575–1583
131. Gadow R, Killinger A, Rauch J (2008) New results in high velocity suspension flame spraying (HVSFS). Surf Coat Technol 202:4329–4336
132. Li JF, Li L, Stott FH (2004) Combined laser and flame surface coating of refractory ceramics: phase and microstructural characteristics. Thin Solid Films 453–454:67–71
133. Li JF, Li L, Stott FH (2004) Multi-layered surface coatings of refractory ceramics prepared by combined laser and flame spraying. Surf Coat Technol 180–181:500–505
134. Babu MV, Kumar RK, Prabhakar O et al (1996) Simultaneous optimization of flame spraying process parameters for high quality molybdenum coatings using Taguchi methods. Surf Coat Technol 79(276):288
135. Iordanova I, Forcey KS, Gergov B et al (1995) Characterization of flame-sprayed and plasma-sprayed pure metallic and alloyed coatings. Surf Coat Technol 72:23–29
136. Killinger A, Kuhn M, Gadow R et al (2006) High-velocity suspension flame spraying (HVSFS), a new approach for spraying nanoparticles with hypersonic speed. Surf Coat Technol 201:1922–1929
137. Li J, Liao H, Coddet C (2002) Friction and wear behavior of flame-sprayed PEEK coatings. Wear 252:824–831
138. Karageorgakis NI, Heel A, Graule T et al (2011) Flame spray deposition of nanocrystalline dense $Ce_{0.8}Gd_{0.2}O_{2-\delta}$ thin films: deposition mechanism and microstructural characterization. Solid State Ionics 192:464–471
139. Sugama T, Kawase R, Berndt CC et al (1995) An evaluation of methacrylic acid-modified poly(ethylene) coatings applied by flame spray technology. Prog Org Coat 25:205–216
140. Flame spray coating info and benefits. Flame Spray Coating Co., Fraser, MI. http://www.flamesprayusa.com/flame-spray-coating.php
141. Yan J, Xu J, Ud-Din R et al (2015) Preparation of agglomerated powders for airplasma spraying $MoSi_2$ coating. Ceram Int 41:10547–10556
142. Pu H, Niu Y, Hu C et al (2015) Ablation of vacuum plasma sprayed TaC-based composite coatings. Ceram Int 41:11387–11395
143. Niu Y, Lu D, Huang L et al (2015) Comparison of W–Cu composite coatings fabricated by atmospheric and vacuum plasma spray processes. Vacuum 117:98–103
144. Dong S, Song B, Liao H et al (2015) Deposition of NiCrBSi coatings by atmospheric plasma spraying and dry-ice blasting: microstructure and wear resistance. Surf Coat Technol 268:36–45
145. Wen B, Man Z, Liu Y et al (2014) Supersonic flame ablation resistance of W/ZrC coating deposited on C/SiC composites by atmosphere plasma spraying. Ceram Int 40:11825–11830
146. Hou H, Ning X, Wang Q et al (2015) Anti-ablation behavior of air plasma-sprayed $Mo(Si, Al)_2$ coating. Surf Coat Technol 274:60–67
147. Daroonparvar M, Yajid MAM, Yusof NM et al (2015) Deposition of duplex MAO layer/nanostructured titanium dioxide composite coatings on Mg–1% Ca alloy using a combined technique of air plasma spraying and micro arc oxidation. J Alloy Compd 649:591–605
148. Zhao S, Zhao Y, Zou B et al (2014) Characterization and thermal cycling behavior of $La_2(Zr_{0.7}Ce_{0.3})_2O_7$/8YSZ functionally graded thermal barrier coating prepared by atmospheric plasma spraying. J Alloy Compd 592:109–114
149. Afzal M, Ajmal M, Nusair Khan A et al (2014) Surface modification of air plasma spraying WC–12% Co cermet coating by laser melting technique. Opt Laser Technol 56:202–206

150. Song B, Dong S, Liao H et al (2015) Microstructure and wear resistance of FeAl/Al_2O_3 intermetallic composite coating prepared by atmospheric plasma spraying. Surf Coat Technol 268:24–29
151. Zhao S, Gu L, Zhao Y et al (2013) Thermal cycling behavior and failure mechanism of $La_2(Zr_{0.7}Ce_{0.3})_2O_7$/$Eu^{3+}$-doped 8YSZ thermal barrier coating prepared by atmospheric plasma spraying. J Alloy Compd 580:101–107
152. Wang Y, Wang D, Yan J et al (2013) Preparation and characterization of molybdenum disilicide coating on molybdenum substrate by air plasma spraying. Appl Surf Sci 284:881–888
153. Daroonparvar M, Yajid MAM, Yusof NM et al (2015) Microstructural characterization of air plasma sprayed nanostructure ceramic coatings on Mg–1% Ca alloys (bonded by NiCoCrAlYTa alloy). Ceram Int. doi:10.1016/j.ceramint.2015.08.118
154. Hu YZ, Li CX, Yang GJ et al (2014) Evolution of microstructure during annealing of $Mn_{1.5}Co_{1.5}O_4$ spinel coatings deposited by atmospheric plasma spray. Int J Hydrogen Energy 39:13844–13851
155. Palacio CC, Ageorges H, Vargas F et al (2013) Effect of the mechanical properties on drilling resistance of Al_2O_3–TiO_2 coatings manufactured by atmospheric plasma spraying. Surf Coat Technol 220:144–148
156. Ercenk E, Sen U, Yilmaz S (2012) The effect of SiC addition on the crystallization kinetics of atmospheric plasma—sprayed basalt-based coatings. Ceram Int 38:6549–6556
157. Niu Y, Zheng X, Liu X et al (2012) Influence of powder size on characteristics of air plasma sprayed silicon coatings. Ceram Int 38:5897–5905
158. Han SJ, Chen Y, Sampath S (2014) Role of process conditions on the microstructure, stoichiometry and functional performance of atmospheric plasma sprayed La(Sr)MnO_3 coatings. J Power Sour 259:245–254
159. Zhao Y, Shinmi A, Zhao X et al (2012) Investigation of interfacial properties of atmospheric plasma sprayed thermal barrier coatings with four-point bending and computed tomography technique. Surf Coat Technol 206:4922–4929
160. Yaghtin AH, Salahinejad E, Khosravifard A et al (2015) Corrosive wear behavior of chromium carbide coatings deposited by air plasma spraying. Ceram Int 41:7916–7920
161. Lin CM, Tsai HL, Yang C (2012) Effects of microstructure and properties on parameter optimization of boron carbide coatings prepared using a vacuum plasma-spraying process. Surf Coat Technol 206:2673–2681
162. Pan ZY, Wang Y, Wang CH et al (2012) The effect of SiC particles on thermal shock behavior of Al_2O_3/8YSZ coatings fabricated by atmospheric plasma spraying. Surf Coat Technol 206:2484–2498
163. Chen X, Gu L, Zou B et al (2012) New functionally graded thermal barrier coating system based on $LaMgAl_{11}O_{19}$/YSZ prepared by air plasma spraying. Surf Coat Technol 206:2265–2274
164. Guanhong S, Xiaodong H, Jiuxing J et al (2011) Parametric study of Al and Al_2O_3 ceramic coatings deposited by air plasma spray onto polymer substrate. Appl Surf Sci 257:7864–7870
165. Zou B, Khan ZS, Fan X et al (2013) A new double layer oxidation resistant coating based on Er_2SiO_5/$LaMgAl_{11}O_{19}$ deposited on C/SiC composites by atmospheric plasma spraying. Surf Coat Technol 219:101–108
166. Rayon E, Bonache V, Salvador MD et al (2011) Hardness and Young's modulus distributions in atmospheric plasma sprayed WC–Co coatings using nanoindentation. Surf Coat Technol 205:4192–4197
167. Niu Y, Zheng X, Ji H et al (2010) Microstructure and thermal property of tungsten coatings prepared by vacuum plasma spraying technology. Fusion Eng Des 85:1521–1526
168. Raj SV, Palczer A (2010) Thermal expansion of vacuum plasma sprayed coatings. Mater Sci Eng A 527:2129–2135
169. Chen H, Gou G, Tu M et al (2009) Characteristics of nano particles and their effect on the formation of nanostructures in air plasma spraying WC–17Co coating. Surf Coat Technol 203:1785–1789

170. Yang W, Li M (2009) Effect of remelting process on characterization of air-plasma sprayed $Fe_{67.5}Ni_{23.5}B_9$ alloy coatings onto $_1Cr_{18}Ni_9Ti$ stainless steel. J Mater Process Technol 209:3256–3263
171. Bolleddu V, Racherla V, Bandyopadhyay PP (2014) Microstructural and tribological characterization of air plasma sprayed nanostructured alumina–titania coatings deposited with nitrogen and argon as primary plasma gases. Mater Des 59:252–263
172. Bozorgtabar M, Rahimipour M, Salehi M et al (2011) Structure and photocatalytic activity of TiO_2 coatings deposited by atmospheric plasma spraying. Surf Coat Technol 205:S229–S231
173. Huang W, Fan X, Zhao Y et al (2012) Fabrication of thermal barrier coatings onto polyimide matrix composites via air plasma spray process. Surf Coat Technol 207:421–429
174. Kwon O, Kumar S, Park S et al (2007) Comparison of solid oxide fuel cell anode coatings prepared from different feedstock powders by atmospheric plasma spray method. J Power Sour 171:441–447
175. Wang L, Wang Y, Sun XG et al (2011) Influence of pores on the thermal insulation behavior of thermal barrier coatings prepared by atmospheric plasma spray. Mater Des 32:36–47
176. Zois D, Lekatou A, Vardavoulias M et al (2010) Nanostructured alumina coatings manufactured by air plasma spraying: correlation of properties with the raw powder microstructure. J Alloy Compd 495:611–616
177. Poza P, Grant PS (2006) Microstructure evolution of vacuum plasma sprayed CoNiCrAlY coatings after heat treatment and isothermal oxidation. Surf Coat Technol 201:2887–2896
178. Huang C, Du L, Zhang W (2009) Preparation and characterization of atmospheric plasma-sprayed $NiCr/Cr_3C_2–BaF_2 \cdot CaF_2$ composite coating. Surf Coat Technol 203:3058–3065
179. Guo HB, Kuroda S, Murakami H (2006) Segmented thermal barrier coatings produced by atmospheric plasma spraying hollow powders. Thin Solid Films 506–507:136–139
180. Sarikaya O (2005) Effect of some parameters on microstructure and hardness of alumina coatings prepared by the air plasma spraying process. Surf Coat Technol 190:388–393
181. Singh VP, Sil A, Jayaganthan R (2011) A study on sliding and erosive wear behaviour of atmospheric plasma sprayed conventional and nanostructured alumina coatings. Mater Des 32:584–591
182. Jaeggi C, Frauchiger V, Eitel F et al (2011) The effect of surface alloying of Ti powder for vacuum plasma spraying of open porous titanium coatings. Acta Mater 59:717–725
183. Hester LC, Nicholson RL (1989) Powder coating technology update, EPA-45/3-89-33
184. Michael MJ (1992) Innovation and regulations aid powder coatings. Mod Paint Coat 82:6
185. Pennisi MS About powder coating. In: Finishing dot com, the home page of the finishing industry. http://www.finishing.com/library/pennisi/powder.html
186. Polyester powder coating. In: Highland-galvanizers and colour coaters. http://www.higalv.co.uk/technical-issues-polyester.asp
187. Muhlenkamp Mac (1988) The technology of powder. Mod Paint Coat 78:52–68
188. Barletta M, Tagliaferri V (2006) Influence of process parameters in electrostatic fluidized bed coating. Surf Coat Technol 200:4619–4629
189. How to powder coat. In: Paint and other finishes. http://www.wikihow.com/Powder-Coat
190. What is powder coating. In: Powder Coating Institute. http://www.powdercoating.org/11/Industry/What-is-Powder-Coating
191. Ellicks DF (1994) Environmental compliant thermoplastic powder coating. In: Proceedings of pollution prevention conference on Lowand No-VOC Coating Technologies, U.S. Environmental Protection Agency. Air and Energy Engineering Research Laboratory. Organics Control Branch. Research Triangle Park, NC. EPA-600/R-94-022
192. MP&C (1991) Powder coatings share program with water-borne and higher solids. Mod Paint Coat, 53
193. Kunaver M, Mozetic M, Klanjsek Gunde M (2004) Selective plasma etching of powder coatings. Thin Solid Films 459:115–117
194. P&SF (1992) Powder coatings play major role in microwave antenna manufacture. Plat Surf Finish, 8

195. CMR (1993) Powder coatings. Chem Mark Reporter, 22
196. Cole GE (1993) Automotive takes a powder. Ind Paint Powder 69:24
197. P&II (1991) Tintas Coral's new anti-corrosion powder coating. Paint Ink Int, 40
198. P&SF (1993) For aluminum window/door fabricator: powder coating is finish of choice. Plat Surf Finish, 8
199. Metallurgia (1991) New guidelines for powder coating of steels, p 443
200. AMM (1991) New powder coating needs no priming. Am Metal Market, 4
201. http://electrostaticrefinishers.com/electrostatic-painting/electrostatic-painting-vs-powder-coating/
202. http://www.decc.com/coating_terminology.php
203. https://en.wikipedia.org/wiki/Electrostatic_coating
204. http://www.electrostaticpainting.org/electrostatic-painting-process.html
205. http://www.fluoroprecision.co.uk/ptfe/ptfe-technical-information/ptfe-coating-process
206. http://www.plasticcoatings.co.uk/plastic-coatings-processes/powder-coating.php
207. http://electro-static.com/index.html
208. Miranda R (2014) Surface modification by solid state processing. Elsevier, UK
209. Dini JW (1993) Electrodeposition: the materials science of coatings and substrates. Noyes Publications, USA
210. Tracton AA (2006) Coatings technology handbook, 3rd edn. Taylor and Francis Group, Boca Raton, FL
211. Brewer GEF, Hamilton AD (1972) Paint for electrocoating. In: ASTM Gardner-sward paint testing manual, 13th edn. American Society for Testing and Materials, Philadelphia, pp 486–489 (Section 8.10)
212. Brewer GEF (1973) Electrodeposition of coatings. In: ACS advances in chemistry series. American Chemical Society, Washington, DC, p 119
213. Brewer GEF (1981) ACS symposium on newer developments in electrocoating. Org Coatings Plast Chem 45:92–113
214. Chandler RH (1966) Advances in electrophoretic painting Bi or Tri annual abstracts. R. H. Chandler, Braintree, Essex, England
215. Duffy JI (1982) Electrodeposition processes and equipment. Noyes Data Corp, Park Ridge, NJ
216. Kardomenos PI, Nordstrom JD (1982) Polymer compositions for catholic electrodeposition coatings. J Coat Technol 54:33–41
217. Machu W (1978) Handbook of electropainting technology. Electrochemical Publications Ltd., Scotland
218. Raney MW (1970) Electrodeposition and radiation curing of coatings. Noyes Data Corp, Park Ridge, NJ
219. Wang T, Tan YJ (2006) Understanding Electrodeposition of polyaniline coatings for corrosion prevention applications using the wire bean electrode method. Corros Sci 48:2274–2290
220. Hamid ZA, Ghayad IM (2002) Characteristics of electrodeposition of Ni-polyethylene composite coatings. Mater Lett 53:238–243
221. Cheng Y, Gray KM, David L et al (2012) Characterization of the cathodic electrodeposition of semicrystalline chitosan hydrogel. Mater Lett 87:97–100
222. Pierleoni D, Xia ZY, Christian M et al (2016) Graphene based coatings on polymer films for gas barrier applications. Carbon 96:503–512
223. Herrasti P, Kulak AN, Bavykin DV et al (2011) Electrodeposition of polypyrrole-titanate nanotube composites coatings and their corrosion resistance. Electrochim Acta 56:1323–1328
224. Lin JY, Liao JH, Chou SW (2011) Cathodic electrodeposition of highly porous cobalt sulfide counter electrodes for dye-sensitized solar cells. Electrochim Acta 56:8818–8826
225. Rapecki T, Stojek Z, Donten M (2013) Nucleation of metals on conductive polymers: electrodeposition of silver in thin polypyrrole films. Electrochim Acta 106:264–271

226. Moral Vico J, Carretero NM, Perez E et al (2013) Dynamic electrodeposition of aminoacid-polypyrrole on aminoacid-PEDOT substrates: conducting polymer bilayers as electrodes in neutral systems. Electrochim Acta 111:250–260
227. Chen J, Dong P, Di D et al (2013) Controllable fabrication of 2D colloidal crystal films with polystyrene nanospheres of various diameters by spin coating. Appl Surf Sci 270:6–15
228. Yang Y, Li Y, Liu H et al (2014) Electrodeposition of gold nanoparticles onto an etched stainless steel wire followed by a self-assembled monolayer of octanedithiol as a fiber coating for selective solid-phase microextraction. J Chromatogr A 1372:25–33
229. Li Y, Xu H (2015) Development of a novel graphene/polyaniline electrodeposited coating for on-line solid phase microectraction of aldehydes in human exhaled breath condensate. J Chromatogr A 1395:23–31
230. Ranjbar Z, Moradian S, Rastegar S (2003) Formation of a percolating cluster in films prepared by cathodic electrodeposition of a mixture of lower and higher molecular weight epoxy-amine adducts. J Colloid Interface Sci 264:420–430
231. Lammel P, Torun B, Kleber C et al (2013) In-situ AFM study of the electrodeposition of copper on plasma modified carbon fiber-reinforced polymer surfaces. Surf Coat Technol 221:22–28
232. Wang Z, Zhang X, Gu J et al (2014) Electrodeposition of alginate/chitosan layer-by-layer composite coatings on titanium substrates. Carbohydr Polym 103:38–45
233. Ngounou B, Aliyev EH, Guschin DA et al (2007) Parallel synthesis of libraries of anodic and cathodic functionalized electrodeposited paints as immobilization matrix for amperometric biosensors. Bioelectrochemistry 71:81–90
234. Ejenstam L, Tuominen M, Haapanen J et al (2015) Long-term corrosion protective by a thin nanocomposites coating. Appl Surf Sci. doi:10.1016/j.apsusc.2015.09.238
235. Zainoldin Z, Harun MK, Bahron H et al (2012) Passive thin film coating through electrodeposition of salicylideneaniline. APCBEE Procedia 3:104–109
236. Balaji R, Pushpavanam M, Kumar KY et al (2006) Electrodeposition of bronze–PTFE composite coatings and study on their tribological characteristics. Surf Coat Technol 201:3205–3211
237. Wolfs M, Darmanin T, Guittard F (2014) Effect of hydrocarbon chain branching in the elaboration of superhydrophobic materials by electrodeposition of conductive polymers. Surf Coat Technol 259:594–598
238. Lacovetta D, Tam J, Erb U (2015) Synthesis, structure and properties of superhydrophobic nickel–PTFE nanocomposites coatings made by electrodeposition. Surf Coat Technol 279:134–141
239. Rastegar S, Ranjbar Z (2006) Cathodically electro-deposited carbon-black-epoxy composite films as primer in two-coat electro-deposition. Prog Org Coat 56:234–239
240. Wang T, Qi S, Ren B et al (2007) Preparation and surface characteristics of low-temperature curing fluorinated cathodic electrodeposition coating. Prog Org Coat 60:132–139
241. Bucko M, Miskovic Stankovic V, Rogan J et al (2015) The protective properties of epoxy coating electrodeposited on Zn–Mn alloy substrate. Prog Org Coat 79:8–16
242. Gonzalez MB, Saidman SB (2015) Electrodeposition of bilayered polypyrrole on 316 L stainless steel for corrosion prevention. Prog Org Coat 78:21–27
243. Hollander A, Amberg Schwab S, Miesbauer O et al (2015) Process control for thin organic coatings using fluorescence dyes. Prog Org Coat 88:71–74
244. Xu J, Zhang Y, Zhang D et al (2015) Electrosynthesis of PAni/PPy coatings doped by phosphotungstate on mild steel and their corrosive resistances. Prog Org Coat 88:84–91
245. Bass M, Freger V (2015) Facile evaluation of coating thickness on membrane using ATR-FTIR. J Membr Sci 492:348–354
246. Mollahosseini A, Noroozian E (2009) Electrodeposition of a highly adherent and thermally stable polypyrrole coating on steel from aqueous polyphosphate solution. Synth Met 159:1247–1254
247. Merisalu M, Kahro T, Kozlova J et al (2015) Graphene-polypyrrole thin hybrid corrosion resistant coatings for copper. Synth Met 200:16–23

248. Zhitomirsky I, Petric A (2000) Cathodic electrodeposition of polymer films and organoceramic films. Mater Sci Eng B 78:125–130
249. Ranjbar Z, Moradian S (2003) Characteristics and deposition behavior of epoxy-amine adducts in cathodic electrodeposition as a function of the degree of neutralization. Colloids Surf A 219:147–159
250. Chen Q, Li W, Goudouri OM et al (2015) Electrophoretic deposition of antibiotic loaded PHBV microsphere-alginate composite coating with controlled delivery potential. Colloids Surf B 130:199–206
251. Wang Y, Guo X, Pan R et al (2015) Electrodeposition of chitosan/gelatin/nanosilver: a new method for constructing biopolymer/nanoparticles composite films with conductivity and antibacterial activity. Mater Sci Eng C 53:222–228
252. Yuan HJ, Lu CX, Zhang SC et al (2015) Preparation and characterization of a polyimide coating on the surface of carbon fibers. New Carbon Mater 30:115–121
253. Mo SJ, Zhang J, Liang D et al (2013) Study on pyrolysis characteristics of cross linked polyethylene material cable. Procedia Eng 52:588–592
254. Bichler C, Kerbstadt T, Landowski HC et al (1999) Plasma-modified interfaces between polypropylene films and vacuum roll-to-roll coated thin barrier layers. Surf Coat Technol 112:373–378
255. Hao S, Wu J, Lin J et al (2012) Modification of photocathode of dye-sensitized nanocrystalline solar cell with platinum by vacuum coating, thermal decomposition and electroplating. Compos Interfaces 13:899–909
256. Chapter 4: Composites. In: Advanced materials in Japan. Elsevier Advanced Technology, UK, pp 73–86
257. Dow lowers filtration costs with solka-floc/dicalite precoating. In: Dicalite technical report, California, p 18A (Ind Eng Chem)
258. Landau LD, Levich BG (1942) Acta Physiochim 17:42–54
259. Strawbridge I, James PF (1986) J Non-Cryst Solids 82:366–372
260. http://www.apexicindia.com/en/technologies/dip-coating-technology8
261. http://www.biolinscientific.com/technology/dip-coating/10
262. Brinker CJ, Hurd AJ, Ward KJ (1988) In: Mackenzie JD, Ulrich DR (eds) Ultrastructure processing of advanced ceramics. Wiley, NY, p 223
263. Stern O (1924) Elektrochem, 508
264. Bravo J, Zhai L, Wu Z et al (2007) Transparent superhydrophobic films based on silica nanoparticles. Langmuir 23:7293–7298
265. Li H, Chen C, Zhang S et al (2012) The use of layer by layer assembly coatings of hyaluronic acid and cationized gelatin to improve the biocompatibility of poly(ethylene terephthalate) artificial ligaments for reconstruction of the anterior cruciate ligament. Acta Biometerialia 8:4007–4019
266. Schroder H (1969) Physics of thin films, vol 5. Academic Press, NY, London, pp 87–141
267. Dislich H (1971) Angew Chem Int 6:363
268. da Conceicao TF, Scharnagl N, Dietzel W et al (2011) Corrosion protection of magnesium AZ31 alloy using poly(ether imide)[PEI] coatings prepared by the dip coating method: influence of solvent and substrate pre-treatment. Corros Sci 53:338–346
269. Fang HW, Li KY, Su TL et al (2008) Dip coating assisted polylactic acid deposition on steel surface: film thickness affected by drag force and gravity. Mater Lett 62:3739–3741
270. Li X, Li Q, Chen GX (2014) Alkali metal surfactant-facilitated formation of thick boron nitride layers on carbon nanotubes by dip coating. Mater Lett 134:38–41
271. Yimsiri P, Mackley MR (2006) Spin and dip coating of light emitting polymer solutions: matching experiment with modeling. Chem Eng Sci 61:3496–3505
272. Benkreira H, Khan MI (2008) Air entrainment in dip coating under reduced air pressures. Chem Eng Sci 63:448–459
273. Cengiz U, Erbil HY (2014) Superhydrophobic perfluoropolymer surfaces having heterogeneous roughness created by dip-coating from solutions containing a nonsolvent. Appl Surf Sci 292:591–597

274. Contreras CB, Chagas G, Strumia MC et al (2014) Permanent superhydrophobic polypropylene nanocomposites coatings by a simple one-step dipping process. Appl Surf Sci 307:234–240
275. Rioboo R, Voue M, Vaillant A et al (2008) Langmuir 24:9508
276. Gao L, He J (2013) A facile dip-coating approach based on three silica sols to fabrication of broadband antireflective superhydrophobic coatings. J Colloid Interface Sci 400:24–30
277. Xu QF, Wang JN, Sanderson KD (2010) ACS Nano 4:2201
278. Li X, Du X, He J (2010) Langmuir 26:13528
279. Xu L, He J (2012) Langmuir 28:7512
280. Camargo KC, Michels AF, Rodembusch FS et al (2012) Chem Commun 48:4992
281. Xu L, Gao L, He J (2012) RSC Adv 2:12764
282. Catauro M, Papale F, Bollino F (2015) Characterization and biological properties of TiO_2/PCL hybrid layers prepared via sol-gel dip coating for surface modification of titanium implants. J Non-Cryst Solids 415:9–15
283. Hao Y, Sano R, Shimomura A et al (2014) Reorganization of the surface geometry of hollow-fiber membranes using dip-coating and vapor-induced phase separation. J Membr Sci 460:229–240
284. Kohoutek T, Wagner T, Orava J et al (2004) Amorphous films of Ag-As-S system prepared by spin-coating technique, preparation techniques and films physico-chemical properties. Vacuum 76:191–194
285. Flack WW, Soong DS, Bell AT et al (1984) A mathematical model for spin coating of polymer resists. J Appl Phys 56:1199–1206
286. Dong Q, Zhou Y, Pei J et al (2010) All-spin coating vacuum free processed semi transparent inverted polymer solar cells with PEDOT: PSS anode and PAH-D interfacial layer. Org Electron 11:1327–1331
287. Tsai ML, Liu CY, Hsu MA et al (2003) White light emission from single component polymers fabricated by spin coating. Appl Phys Lett 82:550–552
288. Jiang P, McFarland MJ (2004) Large-scale fabrication of wafer-size colloidal crystals, macroporous polymers and nanocomposites by spin-coating. J Am Chem Soc 42:13778–13786
289. Barletta M, Gisario A, Rubino G (2011) Scratch response of high-performance thermoset and thermoplastic powders deposited by the electrostatic spray and hot dipping fluidized bed coating methods: the role of the contact condition. Surf Coat Technol 205:5186–5198
290. Gong G, Wang J, Nagasawa H et al (2013) Sol-gel spin coating process to fabricate a new type of uniform and thin organosilica coating on polysulfones film. Mater Lett 109:130–133
291. Dimitriou MD, Sundaram HS, Cho Y et al (2012) Amphiphilic block copolymer surface composition: effects of spin coating versus spray coating. Polymer 53:1321–1327
292. Soz CK, Yilgor E, Yilgor I (2015) Influence of the average surface roughness on the formation of superhydrophobic polymer surfaces through spin-coating with hydrophobic fumed silica. Polymer 62:118–128
293. Dario AF, Macia HB, Petri DFS (2012) Nanostructures on spin-coated polymer films controlled by solvent composition and polymer molecular weight. Thin Solid Films 524:185–190
294. Iyore OD, Roodenko K, Winkler PS et al (2013) Comparison of neat and photo-crosslinked polyvinylidene fluoride-co-hexafluoropropylene thin film dielectrics formed by spin coating. Thin Solid Films 548:597–602
295. Herrera MA, Mathew AP, Oksman K (2014) Gas permeability and selectivity of cellulose nanocrystals films (layers) deposited by spin coating. Carbohydr Polym 112:494–501
296. Syed JA, Lu H, Tang S et al (2015) Enhanced corrosion protective PANI-PAA/PEI multilayer composite coatings for 316SS by spin coating technique. Appl Surf Sci 325:160–169
297. Kim BJ, Han SH, Park JS (2015) Properties of CNTs coated by PEDOT: PSS films via spin coating and electrophoretic deposition methods for flexible transparent electrodes. Surf Coat Technol 271:22–26

298. Miao JT, Sun R, Yan BL et al (2014) Adjustable third order nonlinear optical properties of the spin coating phenoxazinium–PMMA films. Mater Chem Phys 147:232–237
299. Wu X, Wyman I, Zhang G et al (2015) Preparation of superamphiphobic polymer-based coatings via spray and dip coating strategies. Prog Org Coat. doi:10.1016/j.porgcoat.2015.08.008
300. Jasinski P, Molin S, Gazda M et al (2009) Applications of spin coating of polymer precursor and slurry suspensions for solid oxide fuel cell fabrication. J Power Sour 194:10–15
301. Chang H, Wang P, Li H et al (2013) Solvent vapor assisted spin coating: a simple method to directly achieve high mobility from P3HT based thin film transistors. Synth Met 184:1–4
302. Xu L, Yamamoto A (2012) Characteristics and cytocompatibility of biodegradable polymer film on magnesium by spin coating. Colloids Surf B 93:67–74
303. Jayamurugan P, Ponnuswamy V, Rao YVS et al (2015) Influence of spin coating rate on the thickness, surface modification and optical properties of water dispersed PPy composite thin films. Mater Sci Semicond Process 39:205–210
304. Seah CM, Chai SP, Ichikawa S et al (2013) Control of iron nanoparticles size by manipulating PEG-ethanol colloidal solutions and spin coating parameters for the growth of single walled carbon nanotubes. Particuology 11:394–400
305. Sabitha C, Joe IH (2015) Synthesis and characterization of PMMA adherent ZnS thin films by spin coating method. Mater Today Proc 2:1046–1050
306. Kobayashi Y, Oda A (2014) Fabrication of barium titanate nanoparticles/poly (methylmethacrylate) composite films by a combination of deposition process and spin coating technique. J Mater Res Technol 3:290–295
307. http://www.ytca.com/spin_coating
308. Youssef HA, Ali ZI, Zahran AH (2001) Electron beam structure modification of poly(vinyl chloride)-wire coating. Polym Degrad Stab 74:213–218
309. Matallah H, Townsend P, Webster MF (2000) Viscoelastic multi-mode simulations of wire coating. J Nonnewton Fluid Mech 90:217–241
310. Shah RA, Islam S, Siddiqui AM (2013) Exact solution of a differential equation arising in the wire coating analysis of an unsteady second grade fluid. Math Comput Model 57:1284–1288
311. White LK (1986) Spun-on contour coating characterizations for advanced optical lithography. Polym Eng Sci 26:1135–1139
312. Owens MS, Vinjamur M, Scriven LE et al (2011) Misting of non-newtonian liquids in forward roll coating. J Nonnewton Fluid Mech 166:1123–1128
313. Echendu SOS, Tamaddon Jahromi HR, Webster MF (2014) Modelling polymeric flows in reverse roll coating process with dynamic wetting lines and air-entrainment: FENE and PTT solutions. J Nonnewton Fluid Mech 214:38–56

Part II
Applications of Multicomponent Product

Chapter 8
Multilayer Polymer Films

A.R. Ajitha, M.K. Aswathi, Hanna J. Maria, Joanna Izdebska and Sabu Thomas

8.1 The Significance of Multilayer Polymer Films

A single polymer cannot meet all the required properties such as high impact, high stiffness, good heat sealability, stretchability, optical properties, barrier properties etc. of many of the end use polymer products. To compensate these properties a multilayer structure offers a good blend of properties. The whole is more than the sum of its parts properties. Multilayer films could be a natural choice for both rigid and flexible materials. Plastic films compose a broad category of materials that can be relatively simple or complex depending on the demands of a particular product. Multilayer polymer films usually composed of two or more different compositions and possess a thickness of up to about $7.5–8 \times 10^{-4}$ m. The layers are at least partially contiguous and preferably, but optionally, coextensive.

Multi-layer polymer has the advantage that through extruding two or more layers at the same time, it provides a method with fewer steps to produce multilayer films and thus saves time. While it takes almost four hours to produce 10,000 m of a three layer multilayer film, for single layer blown film process, only two and a half hours is required. Furthermore, the film produced from the single layer process would require an extra step

A.R. Ajitha · M.K. Aswathi · H.J. Maria · S. Thomas (✉)
International and Inter University Centre for Nanoscience and Nanotechnology,
Mahatma Gandhi University, Kottayam 686 560, Kerala, India
e-mail: sabuchathukulam@yahoo.co.uk

J. Izdebska
Department of Printing Technology, Warsaw University of Technology,
Konwiktorska, 200-217 Warsaw, Poland

S. Thomas
School of Chemical Sciences, Mahatma Gandhi University, Kottayam 686 560, Kerala, India

J.K. Kim et al. (eds.), *Multicomponent Polymeric Materials*,
Springer Series in Materials Science 223, DOI 10.1007/978-94-017-7324-9_8

to glue the layers together using some sort of adhesive. Making multilayer films is the least expensive means of producing layered films. In multi layer films the three layer film even if it is of the same polymer can increase the tear, dart and drop strengths by as high as 25–30% over those of an equivalent thick film made in single layer. This improvement allows the film to be down gauged (i.e. lower thickness) while still maintaining the product strength and abuse resistance and thus results in raw material savings up to 35–40%. It enhances the barrier properties significantly for use in more complex and demanding applications. The high strength, improved optics and enhanced sealability can all be tailored for a particular application. It is also reported that the ability to combine the superior properties of different polymers, allows the use of higher loading of master batch without affecting sealability, and this results in lower production cost of the film composite. The output rates of individual layers in a multilayer die are very much lower than for equivalent size of single layer die. This results in reduced shear rate and longer resistance times in the die spirals, leading to better homogeneity and additional relaxation of melt. Oxygen and moisture absorption are the two elements which cause lot of damage like loss of nutritive value, bitter taste, loss of crispness in baked/dried foods, hydraulic rancidity etc. All these can be eliminated by using a multilayer film structure which gives excellent moisture and oxygen barrier properties. Also developments in packaging processes and technology based on the reduction of packaging mass and amount of waste are possible with multilayer materials [1].

8.2 Methods of Preparation

There are many ways of combining different materials to form multilayer films. A brief discussion of the most important methods of preparation and their details are given in this section. One of the recent techniques used in multilayer polymer films production is layer by layer assembly. Other major methods include co-extrusion, lamination, coating, solvent casting etc.

8.2.1 *Layer-by-Layer Assembly*

Layer-by-layer assembly (LbL) has been widely applied as a simple and versatile thin-film fabrication technique [2]. The layer-by-layer deposition technique introduced in the early 1990s by Decher, Möhwald, Mallouk, Lvov and others has attracted an increasing number of researchers. All materials such as polymers, proteins, nanoparticles, living cells, etc. which can be dispersed or dissolved in aqueous solutions and are able to form ionic or hydrogen bonds can be used in multilayer composition create by LbL method. A wild range of functional thin film, polyelectrolyte multilayer, tissue with nanoscale controllable thicknesses, special physical and chemical properties as well confined architectures can be produced. Various conventional and unconventional LbL methods are useful to construction of functional films [2].

Multifunctional films with controlled structure and composition are prepared by alternating exposure of a substrate to components with complementary interactions. Resultant thin films can exhibit various properties that include gas barrier, fire retardant, superhydrophobicity and antifogging. In LbL deposition process, treated substrates were dipped in one of the polymer (say polymer A) solution for 5 min, rinsed with deionized water, and dried with filtered air. This procedure was followed by an identical dipping, rinsing, and drying procedure in second polymer solution (say polymer B). After this initial bilayer was deposited, different numbers of layers were added to make the multilayer films under the same rinsing and drying conditions and are repeated until the desired number of layers was achieved [2, 3].

8.2.2 Extrusion

Blown film extrusion is a technology that is the most common method to make plastic films, especially for the packaging industry. Blown film extrusion involves extruding a tube of molten polymer through a die and inflating to several times its initial diameter to form a thin film bubble which is then collapsed and used as a lay-flat film or can be made into bags. During the cooling step the amorphous, transparent melt crystallizes to form a translucent, hazy, or opaque film and the point where opacity begins in the bubble is known as the frost line. This frost line is controlled by several factors like air flow, film speed, and temperature difference between the film and the surroundings etc.

This procedure consists of four main steps [4]:

1. The polymer materials are successively compacted and melted to form a continuous, viscous liquid and is then extruded, through an annular die.
2. Inject air through a hole in the centre of the die, causes the extruded melt to expand into a bubble and replaces air. A constant pressure is maintained to ensure uniform thickness of the film.
3. The bubble is pulled continually upwards from the die. The film is cooled from the inside using internal bubble cooling or by cooling ring which blows air onto the film while pulling the bubble upwards.
4. After solidification at the frost line, the film moves into a set of nip rollers which collapse the bubble and flatten it into two flat film layers. The puller rolls pull the film onto windup rollers. The film passes through idler rolls during this process to ensure that there is uniform tension in the film. Between the nip rollers and the windup rollers, the film may pass through a treatment centre,

depending on the application. During this stage, the film may be slit to form one or two films, or surface treated.

The efficiency of blown film extrusion can be improved by minimizing the temperature of the polymer melt. By decreasing the melt temperature the rate of the process can be increased, thus yielding higher productivity. A way to maintain the melt temperature at a minimum is to choose an extruder that is matched to the specific processing conditions, such as the material of the melt, pressure, and throughout.

8.2.3 Co-extrusion [5]

In co-extrusion two or more different plastic granules are heated and melted separately and are brought together either before or at the die lips so that the combined plastics are extruded as a single product—a multilayer web of plastic film or sheet. The multilayer films can then be extruded using cast extrusion or blown extrusion. The range of thickness of co-extrusion cast film is 80–320 μm where for blow film is 50–100 μm.

In cast or flat die co-extrusion case, there are three extruders processing plastic granules A, B and C to produce a film or sheet with the structure ABCBA. For example, A could be LDPE, B a tie layer, and C EVOH. In the example shown, melts A and B are first split (so they can each form two separate layers) and then brought together with melt C in a combining adaptor or feed block. The geometry of the feed block can usually be adapted to reflect the desired relative layer thicknesses. The multilayer melt is fed into the die, whose "coat-hanger" design spreads the melt flow to the full width of the die lips. The rheological properties of the resins are chosen so as to maintain laminar flow and an even layer thickness during this stage. The web is cooled in a controlled manner using one or more chill rolls. The thickness of the film is controlled at a number of points: viz. the speed and hence the output of the extruders, the speed of the chill roll or by the separation of the die lips [6].

8.2.4 Co-injection Stretch Blow Moulding [6]

Injection stretch blow moulding **(ISBM)** is mostly two-stage process. Other ISBM methods are single-stage and one-and-a-half-stage. Injection stretch blow moulding was developed to manufacture carbonated drinks bottles. The preform has the shape of a test-tube with a neck moulded with a finished thread. The body is first heated by infrared radiation and then placed in a blow mould. It is then simultaneously stretched using a stretch rod inserted through the neck and blown into the final shape by compressed air. The blow mould is then cooled to freeze the material in the correct shape. The different technologies for the co-injection of different polymers can used:

- sequential injection: polymer for example, poly(ethylene terephthalate) (PET) is first injected into the mould, followed by the barrier resin and then again by more PET and the subsequent blowing is done;
- simultaneous injection: PET and the barrier resin are injected together, and the barrier material is directed to the middle of the preform and further step involves the subsequent blowing.

8.2.5 Lamination

Lamination process is advantageous over co-extrusion in the fact that while co-extrusion method can merely combine only two or more plastics, lamination process can be used to combine two or more materials in web form, including non-plastics such as paper and metal foil (usually aluminium). There are several methods of lamination that can be done to prepare multilayer films.

8.2.5.1 Extrusion Lamination

Here the molten polymer, mostly low density polyethylene (LDPE), is extruded from a flat die with 7–20 μm thickness, onto to the primary web of material on two sides. The three layers are pressed against a chilled drum by an impression roll which solidifies the molten polymer, and bonds the two webs together. While the variation of thickness of the extrudate across the web is controlled by die bolts and from the feedback of a thickness measuring device, the bonding of the extrudate to the webs are influenced by chemical compatibility of the polymer and the webs, the line speed, the thickness of the extrudate and the temperature of the polymer. These types of laminates are particularly applicable for vacuum packaging of meat, fish and cheese.

8.2.5.2 Adhesive Lamination [6]

Adhesive lamination is used for making plastic-film-based structures. There are several methods such as:

- wet lamination with one or two composites solvent based adhesives,
- dry lamination,
- with non-solvents adhesives,
- with wax and hot melted adhesives.

Dry bond lamination is one type of adhesive lamination. Here, a coating of solvent- or water-based adhesive is applied to the primary web. The coated web is passed through an oven where hot air is blown onto the web to dry off the solvent or water and raise its temperature. The warm tacky adhesive at the end of the oven is then bonded to the secondary web through a heated nip roll. The resultant laminate is then cooled on a chill roll to increase the adhesive viscosity resulting in immediate bonding (called "green tack"), and then wound under pressure to minimize the chance of the webs separating. For stronger bonds, reactive adhesives are used while in the roll resulting in cross-linking reactions which increases the molecular weight of the adhesive. This causes an increase in the bond strength and the resistance of the laminate to temperature and chemicals. The quality of the bond depends on a number of factors: adhesive selection, adhesive dry weight and correct machine conditions. A water-based adhesive, such as a starch or protein dispersion, is applied to one web and the other web is immediately nipped in. The laminate can then be passed through an oven and the water dried off by passing through the permeable paper.

In wax and hot-melt lamination, a layer of molten wax or hot melt is applied between two webs and then cooled to bond them together. The method of application depends on the viscosity of the laminate and includes roller systems, gravure and slot dies. The wax remains heat-sensitive and hence such techniques are ideally suited to pack types where the material is folded or glued rather than heat sealed.

8.2.5.3 Heat-Welded Lamination

Involves heating and compression of the two films of which at least one has heat sealable properties. Laminates is foamed by increasing the temperature. The total resistance is achieved by cooling the multilayer structure on a cold roll [7].

8.2.6 Coating

Multilayer films can also be produced using coating techniques. There are different types of coating techniques like:

- extrusion coating,
- solvent based coatings,
- wax and hot melt coatings,
- vacuum coating etc.

In extrusion coating the extrudate is simply applied onto a single primary film one or both sides, which results in a two-layer as opposed to a three-layer structure. It can be used only for polymers with similar processing condition or much often for metallic and paper substrates. The most common example is paper/polyethylene (PE) or board/PE. While wax and hot melt coatings use the same techniques as for lamination solvent based coatings are similar to those used particularly for adhesive lamination. Vacuum coating is a fundamentally different process and here a source material is evaporated in a high vacuum and deposit the vapour onto a moving web. The resultant metalized films have a metallic appearance and provide a light, moisture and gas barrier. Analogously to vacuum metallization silicon oxide (SiO_2)

coating on film is applied. The layer is very thin and provides high barrier characteristics and what more is transparent. Also some combination of SiO and MgO coatings and PVOH/EVOH copolymer coatings are used for increasing barrier properties of films.

8.2.7 Metallization

Metallization is a method of application of a thin metal layer on a plastic film substrate or rarely on paper. Heated metal is evaporated under the vacuum. Aluminum is the most common used metal, but in some cases also nickel and chromium are deposited. The thickness of layer is very thin, about 0.5 μm. This metal coating is much thinner then the range of metal foil used in lamination processes, but allows to achieve good gas and moisture barrier. The metallization process is carried out mostly in order to achieve increased barrier characteristics and occasionally for foil appearance and opaque. Metalized polymeric films are good alternative in packaging for lamination with aluminum foil [8].

8.2.8 Thermal Spray Processing of Polymers ("Gun")

Thermal spraying of polymers is as an effective method to produce protective barrier coatings because of the ability to apply coatings of polymers onto a wide variety of materials. The polymers that can be sprayed and can be used for this method includes: PE, PMMA, EMMA, PEEK, PPS, LCP, nylon, phenolic epoxy, Tefzel, post consumer commingled polymer, etc. Here in this method the polymer powder is injected into a heat source (flame or plasma) and transported to a preheated substrate. The thickness of the coating is governed by the number of repeated passes of the spray gun across the substrate. A large particle size or molecular weight distribution may facilitate the formation of numerous heterogeneities within the microstructure by creating voids, trapped gasses, unmelted particles, splats, and pyrolized material.

8.2.9 Spin Coating

Spin coating is another relatively simple method like solution mixing. Here the polymer is dissolved in a volatile solvent and the solution is spinned. Spin coating is the preferred method for application of thin, uniform films to flat substrates. Here an excess amount of polymer solution is placed on top of a substrate which is then rotated at high speed at an angular velocity to spread the fluid by centrifugal force. Rotation is continued until the desired film thickness is achieved. The solvent is usually volatile, providing for its simultaneous evaporation. This method is widely used for preparing films in the area of photoresists, lithography resists, and protective coatings.

8.2.10 Solvent Casting, Painting

It is one of the most simple technique where the multilayer film can be prepared without much sophisticated techniques. Here the polymer solution is deposited on a substrate—drying—solid film for example the solution is simply painted or brushed on a substrate (wall) to get a film with thickness in the micro range depending on the viscosity of the fluid, evaporation rate of the solvent. This process is of large technical importance when it comes to coatings of houses, bridges to protect against rust, water etc. But the thickness can be sometimes inhomogeneous.

8.3 Characterization

Characterization is an essential part of any materials research. Polymer multi-layer films are used in a variety of industries hence the characterization is also oriented based on the applications. Barrier properties, thermal and morphological characteristics, free volume, mechanical properties and toxicity studies, are critical characterisation techniques that help to determine the properties of a multilayer polymer films.

8.3.1 Barrier Properties

The barrier property of multilayer film is very important as packing industry is one of the major industries that use multilayer films. "Barrier property" refers to a material's capability to resist the diffusion of a specific species (molecule, atom or ion) into and through the material [9]. To be a good gas and vapor barrier, the material needs to be pore-free. Multilayered structure is beneficial as the structural characteristics of these films can create a tortuous path, decrease in the number of pores and can decrease the permeation rate. The packaging materials are exposed to wide range of conditions and therefore the analysis of each factor is relevant. For example, for food packaging we need to protect the food especially from oxygen and water vapour. The intrinsic structure of the polymer such as degree of crystallinity, crystalline/amorphous phase ratio, nature of polymer, thermal and mechanical treatment before and after food contact, chemical groups present into the polymer (polar or not), degree of cross linking, and glass transition temperature (Tg) are interrelated to permeation rate. Therefore it is important to study the

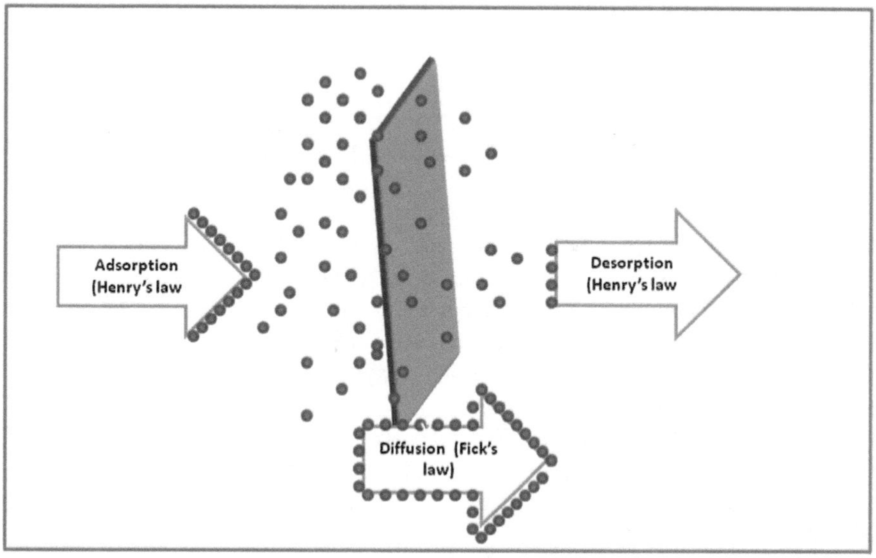

Fig. 8.1 The general mechanism of permeation [10]

relationship of permeation rate and the structural characteristics of multilayer polymer system. In barrier properties we need to analyse the permeation of oxygen, carbon dioxide, water vapour etc. In multilayer films the diffusion of gases/solvents/vapour can be generally considered to taking place through mechanism depicted as in Fig. 8.1.

8.3.1.1 Oxygen Transmission Rate (OTR)

Permeability characteristics of multilayer film to oxygen are important mainly for food packaging application. Various types and forms of packaging materials are available for thermally processed shelf-stable foods. Polymers with inherent low gas permeability like ethylene vinyl alcohol (EVOH), nylon, polyethylene terephthalate (PET), etc., are considered as suitable candidates as packaging materials. The functionality and properties of these polymers can be further enhanced by combining different polymer layers to form multilayer structures where each layer contributes to a specific function. Oxygen permeation in food packaging is one of most important factors as it can severely affect the sensory properties of lipid-containing foods [11]. The oxygen barrier is quantified by the oxygen permeability coefficients (OPCs) which indicates the amount of oxygen that permeates per unit of area and time in a packaging materials (kg m/(m^2 s Pa)) and is related to Oxygen Transmission Rate (OTR) [10].

$$OPC = OTR \times l/\Delta P \quad (8.1)$$

where l is the thickness of the film, and ΔP is the difference between oxygen partial pressure across the film [Pa]. ΔP = p1 − p2, where p1 is the oxygen partial pressure at the temperature test on the test side, and p2 is equal to zero on the detector side.

The allowable permeation of O_2 differs based on the particular packaging application. Therefore, it is important to study the interaction between food-processing techniques, the packaging materials used, storage conditions and their interrelationships which forms the basis of the research in the field. Permission of oxygen in multilayer films is influenced by a lot of factors. There are a many studies which reported an improvement in oxygen barrier properties by incorporating fillers or by some modifications.

Yang et al. [12] studied the OTR of polyethylenimine (PEI)/graphene oxide (GO) films on poly(ethylene terephthalate), PET and reported that on increasing the number of bilayers to 20 and 30 there was a decrease in OTR from 8.48 of the uncoated PET to 0.46 and 0.27 cc/m^2/atm/day (see Fig. 8.2). They also reported that the OTR of PEI/GO was observed to be better than polyethylenimine (PEI)/montmorillonite (MMT) by a factor of 20 and explained to be due to the highest aspect ratio of graphene.

Tashiro et al. [13] have compared the oxygen transmission rate results of silane-treated and diamond-like carbon (DLC)-coated PP film and reported that the samples with the organosilane interlayer fabricated by the silane-coupling agents with amino groups showed very low OTR which was comparable to high gas-barrier polymer like PET. On analysing the OTR results of pure PP, DLC-coated PP, and DLC-coated PP pre-treated by silane-coupling agents with amino groups, a significant improvement in the gas barrier property could be observed for the DLC coated samples (Fig. 8.3). The OTR of different polymer substrates like PP, LDPE, and HDPE also indicated that a high gas-barrier coating could be successfully produced on polymers regardless of their types.

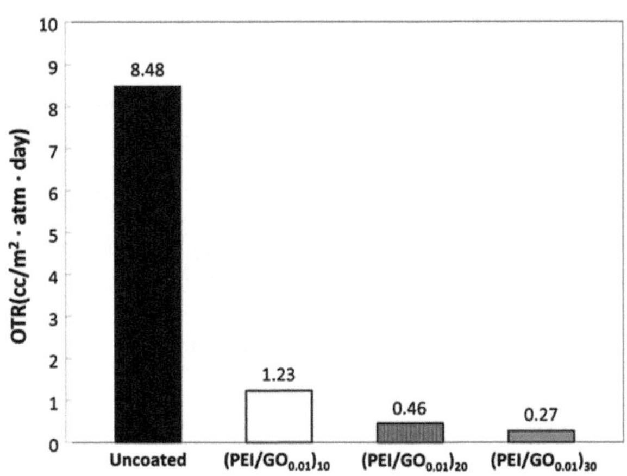

Fig. 8.2 Oxygen transmission rate of PEI/GO assemblies on PET, measured at 23 °C and 0% RH, made with 0.01 wt% GO suspension [12]

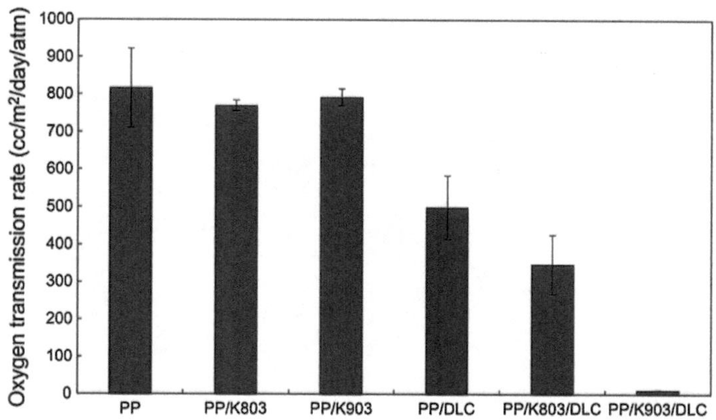

Fig. 8.3 Oxygen transmission rate of silane-treated and DLC-PP coated [13]

The study by Dhawan [14] also reported a 3 fold improvement in OTR when the number of bilayers was changed from 6 to 8. In their study on ethylene vinyl alcohol (EVOH) films they studied the influence of Pressure-assisted thermal processing (PATP) on two state-of-the-art multilayer and reported a reduction in thickness or crosslinking resulting in an increase of OTR. They also reported that the thermal stabilization influenced the OTR of multilayer films. The OTR for the EVOH films observed immediately after the two PATP processing treatments for 3 and 5 min there was a 2.5-fold and 5-fold increase, respectively in the OTR of film. The ethyl content of EVOH was also found to influence the OTR which was in agreement with the study conducted by López-Rubio et al. [15], who observed slightly better barrier properties after high pressure processing for the monolayer EVOH copolymer with lower ethylene content (Fig. 8.4).

Fig. 8.4 Oxygen transmission rate of films A and B as influenced by the two PATS conditions [15]

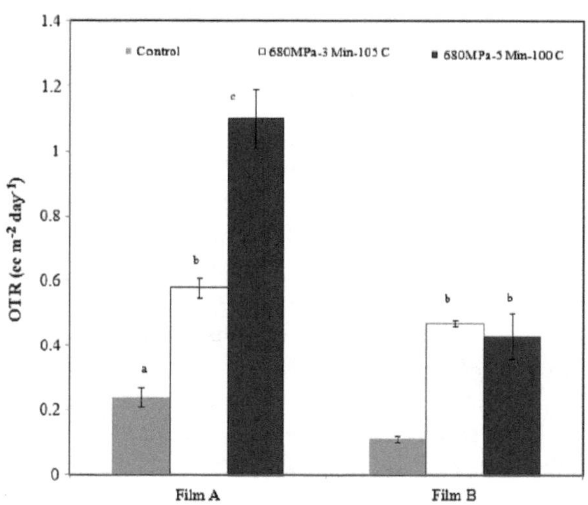

Table 8.1 The OTR values for plain and Al_2O_3-coated plastic samples

Substrate	OTR[a]		
	Polymer	$+Al_2O_3$ batch	$+Al_2O_3$ R2R
CEL	4.0 ± 0.1	2.6 ± 2.3	0.3 ± 0.2
PLA	470 ± 1	0.4 ± 0.1	47 ± 6[c]
PI	30 ± 0.1	<0.01[b]	2.2 ± 1.9[c]

The Al_2O_3-coating has fabricated by using both, batch and R2R ALD processes [7]
[a]$cm^3/m^2/10^5 Pa/day$
[b]Result was under the detection limit
[c]Small sample size [9]

The preparation method and coating also was reported to influence the OTR. Hirvikorpi et al. [9] on their studies on barrier properties of plastic films coated with an Al_2O_3 layer by roll-to-roll atomic layer deposition have reported the positive effect of the Al_2O_3 coating on the barrier properties of plastic films (Table 8.1). Al_2O_3 coating was deposited on one side in batch processing and on both sides of the substrates in the roll-to-roll (R2R) process. Independently of the deposition process used, the Al_2O_3 coating improved both the oxygen and water vapour barrier properties of the polymeric samples.

8.3.1.2 Water Vapour Permeability

Water vapour permeation is another equally important parameter studied in packaging applications, because the permeation of moisture from the internal or external environment through the polymer package wall, results in possible negative changes in product quality and shelf-life. The water vapour barrier properties for the packaged food product, whose physical and chemical deteriorations are related to its equilibrium moisture content, are of great importance for maintaining or extending its shelf-life. It is quantified by the water vapour permeability coefficients (WVPC) which indicates the amount of water vapour that permeates per unit of area and time in a packaging material (kg m/(m^2 s Pa)). The water vapour transmission rate (WVTR) is expressed in cc/m^2 s (org/m^2 day). The WVTR depends on different parameters and can be influenced by processing. Dhawan [14] studied the WVTR of the multilayer EVOH films before (control) and after thermal treatment and reported a significant increase ($P < 0.05$) in the WVTR after processing. Cocca et al. [16] reported the influence of crystal polymorphism on the barrier properties of poly(L-lactic acid) (PLA) film and found that the polymorphic structure significantly affects the water vapour permeability, they related this to the tighter molecular packing chain segments, together with the strong coupling with the amorphous portions of the macromolecules, which reduce the diffusion of the adsorbed water vapour. Kalendova et al. [17] on their work on linear low-density polyethylene (LLDPE) and poly vinyl chloride (PVC) nanocomposites reported an improvement of water vapour barrier properties for most of the PVC nanocomposites. The effect of various parameters like nature of filler, filler concentration etc. were observed. They reported that the maximum enhancement was shown by the nanocomposite with 30B montmorillonite (MMT) and the WVTR was found to be decreasing with filler content.

8.3.2 Morphological Characterization

In multilayer polymer films the morphological analysis plays an important role as it helps in detecting the location and size of packaging defects, the structural defects, the presence and dispersion of fillers etc. Some of the major methods used for morphological analysis includes Scanning Emission Microscopy (SEM), Transmission Electron Microscopy (TEM), Atomic Force Microscopy (AFM), etc. Other techniques such as X-ray diffraction (XRD) and Positron Annihilation Lifetime Spectroscopy (PALS) could help to explain the morphological and free volume characteristics of the polymeric film.

Hirvikorpi et al. [9] reported the effect of Al_2O_3-coating on silicon wafer and studied the surface topography of uncoated and coated plastic samples using non-contact mode atomic force microscopy. AFM images of uncoated and atomic layer deposition (ALD) coated samples revealed that the ALD coating grow as particle-like structure on the polymer substrates [18]. The surfaces of polylactide substrate with and without ALD-coating are presented in Fig. 8.5.

A similar study was done by Howells et al. [19] in their studies on SiO_x gas barrier coatings on polyester films. The high resolution microstructural characterisations were done by AFM imaging (Fig. 8.6). The images showed a rougher surface with discrete 20–30 nm features on the heat stabilised poly(ethylene terephthalate) (HS PET) film compared to the plain PET. This proved an increased degree of surface crystallisation in the coated film.

Porosity and membrane structures of poly(ether-block-amide)/(polypropylene + $CaCO_3$), PEBA/(PP + $CaCO_3$) and PEBA/β-PP, produced through the use of multilayer co-extrusion and axial orientation entirely without the aid of solvents, were studied through the use of scanning electron microscopy (SEM). Both systems, PEBA/(PP + $CaCO_3$) and PEBA/β-PP, maintained a CO_2/O_2 selectivity of at least 7, which is approximately double that of commercial materials being used for modified atmosphere packaging [5] (Fig. 8.7).

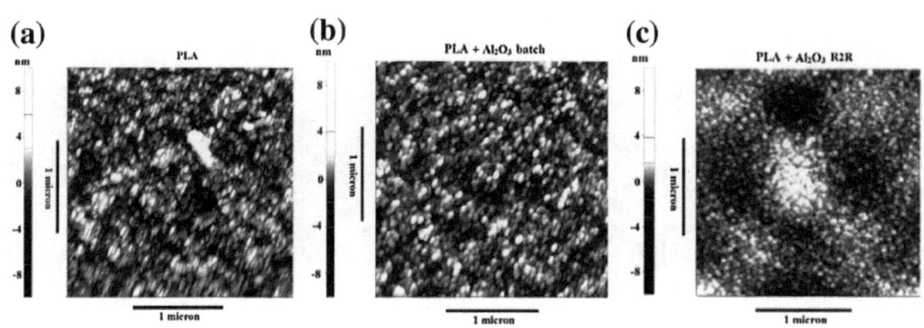

Fig. 8.5 The surfaces of polylactide substrate with and without ALD-coating [9]

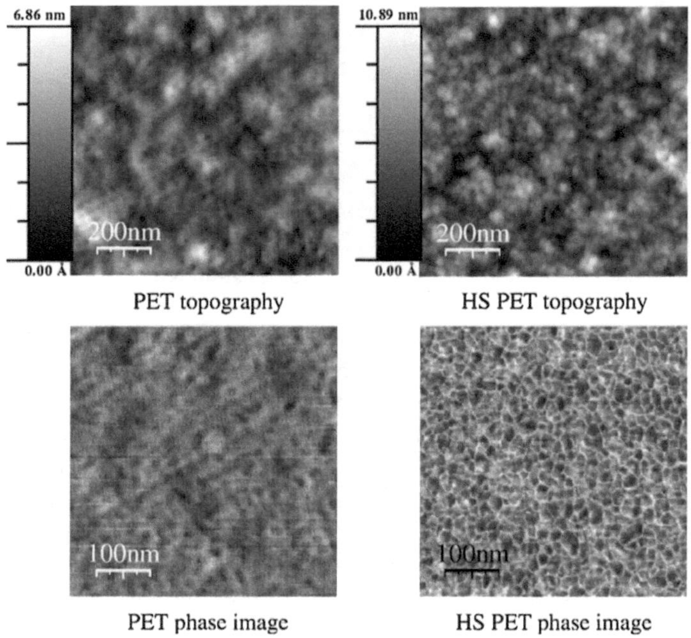

Fig. 8.6 AFM topography and phase images of (*left*) PET and (*right*) HS PET [19]

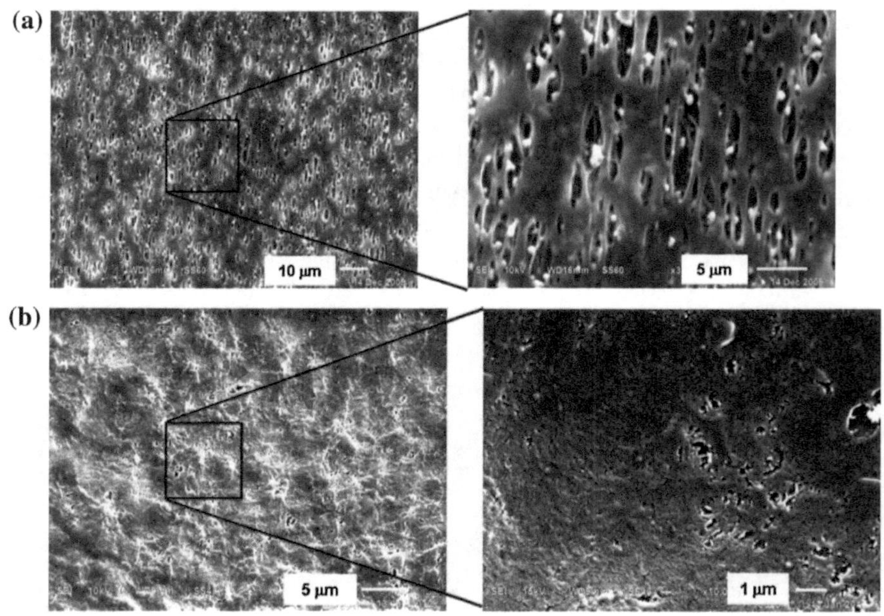

Fig. 8.7 Scanning electron microscopy images of, PEBA/(PP + $CaCO_3$) and PEBA/b-PP [5]. **a** Porous PP + $CaCO_3$ created through uniaxial orientation at 23 °C. Elongated pores with sizes of 0.5–3 µm × 3–7 µm are produced from cavitation around the $CaCO_3$ particles. **b** Porous β-PP created through biaxial orientation at 100 °C to a draw ratio of 2 × 2. Pore sizes of less than 300 nm were produced

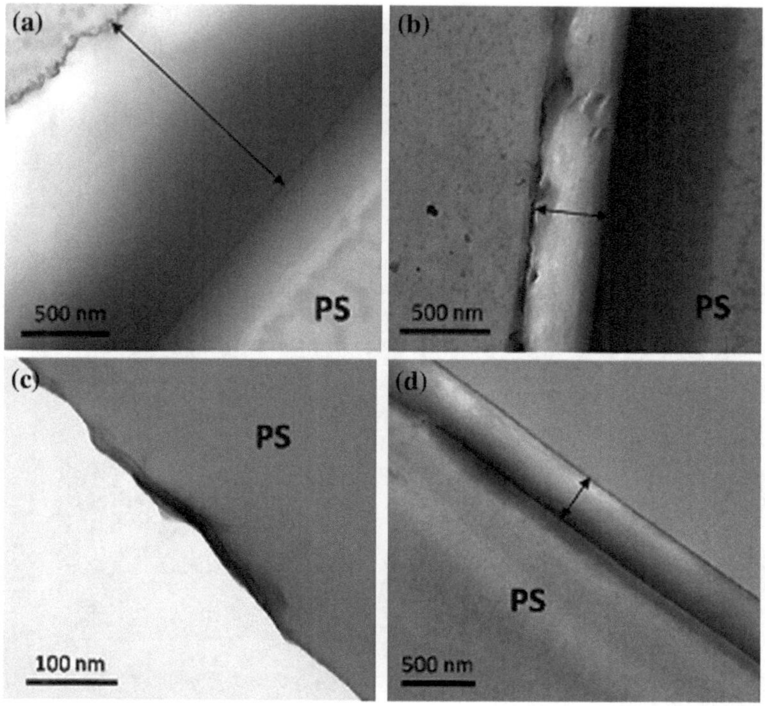

Fig. 8.8 TEM cross-sectional images of $(PEI/PAA)_{20}$ on PS made with various pH combinations: PEI_{10}/PAA_4 (**a**), PEI_8/PAA_6 (**b**), PEI_7/PAA_7 (**c**), and PEI_4/PAA_4 (**d**). The *double arrow bars* highlight the thickness of the films [20]

From TEM and AFM images of thin film assemblies of branched polyethylenimine (PEI) and poly(acrylic acid) (PAA), deposited using the layer-by-layer technique, Yang et al. [20] have observed that altering the pH of PEI and PAA results in large thickness variations (Fig. 8.8). They reported that the surface morphology of PEI/PAA assemblies can be controlled by the pH (Fig. 8.9) and observed that higher charged polymer combinations have soother surfaces and less charged combinations have rougher surfaces. It was reported to be a promising alternative to current polymeric membranes, flexible electronics, and food packaging materials.

The effect of deposition times in improving the gas barrier properties of clay–polymer multilayer thin films was studied by Xiang et al. [3]. The larger overall thickness showed by the TEM images proved larger clay spacing. This increased path facilitated the perpendicular wiggling of gas molecules with respect to the diffusion direction thereby increasing the diffusion path and gas barrier behaviour (Fig. 8.10).

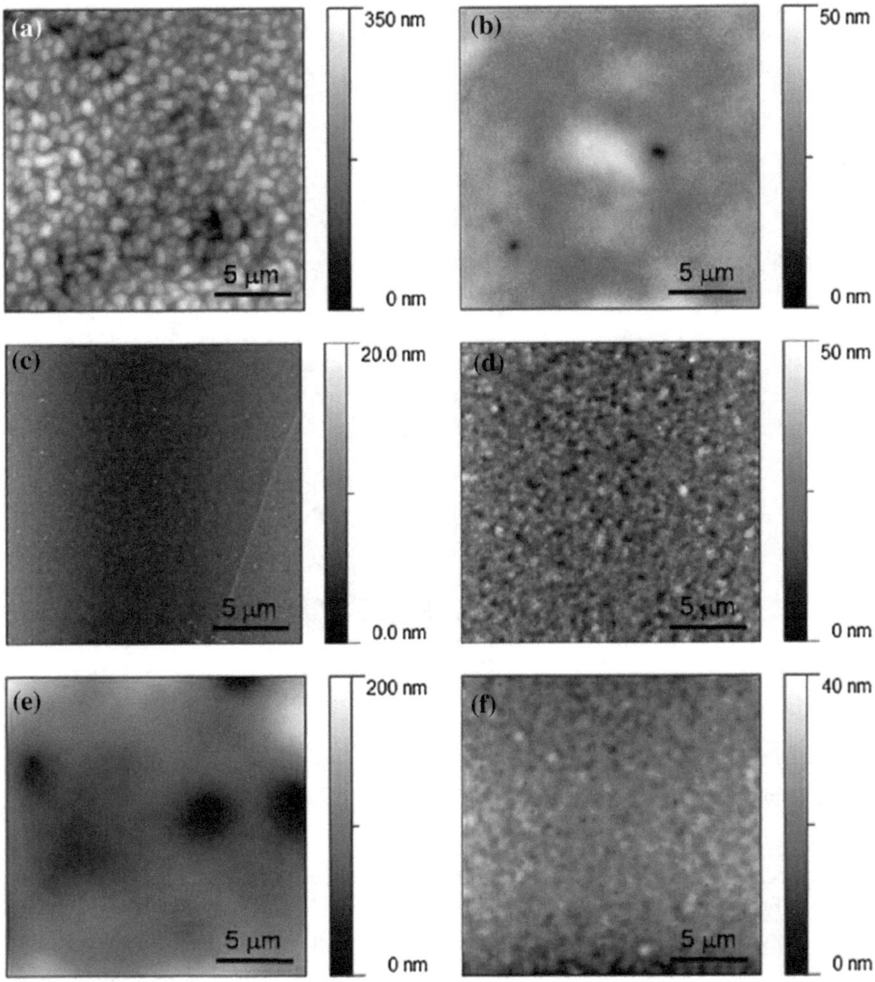

Fig. 8.9 Height images of $(PEI/PAA)_{10}$ on Si wafers: PEI_{10}/PAA_4 (**a**), PEI_8/PAA_6 (**b**), PEI_7/PAA_7 (**c**), and PEI_4/PAA_4 (**d**). Two PEIterminated surfaces, $(PEI_{10}/PAA_4)_{10.5}$ (**e**) and $(PEI_4/PAA_4)_{10.5}$ (**f**) [20]

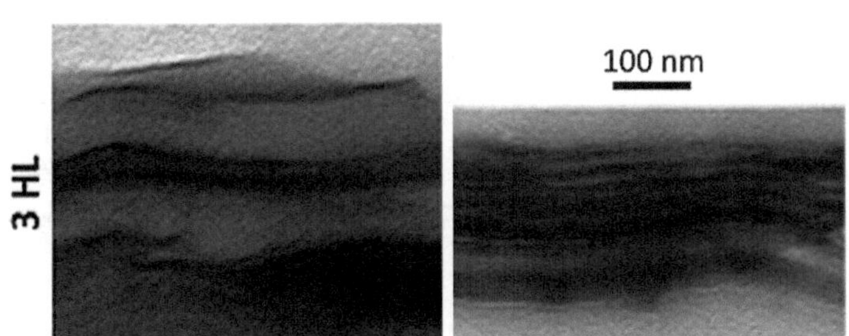

Fig. 8.10 TEM cross-sectional images of three hexalayer films deposited using 5 s and 1 min polyelectrolyte exposure times [3]

8.3.3 Mechanical Study

Mechanical integrity of the multilayer films is very important in most of the application as it has to undergo high pressure treatment during its processing or in its life time. These can cause damage and can lead to integrity defects causing its quality. A decrease in its barrier properties are reported in many cases when the film is subjected to high pressure. Mechanical properties like tensile, flexibility, compression etc. are some major test used in this area. A suitable multilayer structure used to pre-package high pressure-treated food must have sufficient flexibility, resilience and resistance to delamination during the compression process. Caner et al. [21] have studied the effects of High-Pressure Processing (HPP) on the mechanical properties of several high-barrier films by measuring their tensile characteristics. Howells et al. [19] assessed the mechanical properties of poly ethylene terephthalate (PET), SiO_x coated PET films and heat-stabilised (HS) PET film and it is found that the coating displays improved mechanical properties. The HS PET film is found to have inferior mechanical properties compared to the plain PET. They have also reported the effect of substrate in influencing the properties of the film. On comparing the mechanical properties of plain PET, filled PET (F-PET), heat-stabilised PET (HS PET), heat-stabilised PET with an additional primer layer (HS PET_P) and poly (ethylene naphthalate) (PEN) it was found that the addition of the primer layer to the HS PET to increased the $\tau*$ (intrinsic interfacial shear strength) to above 110 MPa and was explained to be due to the increased functionality of the primer surface. The effects of the substrate surface mechanical properties on the $\tau*$ are given in Fig. 8.11.

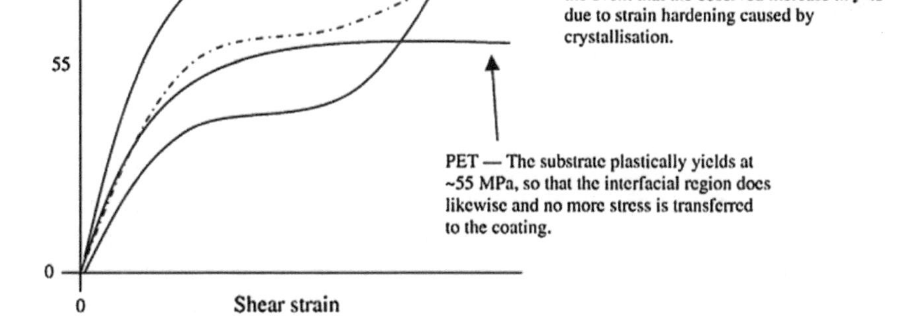

Fig. 8.11 The mechanical behaviour of the polyester substrates under shear, in the vicinity of the interface with the oxide coating [19]

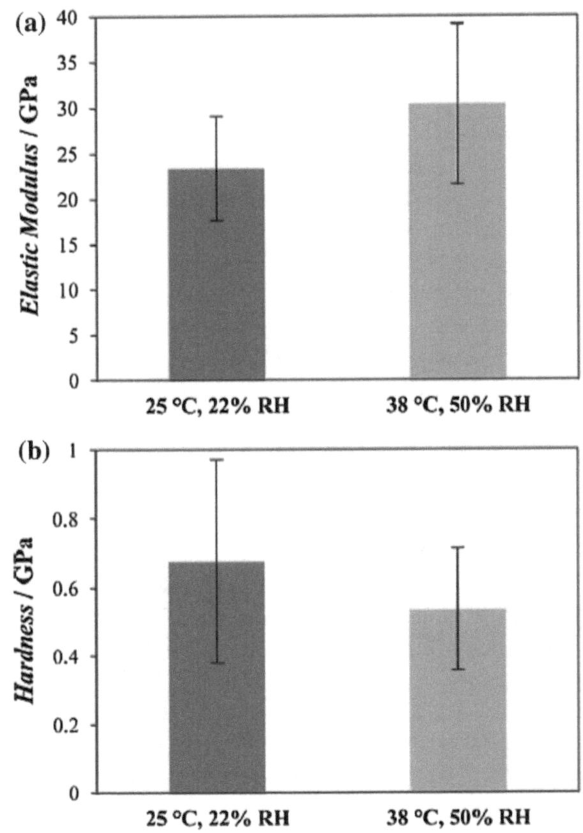

Fig. 8.12 **a** Average elastic modulus and **b** hardness of 10 bilayer PEI/PAA [22]

Kim et al. [22] have studied the compactness of polyethylenimine (PEI)/poly (acrylic acid) (PAA) system from its mechanical behaviour. The compactness of this polymer film can be inferred from its mechanical behaviour. The elastic modulus and hardness was observed to be 23.4 ± 5.7 GPa of 0.7 ± 0.3 GPa at 22% relative humidity (RH) respectively a value which higher than most known all-polymer (Fig. 8.12).

8.3.4 X-Ray Diffraction

Basically X-ray diffraction produces a three-dimensional distribution of intensities that reflects the three-dimensional nature of the material under investigation. The microstructure and optical properties of Ag/Indium tin oxide (ITO)/Ag multilayer films were systematically investigated by X-ray diffraction (Fig. 8.13). It was noted that the multilayer films have a polycrystalline structure. As the thickness of the Ag surface layer increases, the preferred (111) intensified [23].

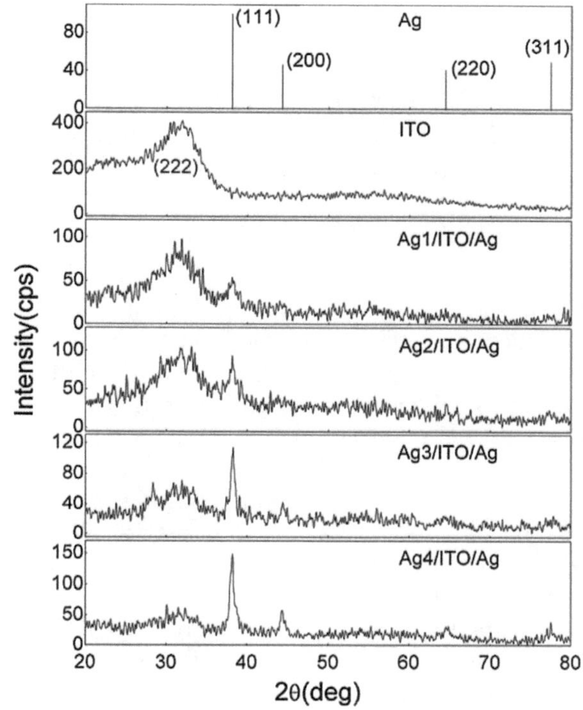

Fig. 8.13 XRD pattern of Ag, ITO and Ag/ITO/Ag multilayer films [23]

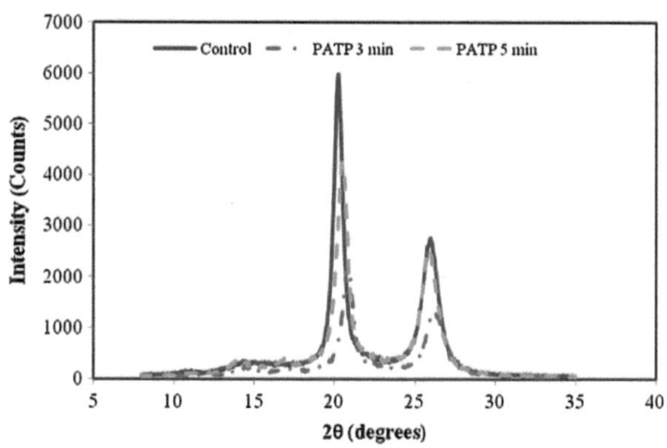

Fig. 8.14 X-ray diffraction pattern before and after PATS treatment [27]

In a study done by Dhawan et al. to evaluate the impact of pressure-assisted thermal processing (PATP) on two state-of-the-art multilayer EVOH based packaging materials the XRD diffractograms showed a decrease in peak intensities after both of the PATP treatments, leading to a decrease in the overall crystallinity (Fig. 8.14). This decrease is reflected in the loss of the film's gas barrier property after the PATP treatments, as a decrease in crystallinity results in a loss of orderliness in the polymeric chains, in turn causing a decrease in tortuous path for the gas to travel though the film [13].

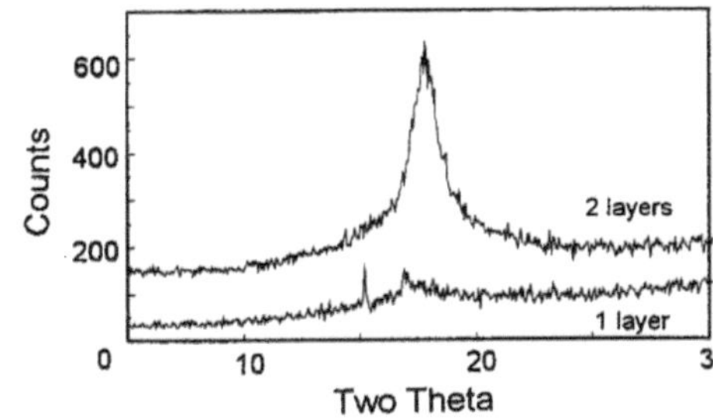

Fig. 8.15 Grazing incidence X-ray diffraction pattern of one and two layers of polymer. For clarity the pattern for two layers of polymer has been displaced vertically by 100 counts [25]

Inhomogeneities in materials can be explored from an entirely different perspective using grazing incidence diffraction (GID), also known as glancing-angle diffraction or surface-enhanced scattering. GID technique is useful in many areas that deal with surfaces and interfaces including paints and coatings, adhesives, polymer-based electronic devices, and biocompatible materials. GID is currently used extensively to study nanostructured surfaces and the structure at air polymer and polymer-substrate interfaces in polymer films deposited onto a substrate [24]. Hammond et al. [25] have reported a grazing incidence X-ray diffraction (GIXD) (Fig. 8.15) study of Langmuir-Blodgett (LB) multilayers prepared from a poly (4-vinylpyridine) quaternized with n-docosyl bromide from which an in-plane packing structure of the polymer has been determined and from which the extent of interdigitation and stresses within the film may be inferred.

8.3.5 *Fourier Transform Infrared Spectrometer*

The chemical composition of the multilayer films, their coatings and nature of interaction between them can measured by Fourier transform infrared spectrometer (FTIR). Cross-linking mechanism of PEI and GA was studied by Yang et al. [20] using FTIR spectra (Fig. 8.16). In their work on LBL assembled polyethylenimine PEI/poly(acrylic acid) (PAA) system deposits over alumina coated porous stainless

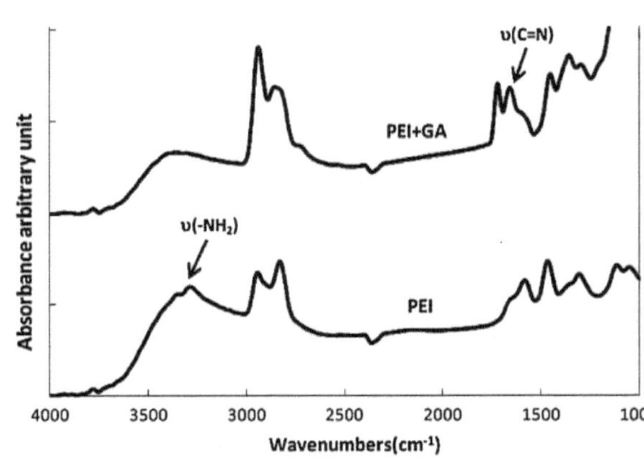

Fig. 8.16 FTIR spectra of PEI and PEI cross-linked with glutaraldehyde (GA) [20]

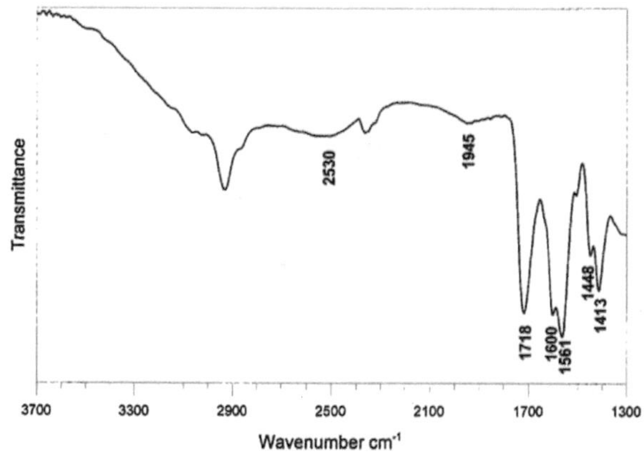

Fig. 8.17 IR spetrum of multilayer film of poly (4-Vinylpyridine) and poly (acrylic acid) [26]

steel (PSS) tube, the superior elastic modulus and hardness of the film were explained using FTIR. The FTIR spectrum showed an ionic crosslinking between the polycation and polyanion.

Investigation of multilayer film of poly (4-Vinylpyridine) and poly(acrylic acid) by Wang et al. [26] showed that the multilayer film was not assembled via electrostatic attraction but hydrogen bonding. The nature of the interaction between the polymers was established by IR spectroscopy. The presence of strong hydrogen bonding between PAA and PVP in adjacent layers was confirmed from the O–H stretching vibration (Fig. 8.17).

8.3.6 Thermal Analysis

Thermal analytical technique is also another valuable analytical tool in the characterisation of multilayer polymer films. An important quality parameter for these multilayer films is the total film thickness, as well as thickness of the individual component layers. Thermomechanical analysis (TMA), because of its ability to measure small dimensional changes, is ideal for evaluating thin films. TMA provides an alternative to spectral techniques, particularly in situations where the presence of inks, adhesives or lacquers might interfere with the measurement. The results for a multilayer packaging film comprised of a metallic substrate coated on both sides with polymer is displayed in Fig. 8.18. The TMA curve shows two sharp negative dimensional changes (penetrations) at 103 and 258 °C. These two penetrations indicate that the film is probably a three layer film containing two polymer layers and the metal foil layer [27].

Other thermal analysis techniques can be used to measure additional multilayer film properties. These include compositional analysis by thermogravimetric analysis (TGA) and softening and crystallinity by differential scanning calorimetry (DSC). The thermal characteristics of the EVOH films after PATP were studied by

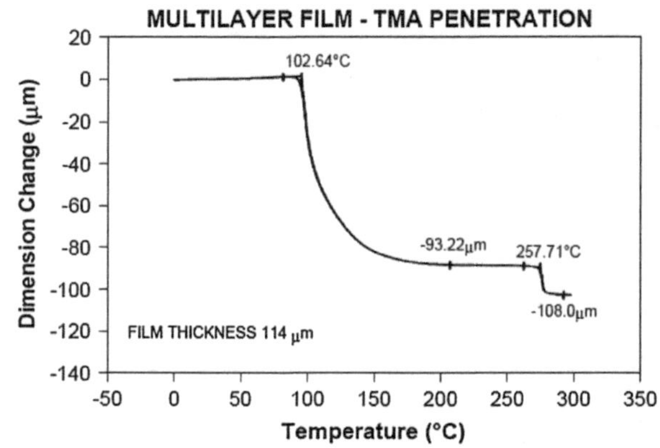

Fig. 8.18 TMA curve of multilayer film [27]

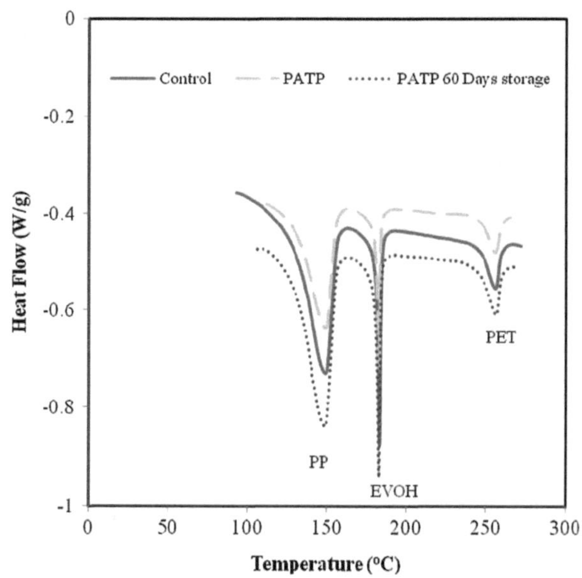

Fig. 8.19 The total melting enthalpy of film after PATS (680 MPa for 5 min at 100 °C) during a storage period of 60 days at room temperature. The DSC scan rate ranged from 20 to 300 °C at a rate of 10 °C/min [28]

Dhawan [14] using DSC experiments (Fig. 8.19). Table 8.2 summarizes the melting temperature and enthalpy of melting for the two films before and after the combined temperature and pressure treatment. The PATP processes had no influence on the melting temperature (Tm) and melting enthalpy (ΔH) of the EVOH layer in both films A and B.

Yang et al. [20] analysed the Tg (measured by DSC) of neat PEI, PAA, and a $(PEI^{10}/PAA^{4})^{200}$ free-standing film. PEI has a Tg of −25.6 °C, and PAA has a Tg of 89.4 °C. Tg of the LbL film was found to be higher than the linear additive combination of neat polymer Tg's, indicating that the polymer chains are more restrained inside of this network and they attributed this to intermolecular hydrogen bonding between these blended polymers which decreases the mobility of each

Table 8.2 Melting temperature and enthalpy of melting for the EVOH layer in films A and B, untreated, and after PATS [14]

Film	Treatment	T_m (°C)	ΔH (J/g)
A (32 mol% ethylene)	Control	182.8 ± 0.3	6.5 ± 0.2
	680 MPa, 3 min, 105 °C	182.2 ± 0.1	6.7 ± 0.1
	680 MPa, 5 min, 100 °C	182.6 ± 0.3	5.8 ± 0.8
B (27 mol% ethylene)	Control	186.1 ± 0.2	4 ± 0.6
	680 MPa, 3 min, 105 °C	186.1 ± 0.1	4.1 ± 0.8
	680 MPa, 5 min, 100 °C	186.1 ± 0.1	4.4 ± 0.5

T_m melting temperature; ΔH enthalpy of melting. Values are means ±1 standard deviation. Means with different letters within a column are significantly different ($P < 0.05$)

polymer, forming a more compact structure with a smaller free volume than normal polymeric membranes.

8.4 Application

Multilayer films are emerging as a new material with various applications. It has gained a good appreciation in recent years and this new resource material could find use in automotive components like fenders doors, hoods, bumpers etc. The application of multilayer films in packaging has benefited the industry economically, functionally and also could impose some aesthetic benefits of replacing its monolayer counterpart.

Multilayer packaging allows vapour barrier layers and moisture locks, that ensure freshness and stability of the contents. By limiting the use of pigments and fillers to only one or two layers can benefit the tensile strength and impact strength of the material which can benefits the automotive industry. The multilayer technology also allows the application of a clear layer that enhances, chroma, intensity of colour and distinctness of image. Many different applications of multilayer films have been suggested and a variety of applications have, indeed, been realized.

8.4.1 Food Packaging

The wide range of food production and the increased transport of food products to different areas create a need for packaging that can extend the shelf life, especially for fresh and chilled products. The fewer but larger manufacturers and retailers have driven down packaging costs by using better pack design, thinner materials and faster, more automated packaging processes. The changing requirements for packaging have become more and more difficult to achieve with single materials. To achieve the required performance the properties of different materials need to be combined. The use of multilayer materials should combine different functional requirements. One particularly important feature for good food packaging is collation that requires good strength and heat sealability. Along with this there are certain criteria that should be possessed by the food packaging materials. So the multilayer film that is used for food packaging should posses:

1. preserving capacity,
2. moisture control,
3. oxygen barrier,
4. light transmittance,
5. retention of flavours,
6. suitability for diverse machinery packaging processes,
7. compatibility,
8. aesthetics, communication and display.

The multilayer film used for packaging must help to preserve the food, to it protecting it from external influences and thereby increasing the shelf life. The food should be protected from moisture, oxygen and other gases like other gases, like carbon dioxide and nitrogen.

The moisture content of many foods must be controlled. While certain food products absorb moisture from the atmosphere to become soft some will lose moisture to become dry. Therefore, packaging materials are should provide a barrier to the transmission of water vapour in or out of the pack. The permeabilities to moisture, is expressed as water vapour transmission rates (WVTR) with units of $g/m^2/day$.

Through permeation of oxygen, food products can be oxidized and can lose its flavor and nutritions. Foods such as fresh meat and fish, the ratio of atmospheric gases that is permeating the package should be altered in order to reduce microbial activity and spoilage. Therefore, the packaging material must provide a barrier to gas transmission. Generally, oxygen transmission rates (OTR) are used to measure permeability and are expressed as $cc/m^2/day/bar$. Other gases, e.g. carbon dioxide and nitrogen, have different transmission rates but these are often in proportion to the OTR (with CO_2 typically migrating through a material faster than O_2, and N_2 slower).

It is important that the packaging materials should provide a physical light barrier to prevent light-induced changes in flavour or nutritional quality. Hence the multilayer film that is used for food packaging application should prevent visible or UV light which can be an initiator for oxidative reactions.

It is important that the packaging materials should be able to lock the flavor and odour. Also it should prevent the absorption of chemicals and unwanted odours from the outside environment.

Next to the barrier properties of the final packaging, the mechanical properties of the materials are important for processing and also during the use of the products derived from these materials. As the food is packed with high speed automated machinery the multilayer film should also posses good mechanical strength to:

- prevent stretching or snapping it should possess good tensile strength and should provide a sealing medium that is heat resistance,
- maintain the shape of packaging the material should possess rigidity, pliability or dead-fold,
- the material should allow it to be gripped firmly on the machine part.

The sealing of the food packing material is also important as in many cases it may need to prevent microbial contamination and also the food should be kept as clean and hygienic as it was when packed. It is vital that both the packaging material and the seals prevent micro-organisms and other contaminants from entering into the pack.

For food packaging applications it is important that the materials used should be compatible with the food stored. It should not undergo any chemical interactions with the food and also should not transfer any toxic chemicals to the food. The package must be designed so as to ensure the safety of the eventual consumer of the packed food and to comply with all food contact and other legal requirements mentioned, barrier to moisture and gases such as oxygen is needed for many packs.

The appearance of a pack is important to attract the consumer and to present a quality image that can aid in its promotion. One of added advantage of using multilayer films in food packaging application is that for multilayer films only the outer layer needs to printed and it also give economic benefits. It also allows long lasting printing which protects the ink from abrasion and helps to provide high gloss or matt finishes.

Some of the major polymers used for food packaging includes: polyethylenes (PE), polypropylenes (PP), poly(ethylene-vinyl acetates) (EVA), polystyrenes (PS), polyvinylchlorides (PVC), poly(ethyleneterephthalates) (PET), polyvinylacetates (PVAc), polycarbonates (PC), polyamides (PA), polyvinylalcohls (PVOH), polyvinylidenechlorides (PVDC) etc. Multilayer films made by combining these polymers using different methods of preparation are being extensively used. However, recent trends in packaging industry is to make use of biodegradable polymers like poly(lactic acid) (PLA), starch, cellulose, polycaprolactone (PCL) etc. The degradation of synthetic polymer films can also be accelerated by incorporating biofillers. The principle function of packaging like retardation of the deterioration of food stuffs, extension of shelf life and maintenance of quality and safety of packaged food etc. fulfilled by biodegradable polymers without causing any threat to the environment. Biodegradable polymer materials will thus reduce the need for synthetic polymer production thus reducing pollution.

8.4.2 Agricultural Application

The first use of plastic film in agriculture was to make a cheaper version of a glasshouse. The use of multilayer films in agricultural field in recent years have influenced positively in different sectors of agriculture like water management, cultivation, handling of fertilizers. A huge economic benefit also could be attained by using plastic films in agricultural field. Multilayer films give an added advantage that it offers scope for incorporation of different type of functional additives for different layers and can be made to meet a wide range of requirements in agricultural fields. Greenhouses, tunnels, direct covers, windbreaks, mulching films, and protective nets against birds are all examples of different areas where multilayer films can be used in the agricultural field. The requirements of these films depend on the application it has in each filed.

8.4.2.1 Mulching

Mulching is one of the important application that uses multilayer films as it plays valuable role in agricultural fields. Mulching is placing thin plastic film placed over the ground, poking holes at regular intervals for seeds to be planted in, or placing it directly over plants in the beginning stages of growth. The main functions of plastic mulch are to improve cultivation conditions by insulating and maintaining a consistent temperature and humidity of the soil by preventing evaporation of moisture from the soil, and to prevent erosion. Polymer films are used to create microclimates in the form of mulching. They are also used in soil sterilization by fumigation, and also in the handling of fertilizers and their distribution in the soil in association with water. The improvement in cultivation conditions is effected by acting as binding agents for soils, by avoiding of soil compaction, by preventing leaching of nutrients, by controlling the climate, increased plant growth by carbon dioxide retention under the film, soil protection from erosion and leaching of nutrients, action as thermal insulation for the roots in cold climates (in winter), protection from frost and the action of torrential rain, saving and retention of irrigation water, reduction of evaporation by insulating soil surface against direct solar radiation and by obstructing vapour diffusion, suppression of transpiration losses without reduction in photosynthesis etc.

The use of polymer film for mulching is particularly beneficial in areas with water deficiency and in sandy soils with rapid drainage. The advantages of plastic film mulching over traditional mulching are in its light weight, that it covers a much greater area per volume than natural mulches, its being amenable to mechanized installation, and its lower cost. Use of multilayer films further benefits it by tuning the properties based on the particular requirement.

The plastics used for mulching soil surfaces are of various types. Mulch films are classified into the following types:

- transparent film mulching,
- black and coloured film mulching,
- white film mulching,
- photo/biodegradable film mulching.

The use of transparent film increases the soil temperature during the day according to the season, type of soil, the level of sunshine, and the water content, thereby increasing the activity of the volatile fumigants within the enclosed area. It facilitates rapid heating of the soil and helps in conserving moisture and preventing leaching of nutrients. The film transmits most of the incoming radiation which warms the soil and the moisture droplets that collected on the underside of the film block, much of the radiation is emitted as the soil cools at night.

(a) The use of black plastic mulch eliminates the need for mechanical cultivation. The increase in crop yield by using black PE mulch is based on the elimination of weeds and the avoidance of soil compaction. Here the film retains its position for several years. One example for this is mulch made of rubber from recycled tires that avoids the need for otherwise frequent replacement while using normal films.
(b) The use of white film mulching lowers soil temperature in relation to uncovered soil. This type is used either in regions with high levels of sunshine, where it is required to reduce the transmitted radiation and soil temperature, or in regions of low luminosity, where there is a need to increase the amount of reflected light on the lower and middle leaves.

Photo/biodegradable film mulching is significantly used in agricultural mulch as it is completely degraded in a short time when buried in the soil at the end of the crop season and therefore the disposal of the used films can been overcome by the use of photodegradable film.

8.4.2.2 Multilayer Films for Green Houses

Greenhouses are like intensive-care units. Plants are exposed to the sunlight and can grow in ideal conditions according to their physiological properties. The use of greenhouses indeed provides farmers with the possibility to create the appropriate environmental conditions that plants require for faster and safer growth, to avoid extreme temperatures and protect crops from harmful external conditions. In the past few decades use of plastic films for greenhouse has gained prominence across the globe. The advantages offered by them such as lightweight, translucency, flexibility, toughness, hydrophobic nature and durability, make plastics films the most preferred material for cladding of greenhouses. Different types of plastic materials such as poly (vinyl chloride) (PVC), polycarbonate (PC) and low density polyethylene (LDPE) are commonly used for cladding of greenhouse structures. LDPE remains the most popular of all plastic material because of its features, ease in availability and economics. The major requirements of a film to be used in green houses are durability diffusion thermal effect—thermicity anti-dustproperty and good mechanical strength.

8.4.2.3 Controlled Release of Agricultural Chemicals

Multilayer films can also be used for the controlled release of agricultural chemicals. Controlled release (CR) is a method by which biologically active chemicals are made available to a target species at a specified rate and for a predetermined time. The polymer films can serve primarily to control the rate of delivery, mobility, and period of effectiveness of the chemical component.

8.4.2.4 Polymeric Windbreaks and Protective Nets

The use of films, set around the plants is more effective to create a channel for plant protection against damage by cold weather, excessive insolation, and animals. The purpose of protection is to increase the crop and accelerate maturation, or to extend the cropping season. The main form of protection is achieved through regulating the temperature and moisture levels, and eliminating wind and possible damage from the adverse weather conditions as high temperature, hail, or wind. Such protection can also modify the spectrum of light reaching the plants which modifies their growth. The mechanics of this type of protection primarily involves a covering of film, but netting is used when shading is required to reduce temperature. Windbreaks are a permeable barrier rather than a covering.

8.4.3 Medical Application

Examples of potential and actual applications of these films in medical field include coatings for cell and tissue culture, and artificial red blood cells, drug-delivery applications, encapsulation to protect biologics from the surrounding environment and specific targeting to enhance drug efficacy or reduce side effects. The films could also be utilized for drug reformulation; for example, to extend patent lifetime, which is a major interest of pharmaceutical companies today. The substrates for film assembly in a drug delivery context have included small crystals of hydrophobic drugs and hydrogels impregnated with small-molecule drugs, multilayer films from polymers onto which peptide drugs have been grafted and could be released by a change of environment and polymers that incorporate T-cell antigens.

8.4.4 Optical Devices

Reflective thin films optical, electro optic, telecommunications, solar concentrator and architectural applications, reflectance must either be controlled (reduced or enhanced), or the colour of the object changed (e.g., given a "gold" colour). It is essential in determining the colour, transparency and polarization characteristics of the film. Reflectance (R) is the fraction of incident light reflected from a surface and is an intrinsic optical property of thin films. Combined with absorption, reflectance determines colour and intensity (or energy) of reflected light. Multilayer coatings can be designed for high reflectance at virtually any wavelength or spectral region, and can change the simple glass or plastic windows into a highly reflective mirror, low-e and solar control coatings, astronomical telescope mirrors, laser mirrors, satellite optics, DWDM filters, optical cavities, Dichroic beam splitters, polarizers and electromagnetic shielding are the applied field. Unlike polished metals, multilayer high reflector coatings can have almost perfect reflectivity and are used as laser mirrors [28].

8.5 Conclusion

Multi layer polymer films are of highly attractive in the area of food packaging, agriculture and in medical field. This chapter discusses the major areas in the field of multilayer polymer films. The preparation methods and the characterisation techniques used in multilayer polymer films are discussed in detail. The current trend in multilayer polymer films and their applications in different areas are also discussed. The chapter also discusses the important parameters that needed to be taken care off in the preparation of multilayer polymer films. However, there is still a lot of scope of in the field of multilayer polymer films which can really contribute to the industrial requirements.

References

1. Engelmann S (2012) Advanced thermoforming: methods, machines and materials, applications and automation. Wiley
2. Decher G (2012) Layer-by-layer assembly (putting molecules to work). In: Multilayer thin films: sequential assembly of nanocomposite materials, 2nd edn. pp 1–21
3. Xiang F et al (2013) Improving the gas barrier property of clay–polymer multilayer thin films using shorter deposition times. ACS Appl Mater Interfaces 6(9):6040–6048
4. Vogler J (1984) Small scale recycling of plastics. Intermediate Technology Publication, pp 6–7
5. Armstrong SR et al (2014) Co-extruded polymeric films for gas separation membranes. J Appl Polym Sci 131(2)
6. Dixon J (2011) Packaging materials: 9. Multilayer packaging for food and beverages, ILSI Europe a.i.s.b.l. ISBN 9789078637264
7. Saldivar-Guerra E, Vivaldo-Lima E (2013) Handbook of polymer synthesis, characterization, and processing. Wiley
8. Abdel-Bary E (2003) Handbook of plastic films. Smithers Rapra
9. Hirvikorpi T et al (2014) Barrier properties of plastic films coated with an Al_2O_3 layer by roll-to-toll atomic layer deposition. Thin Solid Films 550:164–169
10. Siracusa V (2012) Food packaging permeability behaviour: a report. Int J Polym Sci

11. Mokwena KK et al (2009) Oxygen transmission of multilayer EVOH films after microwave sterilization. J Food Eng 92(3):291–296
12. Yang Y-H et al (2013) Super gas barrier and selectivity of graphene oxide-polymer multilayer thin films. Adv Mater 25(4):503–508
13. Tashiro H, Nakaya M, Hotta A (2013) Enhancement of the gas barrier property of polymers by DLC coating with organosilane interlayer. Diam Relat Mater 35:7–13
14. Dhawan S (2013) Polymeric-based multilayer food packaging films for pressure-assisted and microwave-assisted thermal sterilization. Ph.D thesis, Washington State University
15. López-Rubio A et al (2005) Effect of high pressure treatments on the properties of EVOH-based food packaging materials. Innovative Food Sci Emerg Technol 6(1):51–58
16. Cocca M et al (2011) Influence of crystal polymorphism on mechanical and barrier properties of poly (l-lactic acid). Eur Polym J 47(5):1073–1080
17. Kalendova A et al (2013) Polymer/clay nanocomposites and their gas barrier properties. Polym Compos 34(9):1418–1424
18. Puurunen RL et al (2004) Island growth in the atomic layer deposition of zirconium oxide and aluminum oxide on hydrogen-terminated silicon: growth mode modeling and transmission electron microscopy. J Appl Phys 96(9):4878–4889
19. Howells DG et al (2008) Mechanical properties of SiO_x gas barrier coatings on polyester films. Surf Coat Technol 202(15):3529–3537
20. Yang Y-H et al (2011) Super gas barrier of all-polymer multilayer thin films. Macromolecules 44(6):1450–1459
21. Caner C et al (2003) The use of mechanical analyses, scanning electron microscopy and ultrasonic imaging to study the effects of high-pressure processing on multilayer films. J Sci Food Agric 83(11):1095–1103
22. Kim D et al (2013) Highly size-selective ionically crosslinked multilayer polymer films for light gas separation. Adv Mater
23. Sun Z et al (2013) Microstructure and optical properties of Ag/ITO/Ag multilayer films. Nanoscale Res Lett 8(1):1–5
24. Murthy NS (2004) Recent developments in polymer characterization using X-ray diffraction. Parameters 18:19
25. Hammond DB et al (1998) A grazing incidence X-ray diffraction study of Langmuir-Blodgett films of poly (vinylpyridine) quaternized with n-docosyl bromide. Langmuir 14(20):5896–5899
26. Wang L et al (1999) Investigation into an alternating multilayer film of poly (4-vinylpyridine) and poly (acrylic acid) based on hydrogen bonding. Langmuir 15(4):1360–1363
27. http://www.tainstruments.co.jp/application/pdf/Thermal_Library/Applications_Briefs/TA108.PDF
28. www.materion.com/microelectronics

Chapter 9
Hybrid Systems for Multi-layer Fuel and Air Hoses in Automobiles

Narendra Reddy

Abstract Transportation of fuel and air are critical for the operation of any automobile. Hoses made from rubber have been the primary means of carrying fuel or air although metal tubes have also been used in some instances. Increasing use of alcohol as fuel, stringent environmental regulations on emissions from petroleum fuels and restrictions on the use of several chemicals as additives in fuels have made conventional rubber, plastic or metal hoses unsuitable for fuel or air transportation in modern automobiles. Recent trend has been to develop multilayer hoses using hybrid materials with distinct functionalities and specific to the requirements. In addition to multi-layer rubber hoses, a plethora of polymers such as nylon and its derivatives have been used for developing the multi-layer hoses. Several companies are commercially marketing multilayer hybrid hoses but research to develop new hoses to meet latest specifications is ongoing. Considerable numbers of patents on development of new type of multilayer hoses using different materials or technologies have been reported. Although a large number of patents have been filed on developing hybrid hoses using various materials, there is limited scientific literature on the structure and properties of multilayer hybrid hoses. This chapter covers the various multilayer hybrid hoses developed and those that are commercially available.

9.1 Introduction

Safe and efficient delivery of fuel to the various parts in an automobile is one of the most critical aspects in vehicle design and development. In addition, increasing need and efforts to curb pollution from automotive exhausts have led to stringent norms on use of additives in fuels. Also, the use of biofuels such as ethanol that have considerably high permeability makes conventional metal tubes and rubber hoses unsuitable

N. Reddy (✉)
Center for Emerging Technologies, Jain University, Jain Global Campus, Jakkasandra Post, Ramanagara District, Bengaluru 562112, India
e-mail: nreddy3@outlook.com

J.K. Kim et al. (eds.), *Multicomponent Polymeric Materials*,
Springer Series in Materials Science 223, DOI 10.1007/978-94-017-7324-9_9

Table 9.1 Minimum values required for fuel, oil and emission hose as per SAE standard J 30

Nominal size (mm)	Burst test (MPa)	Hydrostatic proof (MPa)	Change in length (MPa)
4.78	13.8	6.7	3.5
6.35	11.0	5.5	2.8
7.94	8.3	4.1	2.1
9.53	6.2	3.1	1.5

for use. Recently, car manufacturers in the USA, Japan, Korea and the European Union are also demanding conducting automotive fuel lines that allow static discharge and avoid static build-up and potential risks of explosion. For instance, SAE standard J 30 includes standards for fuel, oil or emission hose in coupled and uncoupled applications (http://standards.sae.org/j30_201202/). This standard covers three types of hose construction and provides dimensions and tolerances to perform the test. Some of the parameters tested in the standard include change in length, burst resistance, vacuum collapse, cold flexibility, tensile strength, elongation for tube and cover, dry heat and fuel resistance, ozone and oil resistance (https://law.resource.org/pub/us/cfr/ibr/005/sae.j30.1998.pdf). Acceptable values for the tubes depend on their diameter. Some of the acceptable values for the performance of the fuel hose are listed in Table 9.1.

Even in terms of safety, conventional methods of delivering fuel using metal pipes or rubber hoses have several limitations and pose considerable risks to the user and the environment. For instance, metal and rubber hoses are connected using clamps which increase the chances of leakage. Metals have low resistance and cannot resist large shocks and lead to breakage. Similarly, rubber hoses have high permeability leading to loss of fuel and are also less resistant against environmental conditions. Metal tubes have low flexibility and need longer lengths to deliver the fuel to the designated part. Disposing metal and rubber hoses also causes considerable harm to the environment and longer hoses means more material to be disposed.

More importantly, the mandates to reduce petroleum based emissions have led to considerable increase in the use of biofuels, especially methanol and ethanol. Conventional rubber or metal based hoses are not suited to handle alcohol based fuels due to corrosion and permeation problems. Table 9.2 provides an indication of

Table 9.2 Comparison of the permeability (g/m^2/day) for the polyamide and elastomers based hoses with and without alcohol

Hose material	No alcohol	10% ethanol	15% methanol
NBR, 33% CAN	1190	1030	1190
PA 12 flexible	5.5	24	83
FKM A-200, 66% fluor	0.76	7.5	36
FKM B-200, 68% fluor	0.68	4.1	12

Inner surface of the hoses was exposed to fuels at a temperature of 23 °C
FKM is Viton® *A-200* Fluoroelastomer

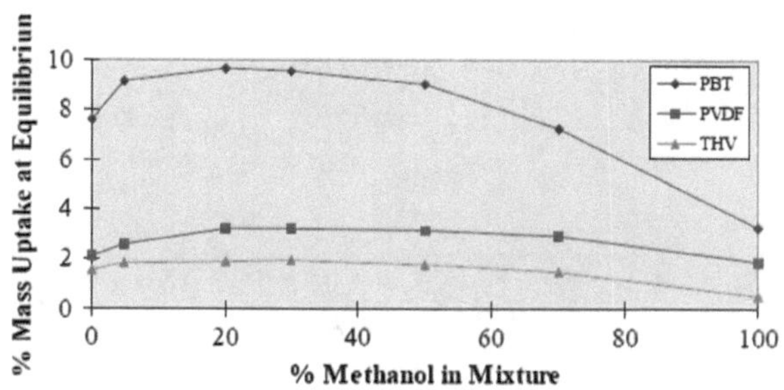

Fig. 9.1 Variation in the % mass uptake of methanol at equilibrium at 60 °C for three polymers commonly used to manufacture hoses [11]

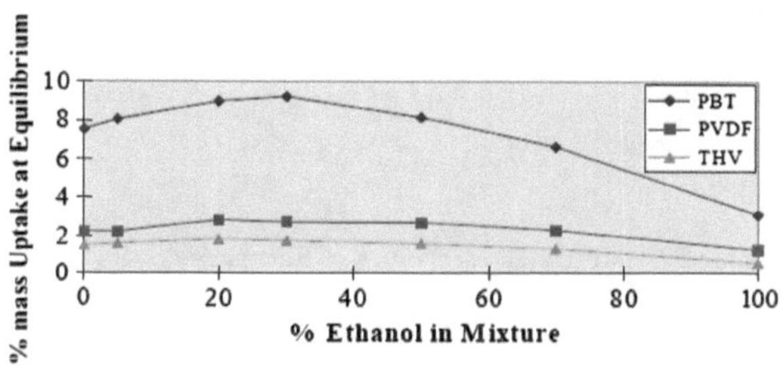

Fig. 9.2 Variation in the % mass uptake of ethanol at equilibrium at 60 °C for three polymers commonly used to manufacture hoses [11]

the increase in permeability due to the addition of ethanol and methanol. Rubber based hoses have extremely high permeability to alcohols compared to polyamide and fluoropolymer based hoses. Addition of alcohol showed substantially higher permeability in both the polyamide and fluoropolymer based hoses. It has also been observed that % mass uptake of methanol and ethanol by poly(vinylidene fluoride) (PVDF), poly(butylene terephthalate) (PBT) and terpolymer of vinylidene fluoride tetrafluoroethylene and hexafluoropropylene (THV) varied with the % of alcohol in the fuel (Figs. 9.1 and 9.2). PBT had the highest uptake and therefore highest permeability followed by PVDF and THV [11]. Although THV showed lower % uptake, THV is considerably expensive and not ideal for use in automotive hose applications. Several materials suggested for use in automotive fuel hoses were tested for the permeation resistance and mechanical properties after immersing in solvents for up to 21 days. Materials show completely different permeability depending on their composition and fuel used for testing (Table 9.3). For instance, Vistamid 7344 (T) had excellent resistance to toluene but very poor resistance to methanol. However, materials such as Kynar, Dyflor and THV 500G had good permeability to all the three solvents studied [10]. It is also interesting to note that the permeability of the hose to the different fuels was not related to the mechanical properties (Table 9.4).

To overcome these limitations, polymeric materials are being used to develop multilayer hybrid fuel and air hoses. Many polymers but predominantly nylon and its derivatives and blends have been used to develop multilayer hybrid hoses. Conventionally, multilayer hoses were developed through lamination but the advent of co-extrusion has made it possible to extruder multiple layers and use different

Table 9.3 Some properties of different polymers used to develop the multilayer hybrid hoses and their performance against various fuels [10]

Material	Function	Equilibrium weight gain (%) after immersion for 21 weeks in various chemicals		
		Methanol	Ethanol	Toluene
Grilamid (N)	Outer layer	1.72	2.20	4.82
MA 4411 (N)	Outer layer	2.64	3.35	5.49
Vestamid 7293 (N)	Outer layer	2.19	2.01	4.95
Vestamid 7344 (T)	Tie layer	12.39	11.55	4.25
Adheflon (T)	Tie layer	6.44	2.89	45.79
Kynar (B)	Barrier layer	0.52	0.19	0.22
Dyflor (B)	Barrier layer	0.68	0.21	0.29
THV 500G (B)	Barrier layer	0.04	0.42	0.26

Table 9.4 Mechanical properties of the materials intended for hose applications after conditioning at 80 °C

Material	Modulus (MPa)	Stress (MPa)	Strain (MPa)
Grilamid (N)	340	34.9	258
MA 4411 (N)	489	46.6	316
Vestamid 7293 (N)	241	40.1	314
Vestamid 7344 (T)	1195	42.5	43
Adheflon (T)	759	25.3	65
Kynar (B)	265	20.9	566
Dyflor (B)	1325	47.8	23
THV 500G (B)	1415	47.5	27

polymers rather than laminating different layers. Extrusion is convenient, faster, inexpensive and offers more flexibility compared to other hose manufacturing techniques. The reader is directed to watch some videos on youtube (https://www.youtube.com/watch?v=VDAqufbmsDg) on multilayer extrusion of automotive hoses to get a better picture of the overall process of hose production.

9.2 Multilayer Hybrid Hoses Made Using Rubber

Rubber was the primary source for all hoses before the commercialization of polymer based hoses. Although rubber provided excellent flexibility and good mechanical properties, the poor chemical resistance and more importantly the low resistance to permeation led to considerable risks when rubber was used as automotive fuel hoses. Low permeation rubber hose have been developed using several techniques. Applying a layer of resin (usually nylon) on the inner surface of the

Table 9.5 Components of a rubber-metal hybrid hose that has low permeability (http://www.maillefer.net/en/)

Component	Parts by weight
Natural rubber	85
Styrene-butadiene rubber	15
Carbon black (FEF)	45
Mineral oil	5
Antioxidant	0.7
Stearic acid	0.5
Zinc oxide	5
Vulcanization accelerator	0.5
Sulfur	2

hose or using different types of rubbers were considered to develop high permeation resistance hoses. Higher costs and poor binding between nylon and rubber made these types of hoses to be unpopular. To overcome these limitations, a layer of aluminum, copper strip or plastic tape was wound around the outer periphery of the resin coated rubber tubing [17]. Manufacturing such as tubing was difficult and penetration of the fuel through the resin was also observed. In further development of the rubber hose, a thin film of silver alloy was coated on the outer surface of the inner tube and then a rubber layer was heat cured to the metal film using an adhesive (isocyanate, phenol resin, chlorinated rubber based) [20]. Since the metal film was plated unlike wrapping in the previous approaches, the films did not lose their barrier resistance and the rubber hose could retain the flexibility and mechanical properties. Composition of a typical low permeable rubber-metal hybrid hose is shown in Table 9.5. When tested with Freon, the permeability of the metal coated hoses was about $0.5–1.5 \times 10^{-4}$ cc/cm^2 h compared to 8–25 for the untreated hose [20].

A rubber based 3 or 4 layer fuel hose was developed by Tokai Rubber Industries, Ltd. In their invention, a three layer hose consisting of inner acrylonitrile-butadiene rubber (NBR), intermediate layer composed of a tetrafluoroethylene-hexafluoropropylene-vinylidene-butadiene rubber (THV) and an outer layer composed of NBR-PVC was developed. A optional fourth layer (outermost) composed of chlorosulfonated polyethylene rubber (CSM) was included. However, the NBR layer had poor adhesion to THV and to improve interlaminar adhesion, 1,8-diazabicycloundecene-7 salt (DBU salt) was added into the layer containing NBR [14]. Addition of the salt provided good adhesion for the ternary hose but poor adhesion was observed in the four layer hose between NBR and THV even with the addition of salt. It was discovered that using diene rubber vulcanized by sulfur and addition of DBU and DBN salt lead to excellent interlaminar adhesion between resin layer and THV. Such a hose could be manufactured by concurrent extrusion or by extruding individual layers and then laminating.

A layered and laminated rubber structure (Fig. 9.3) was developed using a layer of fluororubber and a layer of Teflon superimposed on the fluororubber layer that exhibits excellent inter layer bonding characteristics and at the same time provides

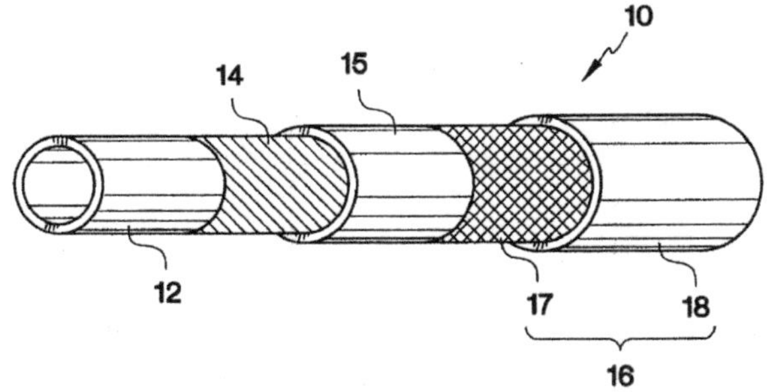

Fig. 9.3 Schematic representation of the layers in a hybrid hose developed using various materials in different configuration [12]

ultra low permeation and excellent bending properties [12]. Inner layer was formed of a veneer of a fluorovinylidene and hexafluoropropylene. Attached to the inner layer [13] is a Teflon coating (0.02–0.05 mm) [15] which has excellent binding with the fluoroelastomer without affecting the bending or durability of the inner layer. Another tie layer [16] is attached between the Teflon and the outer tubing. The outer tubing is composed of a reinforcing single or double layer formed of braided synthetic fibers such as nylon-12 leading to a hybrid hose. Although the development of the hose was described, the properties of the hose were not reported.

9.3 Fluoropolymer Based Hoses

A laminated multi layer hybrid hose consisting of fluoroelastomer having thickness of 0.1–1.5 mm that was bonded to a second layer of non-elastomeric fluoroplastic such as polyvinylidene fluoride and treated to promote adhesion and a third structure made from a non-fluorinated elastomer (1–2.5 mm thick) such as nitrile-butadiene rubber, epichlorohydrin or ethylene copolymers was developed to achieve both flexibility and high permeation resistance [18]. The three layer laminates were cured at 162 °C for 30 min and excellent adhesion was observed between the three layers. A comparison of the 3-laminate hose in terms of permeation (g/m^2/day) to two types of fuel with a fluoroelastomer layer revealed that the laminate had considerably lower permeation even after exposure for 672 h (Table 9.6).

Viton® from Dupont is a fluoroelastomer that has low permeation resistance and provides resistance to most automotive fuels and fuel blends. Further, Viton® has been optimized for fuel hose applications with improved extrusion processing, good adhesion and low permeation. Hoses manufactured using Viton® are marketed under the trade name VTR-9209. Fuel hoses are manufactured with an inner layer of Viton and an outer layer of Vamac® when good fuel resistance and high temperature properties are desired. Hoses manufactured with these two polymers are reported to exceed performance requirements for mechanical strength, fuel aging, hydrocarbon permeation resistance and high temperature performance at an

Table 9.6 Comparison of the permeation ($g/m^2/day$) of the three-layer laminate with a fluoroelastomer layer [18]

Time (h)	Fuel type			
	75% fuel/25% methanol		90% fuel/10% ethanol	
	A	1	A	1
48	–	−0.17	–	−0.06
96	6.7	−0.88	−0.16	−0.93
264	14.7	0.02	2.5	0
336	15.2	0.08	3.8	−0.26
408	19.9	–	5.6	–
672	15.0	0.25	4.0	0.27

optimum price (http://www.dupont.com/products-and-services/plastics-polymers-resins/elastomers/articles/automotive-fuel-hoses.html).

Conventional hoses made using FKM/Textile/ECO (fluoroelastomer/aramid fiber/epichlorohydrin rubber) have good mechanical flexibility and anti vibration properties but have poor permeation resistance. Conversely, plastic tubes coated with fluoroplastic/nylon offer low fuel permeability but are rigid and have poor sealability. To overcome these limitations, a new lamination technology was developed by Dupont using F200 low permeation lamination. This technology utilizes a 3–5 layer construction and meets CARB LEV II and partial zero emission vehicle (PZEV) fuel permeation compliance. Figure 9.4 shows that the F200 technology provided hoses with higher permeation resistance than nitrile rubber/fluoroplastic/NBR rubber tube and THV-ECO (THV inner layer with cover of epichlorohydrin rubber hoses). F200 showed negligible permeation to CE 10 fuel even after exposure for 840 h at 40 °C. Hoses made using F200 technology are reported to be used in commercial vehicles and a production rate of up to 7 m/min have been achieved.

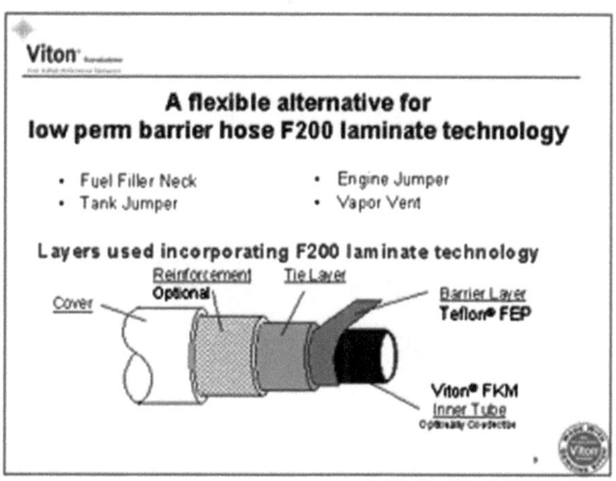

Fig. 9.4 Schematic of the 5-layer lamination technology to develop fuel hoses using F200 (http://www.dupontelastomers.com/autofocus/a2/af2.asp?article=af2_barrier)

A multilayer fuel hose was developed using blends of fluoropolymers where the first fluoropolymer (hexafluoropropylene, vinylidene fluoride and tetrafluoroethylene) had elastomeric properties and the second fluoropolymer had thermoplastic characteristics. In addition to the fluoropolymer layers, additional reinforcing layers made from polyamide fibers, polyester, rayon, glass or cotton were included. Preferably, an outer protective layer consisting of a synthetic elastomer such as styrene-butadiene rubber or high density polyethylene were included [15]. Although the development to the hose was described in detail, the properties of the extruded hose were not reported.

Fluoropolymers are suggested to be ideal for use for fuel transport due to the low permeation and high mechanical properties of the polymers. Typically, several thin layers of the fluoropolymers are used in combination with other materials to develop the multi-layer hoses. However, fluoropolymers have poor bonding between themselves and also with other polymers. Several methods have been used to improve the properties of fluoropolymer based hoses. A combination of fluoropolymers and non-fluoropolymers, treating the surface, blending with other polymers or grafting the fluoropolymers to introduce polar functionality have been studied as possible chances [9]. A perfluorothermoplastic (FEP) was blended in various ratios with a partially fluorinated polymer (PF-3) and extruded at 300–360 °C to form films. The films were tested for permeation resistance for eventual use as hose. Table 9.7 shows that the blend material had considerably lower permeation compared to the neat polymer.

A flexible hose for transporting volatile hydrocarbons was made by Hiroaki et al. which consisted of an inner layer made from polyvinylidenedifluoride (PVDF), an intermediate layer made from thermoplastic polyurethane (TPU) that was extruded over the PVDF layer and finally an outer layer covering the intermediate layer and made from polyvinyl chloride. The hose developed had a maximum permeation rating of 15 g/m^2/day when tested according to SAE standard J1737 [6]. One of the advantages of this hose was that the three layers could be co-extruded without the need for tie layers. Up to 12 layers of tubing could be formed with the desired color and with required logos since the polymers were thermoplastic. Conditions used to produce a three layer hose that had a permeation of 14 g/m^2/day are given in Table 9.8. Hose developed had a tensile modulus of 5.8 MPa and tensile strength of

Table 9.7 Permeation and surface energy of hybrid material made from various ratios of FEP and PF [9]

FEP/PF ratio	Permeation constant (g mm/m^2/day)	Surface energy (mJ/m^2)
90/10	0.78	18.9
80/20	1.06	18.9
70/30	–	20.0
50/50	2.21	20.3
100/0	0.59	17.6
0/100	9.5	21.7

Table 9.8 Raw materials and conditions used to produce the three layer hose

	Inner	Intermediate	Outer
Material	KynarFlex 2500-00	Estane 58213	C-527-D11
Extruder size (in.)	1.25	1.25	2.5
Temperature range (°F)	490–510	370–400	310–330
Extruder screw (rpm)	4–9	2–4	20–32

Table 9.9 Properties of the three layer hose evaluated using various standards

Property	ASTM method	Value or rating
Durometer hardness (Shore A), 15 s	D2240	69
Tensile strength, psi (MPa)	D412	2400 (16.5)
Ultimate elongation (%)	D412	450
Tensile modulus, psi (MPa)	D412	850 (5.8)
Specific gravity	D792	1.29
Operation temperature range (°F (°C))	–	−20 to 180 (−28 to 82)
Color	–	Yellow

16.5 MPa (Table 9.9) and excellent resistance to acids, bases, oils, greases and fuels and also to hydrocarbons but unsatisfactory resistance to ketones [6].

To overcome the limitation of fluoropolymer based multi-layer tubing fabricated by lamination, Dana corporation developed a multi-layer tubing by extrusion using polybutylenenaphthalate (PBN) which could be formed into hoses in the conductive, non-conductive and corrugated configurations. The PBN based tubing had low permeation, electro static discharge and also lower weight and cost. The hose was made from Nylon-12 internal and external layers, a PBN middle layer and two adhesive layers (http://www.prnewswire.com/comp/226839.htm).

9.4 Polyamide Based Multilayer Hybrid Hoses

A typical fuel hose made using polymeric system consists of three to seven layers (Fig. 9.5) depending on polymer system used, the type of vehicle in which the fuel hose will be deployed and the manufacturer. Since fuel hoses have specific requirements necessary to perform at higher pressures, multilayer fuel systems have been developed mainly using PA 12 with barrier materials such as polybutyleneterephthalate (PBT), polybutylenenaphthalate (PBN), polyvinylidenefluoride (PVDF), ethylene perfluoroethylenepropene-copolymer (EFEP), ethylenetetrafluoroethylene-copolymer (ETFE), polyphenylene sulfide (PPS), and ethylene-vinylalcohol-copolymer (EVOH). Further, conductive PA12, EFFE and EFEP are used as conductive layers to reduce static build-up. Although fluorine based materials provide good properties, these materials are expensive and also have high density which

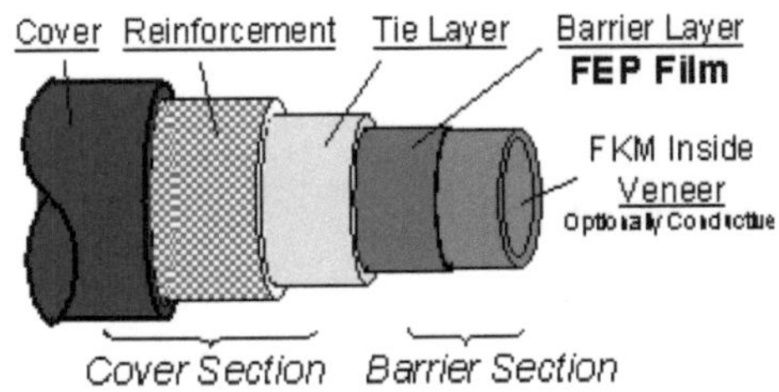

Fig. 9.5 A five layered F-200 hose with low permeation of <7 g/m^2/day (http://www.dupontelastomers.com/autofocus/a2/af2.asp?article=af2_extrusion)

increases the weight of the vehicle and are not ideal for use with PA based hoses. Instead, a novel multilayer tubing has been developed by combining PA 12 with another polyamide (Genestar Polyamide PA 9T) manufactured by Kurraray.

A new generation multilayer tubing system for fuel lines was developed by Kuraray and Ube with the commercial name Ecobesta-9T. This system combines PA 12 with polyamide PA 9T which is a fuel barrier material. Schematic of the hose reveals that the outer layer consists of heat stabilized PA 12, a non-conductive PA 9T as the middle layer and the new PA 9T as the inner layer. Some of the features of the new hose include excellent cold impact and chemical resistance (PA 12), superior adhesive strength with PA 9T which has low permeability to gasoline and low monomer/oligomer elution. Other commercial products developed using the same technology include Sunbesta (PA12/ETFE) and Sunbesta-ZV (Fig. 9.6) which combines the performance of Ecobesta and Subesta (http://k.plasticstoday.com/article/multi-layer-polyamide-system-conductive-fuel-lines).

Dupont has developed a 612 polyamide based resin (Zytel® LCPA series) that contains about 60% of renewable material and has outstanding heat resistance and also chemical and hydrolysis resistance. The resin can be tailored to have high temperature resistance and lower permeability to fuel and gases. These resins are used with traditional reinforcing materials such as glass fibers to form the fuel hoses. Some of the properties of the PA 12 based commercially available hose are given in Table 9.10.

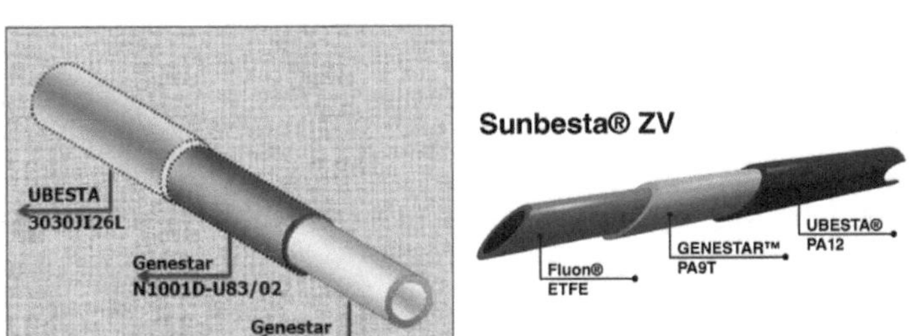

Fig. 9.6 Schematic depiction of the construction of the polyamide based hybrid hoses that are commercially available. http://k.plasticstoday.com/article/multi-layer-polyamide-system-conductive-fuel-lines; http://articles.sae.org/12982/

Table 9.10 Typical performance properties of a commercially available 5-layer PA tube

Property	Acceptable values
Burst strength	
– Room temperature	5.6 MPa
– High temperature	2.1 MPa
Bending	
– Minimum bending radius	Meet QCT-798-2008 requirements
– Resistance to bending deformation	–
– Burst strength after bursting	5.6 MPa
Resistance to ZnCls	
– Appearance	No cracking
– Burst strength, room temperature	4.2 MPa
– Cold-temperature appearance	No cracks in appearance
– Cold temperature impact	4.2 MPa
Permeation resistance	
– Permeation rate	0.05 g/mg/day
– Length change rate	−1 to +1%
– Burst strength, room temperature	5.6 MPa
Yield stress	
– Parallel extrusion direction	24 N/m
– Vertical extrusion direction	25 N/m
Elongation	
– Parallel extrusion direction	170%
– Vertical extrusion direction	170%
Performance to fuel resistance	
– Appearance	No cracking
– Low temperature impact	No cracking
– Burst strength, room temperature	5.6 MPa
– Stress yield at extrusion direction	20 N/m
– Elongation at extrusion direction	150%
– Layer adhesive force	No clear delamination observed

The tube has external diameters from 3 to 16 mm. The barrier layer is composed from ethylene vinyl alcohol (EVOH) and the inner layer is made from PA 12 (http://eternal.en.gasgoo.com/auto-products/1856323.html)

Another company EMS Switzerland has developed a 3 layer hose (EVX 306) containing an outer PA 12 layer of 0.45 mm thickness, a barrier layer made from EVOH of 0.1 mm thickness and an inner PA 6 layer of 0.45 mm thickness. The tube was claimed to have exceptional toughness and corrosion resistance and suitable for fuel lines requiring low permeation and was also suitable for use in automotives running on 100% ethanol.

Although not developed into commercial production, several researchers and companies have patented novel technologies to develop multilayer hoses for fuel

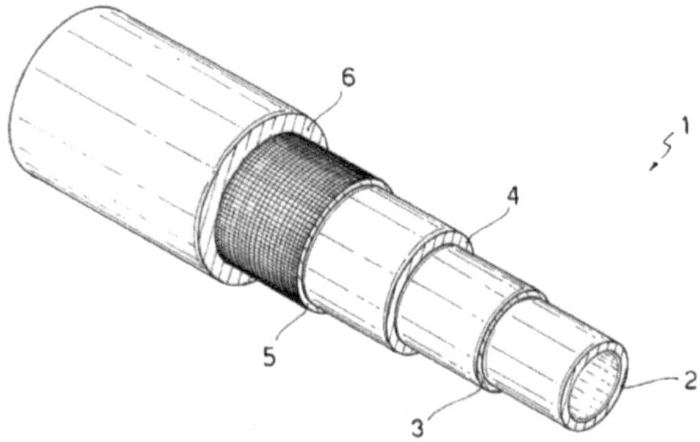

Fig. 9.7 A5-layer hose developed by Bertero and Dimattia according to U.S. patent number 6,435,217 [1]

and air technologies. US patent 6435217 [1] describes the development of a multilayer hybrid hose for conducting fuel, specifically from the vehicle pump and withstand pressure over 100 bar, impermeable to hydrocarbon vapors and can replace conventional metal pipes. The hose was designed to have one inner layer of elastomeric material (acrylonitrile/butadiene and hydrogenated acrylonitrile/butadiene), a barrier layer made from nylon 6/6.6 or mixtures of nylon and polypropylene and a reinforcing layer made from a spun fabric of aramid fibers such as Twaron or Kevlar. Finally, a cover layer composed of chlorinated polyethylene was used (Fig. 9.7).

One of the earliest inventions on a multilayer tubing to be used for fuel supply was done by Brunnhofer where three layers were coextruded into a hose. Outer layer of the hose was made from a polyamide, primarily nylon 12, an inner layer made from nylon 6, 6.6 and nylon 11 and intermediate layer composed of a polyolefin that can block perfusion of alcohol to the outer layer. However, such a hose cracks and oxidizes when surface active agents or oxygen releasing substances pass through the tube. Also, the alcohols and the aromatics in the fluid diffuse through the layers and change the composition of the fuel. Crystallization of the monomers and oligomers from nylon 12 were also observed. To overcome this limitation, a three-layer multipolymer tube was developed. Polyamide (nylon 11 or nylon 12) was used to form a thick outer layer (0.2–0.7 mm), a inner solvent blocking layer (0.1–0.2 mm) used was made of a copolymer of ethylene and vinyl alcohol and a intermediate layer (0.05–0.2 mm) made of about 30–45% polyethylene or polypropylene. With this configuration, the outer layer prevented the tube from stretching and since the inner layer was made of an inert material that did not dissolve in the fuels or fuel additives, the possibility of contamination was eliminated [3].

A multilayer thermoplastic hybrid hose specifically resistant to alcohols in fuels was developed using inner and outer layers made from polyamide and an intermediate layer made from ethylene/vinyl alcohol for fuel hose applications. Although polyamide could provide the required mechanical properties to the hose, the poor permeation resistance, especially to alcoholic fuels made it necessary to have an intermediate layer. The intermediate layer was bonded to the outer and

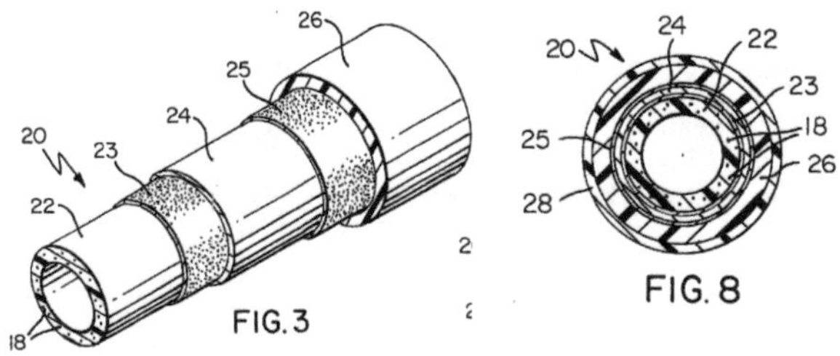

Fig. 9.8 A multilayer hybrid hose developed using metals and polyamide

inner layers using an adhesion promoting layer of polyamides and a compatibilizer. The composite hose (PA6/EVOH/CoPA6/12/PA12) developed in this research had a permeation rate of less than 1 compared to 2 and 3 for commercially used fuel hoses [5]. Permeation level achieved with this tube was within the permissible limit according to the *Sealed Housing Evaporative Determination* (*SHED*) test prescribed in the United States.

Several inventors have also attempted to use metals to resist permeation of fuels in multi-layer polymer based hybrid hoses. In one such invention, a nylon based hose was developed with nylon as the inner tube (22) and an aluminum barrier layer (24), an outer non-conductive nylon tube (26) with two tie layers (23 and 25) (Fig. 9.8) between the nylons and aluminum. In addition, a conductive layer composed of carbon black was also included [16]. Inner and outer nylon tubes were extruded and the aluminum layer was helically wrapped on the inner conductive tube or using electrolytic deposition.

9.5 Fuel Hose for Hydrogen Transport

Fuel-cell based vehicles in which hydrogen and oxygen are electrochemically mixed to generate electricity and fuel vehicles are gaining traction across the world. To operate the fuel cell, it is necessary to transport hydrogen gas which is explosive and has high degree of permeability [8]. Metal hoses made from stainless steel were used for transporting hydrogen to fuel cells. Unfortunately, metal tubes have poor flexibility and since wet hydrogen also flows through the tube, chances of corrosion of the metal and contamination of fuel are possible. To overcome this limitation, a rubber tubing having a high degree of flexibility, impermeability to hydrogen fuel and high electric resistivity and low ion extractability was developed. Such a hose was manufactured using a rubber material consisting of one inner rubber layer, one outer rubber layer composed of a halogenated rubber material, a blend of halogenated and a butyl rubber material or a ethylene-propylene rubber material. Properties (Tables 9.11 and 9.12) of the hose developed in comparison to commercially available hose show that the invented hose had hydrogen gas permeability between 5.4–11 compared to 5.8–11 for the commercial hose. The rubber hose also had higher electrical resistivity.

Table 9.11 Composition (% of weight) and properties of a rubber based hose developed for transporting hydrogen

Component	Type	Samples		
		1	2	3
Rubber	CI-IIR	100	100	–
	EPDM	–	–	100
Vulcanization system	Peroxide vulcanizing agent	4.2	–	4.2
	Resin vulcanizing agent	–	9	–
	Co-crosslinking agent	2	–	2
Filler	Talc	70	70	75
Carbon black		50	50	100
Softener	Paraffin softener	2	5	60
Characteristics				
Hydrogen gas permeability (gas permeability coefficient) (10^{-9} cm^3 cm/cm^2 s cmHg)		6	5.4	11
Ion extractability (electric conductivity) (μS/cm)		15	9.8	14.8
Electric resistivity (volume resistivity) (Ω cm)		5×10^{12}	5×10^{12}	4×10^{11}

9.6 Hoses Made from Hybrid Yarns

In a unique approach, hybrid yarns were used to develop a fluid transfer hose with the yarns and a cover forming the hose. The flexible hose was made using a knit fabric comprising hybrid yarns containing a blend yarns of co-para-aramid fibers and a second yarn of meta-aramid fibers that were twisted together (Fig. 9.9) [2]. High burst pressure and/or improved impulse fatigue resistance was obtained. Hybrid yarns of aramid fibers and meta-aramid fibers and yarns had a twist per inch (TPI) of 5 TPI and the fabric reinforcement had about 50–200 loops per foot. In one such hose, individual yarns having fineness of 400–3000 denier were used to prepare hybrid yarns with deniers from 800 to 30,000. Properties of the hose developed when compared to commercially available hoses showed that the hose made from hybrid yarns had higher impulse fatigue resistance and also passed the field test results (Table 9.13).

Table 9.12 Composition (% of weight) and properties of a commercially available rubber based hose for comparison with the hose developed with properties in Table 9.11

Component	Type	Samples		
		1	2	3
Rubber	CI-IIR	100	–	–
	EPDM	–	100	100
	NBIR	–	–	–
Vulcanization system	Sulfur	1	1	1
	Vulcanization	4	4	2
	Accelerator TT			
	Zinc oxide	–	–	5
Filler	Talc	70	75	–
Carbon black		50	100	60
Softener	Paraffin softener	5	60	–
	Phthalate softener	–	–	10
Characteristics				
Hydrogen gas permeability (gas permeability coefficient) (10^{-9} cm^3 cm/cm^2 s cmHg)		6	5.4	11
Ion extractability (electric conductivity)μS/cm		5.8	11	7.2
		800	820	770
Electric resistivity (volume resistivity) (Ω cm)		2×10^{11}	6×10^{9}	4×10^{8}

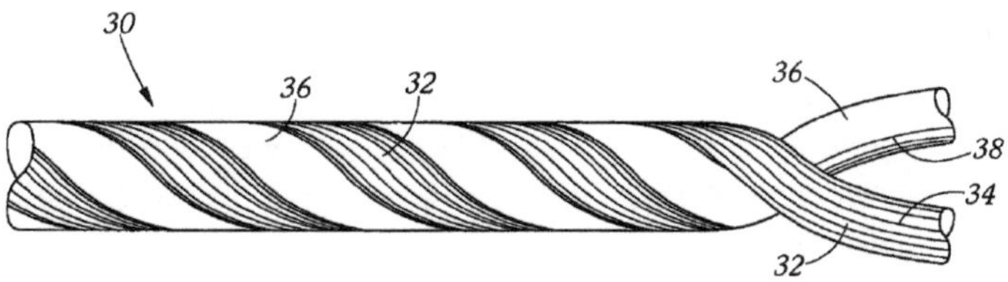

Fig. 9.9 Schematic of the hybrid yarns used to manufacture for transporting fuel

Table 9.13 Properties of hybrid yarns used for automotive fuel line

	Twist (TPI)	Yarn tensile strength, lbf	Burst strength, psi	Pressure 200 °F	Field test	Impact fatigue resistance, cycles
Comparative 1	0	78	118	–	Fail	–
Comparative 2	2.3Z	24	134	1.0	Fail	–
Example	3.0Z	39	226	1.4	Pass	>300 k

9.7 Air Hoses

Latest regulations for automotive greenhouse gas reductions recommend replacing the common coolant R-134a with HFO-1234yf or 2,3,3,3-Tetrafluoropropene. Current hoses used in air conditioning cannot handle the new coolants and therefore newer hoses are being developed. To overcome the limitation with the conventional air hoses, a multilayer hose was developed that consisted of a inner layer made from PPA (semiaromatic polyamide or polyphtalamide or high performance polyamide), a second layer made from an elastomeric material (natural rubber, polychloroprene, acrylonitrile-butadiene (NBR)), a third layer made of reinforcing yarns and finally a fourth layer made of a different type of elastomeric material than that in the second layer [4].

PlastAir manufactures a multilayer polyurethane reinforced air hose that is not only flexible but also durable (http://www.plastair.com/en/polyurethane-reinforced-air-hose/). Working temperature of the hose ranged from 76 to −45 °C. Similarly, Mebra Plastik manufactures a triple layer hose consisting of two layers of polyamide 12 and one middle layer of ether-based polyurethane that was chemically bonded. The hose was highly flexible, resistant to low temperature impact and suited for pneumatic applications (www.mebraplastik.com).

Air brake system used in heavy duty vehicles is activated by pressurized air that is transported through hose. Conventionally, metal tubing was used for air supply but nylon and other polymeric materials are recently used. A multi-layer air hose was developed by Saint-Gobain performance plastics corporation using a first outer layer made from nylon (11,12 or blends) and a second layer made from impact modified nylon 6–12, maleic anhydride modified polyolefin and blends of nylon 6–12 and a third layer made of nylon 6 which has a flexural modulus of at least 40,000 and an elastic modulus of at least 10,000, yield strength of 1000 and Izod impact strength of 0.7 (http://www.google.co.in/patents/US6670004). Up to 12 layers were incorporated to develop the hose to obtain a flexural modulus ranging from 40,000 to 10,000 PSI, elastic modulus of 10,000 to 40,000 PSI and yield strength of 1000 to 2000 PSI.

Charge air hoses are used to connect the hot and cold sides in turbo charged engines and require considerably high performance properties. While one side of the hose could experience temperatures as high as 500 °F, the other side will have temperatures of 350 °F but the hose has to be capable of handling temperatures between −65 and 350 °F (http://www.goodyearrubberproducts.com/files/Parker/ParkerSiliconeCatalog4888/ParkerSiliconeCatalog48881.Page12.pdf). Several companies offer such hoses in different sizes and configurations to meet particular requirements. For instance, Baker Precision manufactures high performance silicone hoses reinforced with 4-ply Nomex and suitable for use in racing cars, commercial trucks and buses and automobiles having turbo diesel engines (http://bakerprecision.com/hps_CAC_hoses.htm). Similarly, molded hoses with fabric reinforcement are manufactured by Continental that have performance from −50 to +260 °C at pressure of up to 250 kPa (http://www.contitech.de/pages/produkte/schlauchleitungen/ladeluft/form_en.html). The hoses were manufactured using chloroprene rubber, ethyl acrylate rubber and acrylate rubber as the base with the fabric reinforcement. A 4-ply hot side hose made from silicone and meta-aramid reinforcement could perform at 500 °F and higher than 50 PSI pressure (www.flexfab.com).

Ethylene acrylic acid elastomer (AEM) are commonly used as air hose due to their high temperature and oil resistance. However, a charge hose should consist of an inner layer that has high heat resistance which is not necessary for the outer layer. In addition, the various layers in the hose should have good bonding to each other. It was suggested that using a peroxide cured rubber layer on the outside reduces cost and provides the desired features. A hose for charge air delivery was constructed using an inner layer (1.2–1.5 mm) made from amine curable acrylic based elastomer, an intermediate layer (1.5–1.8 mm) made from a blend of ethylene propylene dientetrapolymer, outer layer (1.9–2.2 mm) made from a blend of ethylene propylene diene ter polymer and an outer layer made from ethylene propylene diene terpolymer [19]. In a similar attempt, a hose intended to be used as air inlet circuit of motor vehicle engine was designed using copolymers of ethylene and acrylic acid as the inner layer, a fabric layer (aramid) placed between the intermediate layer and the outer layer. The inner layer had a thickness of 1.5 mm, intermediate layer was 0.5 mm and outer layer was 3 mm in thickness. Although the hose developed was reported to meet the requirements, properties of the hose were not reported (Fig. 9.10) [7].

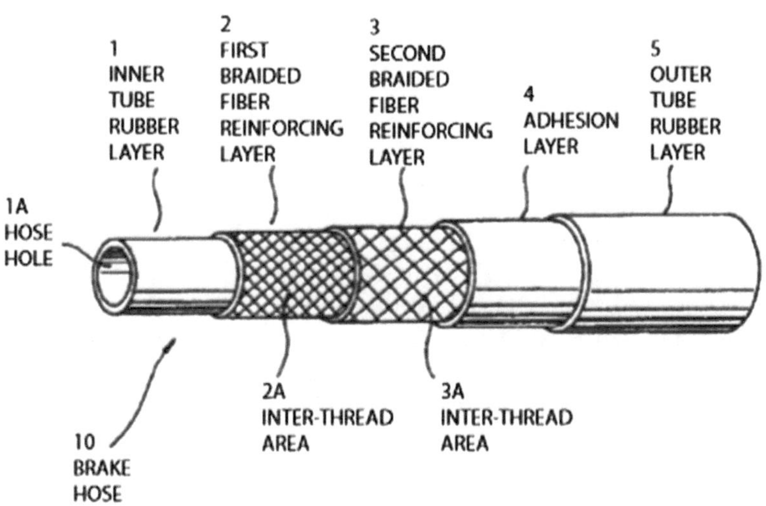

Fig. 9.10 Structure of a multi-layer hybrid air hose

Table 9.14 Some of the common standards used to evaluate the performance of automobile hoses

Standard	Description
ISO 1307:2006	Rubber and plastics hoses—Hose sizes, minimum and maximum inside diameters, and tolerances on cut-to-length hoses
ISO:DIS5774 and ISO 5774:2006	Plastics hoses—Textile-reinforced types for compressed-air applications—Specification
ISO 10619-1:2011	Rubber and plastics hoses and tubing—Measurement of flexibility and stiffness—Part 1: Bending tests at ambient temperature
ISO/DIS 13775-2	Thermoplastic tubing and hoses for automotive use—Part 2: Petroleum-based-fuel applications
BSEN13765:2010	Thermoplastic multi-layer (non-vulcanized) hoses and hose assemblies for the transfer of hydrocarbons, solvents and chemicals —Specification
BS EN 16643 assemblies	Unbondedfluoroplastic lined (e.g. PTFE) hoses and hose assemblies for liquid and gaseous chemicals—Specification

9.8 Standards for Evaluating Multilayer Hoses

The multilayer hybrid hoses developed have to be tested to meet specifications. Since the type of vehicles and fuel and environmental regulations vary from region to region, there are several standards that have been established. Some of the common standards for testing multilayer hybrid hoses are listed in Table 9.14.

9.9 Conclusions

Multilayer hybrid hoses that can perform under extreme conditions and meet most of the stringent environmental and automotive performance requirements have been developed. Nevertheless, newer materials and technologies are being adopted to increase the performance and reduce cost of multilayer hybrid hoses. Most of the research and development on developing multilayer hybrid hoses is being done by industries and very few academic reports are available in this area. Although multilayer hybrid hoses that satisfy requirements are available, current hoses are mostly made from synthetic polymers or rubber. Understanding the possibility of using biopolymers that can degrade in the environment should be considered and efforts should also be made to reduce the weight and cost of the hoses.

Acknowledgments Author thanks the Ministry of Science and Technology, Department of Biotechnology for financial support through the Ramalingaswami Fellowship. The Center for Emerging Technologies at Jain University is also acknowledged for their support and encouragement to complete this work.

References

1. Bertero L, Dimattia D (2002) Multilayer fuel pipe. United States Patent # 6435217
2. Branch M et al (2009) Fluid transfer hose reinforced with hybrid yarn. United States Patent Number 7572745
3. Brunnhofer E (1992) Three-layer fuel line hose. United States Patent # 5167259
4. Carrano A, Zanardi M (2014) Multilayer hose for vehicle air conditioning system. World Patent # WO/2014/06100
5. Flepp A, Hoffman M (2003) Thermoplastic multilayer composites. United States Patent # 6555243
6. Garver WE, Colton MF (2012) Multi-layered fuel tubing. United States Patent # 8092881
7. Houee S (2014) Pipe which is intended in particular for an air intake circuit in a motor vehicle engine, WO2004029493
8. Ikemoto A et al (2004) Hydrogen fuel transporting hose for fuel cell powered vehicle. United States Patent # 6745796
9. Jing N et al (2005) Fluoropolymer blends and multilayer articles. United States Patent # 6849314
10. McNally GM, McCourt MP, Murphy WR (1999) The effect of fuel components on the mechanical properties and glass transition temperatures of polymers used in multi-layer fuel line tubing. SAE technical paper series, international congress and exhibition, Detroit, Michigan, 1–4 March 1999
11. McNally TGM et al (1998) Transport and sorption properties of various barrier materials used in multilayer fuel line tubing on immersion in fuel components. In: The 1998 56th Annual Technical Conference, ANTEC. Part 3(of 3), pp 2642–2646
12. Plumley S (1996) Hose for an automotive fuel line. United States Patent # 5507320
13. Ries H (1995) Advanced multilayer tubing to cut petrol emissions. Brit Plast Rubber 2:25–29
14. Sakazaki K, Mori H, Iio S (2012) Fuel hose. United States Patent # 8329274
15. Shifman J (2007) Flexible hose having reduced fuel vapor permeability and method of manufacturing such hose. United States Patent # 7228877
16. Smith CW et al (2003) Low permeation nylon tube with aluminum barrier layer. United States Patent # 6652939
17. Spurgat JC (1995) Low permeation fuel hose. United States Patent # 5398729
18. Stevens RD (1994) Fluoroelastomer laminated. United States Patent # 5320888
19. Unal G (2014) A multilayer hose. World Patent # 2014002009 A3
20. Yoshikawa M, Hideyuki K, Fukuura Y, Naito K (1995) Low permeable rubber hose. United States Patent # 5476121

Chapter 10
Multi Layer Pipes

R. Rajasekar, K.S.K. Sasi Kumar and P. Sathish Kumar

10.1 Introduction

Multi-layer pipe was first developed around 20 years ago. Since its entry, it has been utilized in countless installations across the globe and has recognized an excellent track record of reliability. During this period, it was in relatively limited supply in the UK, and carried a high price tag due to its cutting-edge design and the complicated manufacturing processes involved. The Multi-layer Pipe System (MLP) is a progressive, very hard wearing, and easy to work with system. It is manufactured using a well-engineered combination of toughened Pex-b plastic and overlay welded aluminum. This composite of high quality materials proves that multi-layer pipe has same strength and stability to metal pipes such as copper; but also MLP are lightweight, easy to cut, shape and install like plastic pipes. It is a tried and tested product designed to give plumbers and installers the perfect balance of dependability and ease of installation. Homogeneous (one-layer) pipes are damaged by cracks induced at the free surface and propagating through the pipe wall. This type of pipe failure occurs often, especially in cases of below ground plastic pipes where the outside surface cracks are created due to scratches formed during pipe installation and mounting. Plastic pipes are then cracked in a brittle-like manner in many instances. Multi-layer (composite) pipes are preferred to overcome this type of pipe damage. Basically, multi-layer composite pipes comprise a main (functional) pipe and one or two (inner and outer) protective layers (Figs. 10.1 and 10.2). The purpose of multi-layer expansion is mainly to guard the main part of the pipes from peripheral damage made during their implanting and manipulation. To accomplish this, the

P. Sathish Kumar
Department of Mining Engineering, Indian Institute of Technology Kharagpur, Kharagpur 721 302, West Bengal, India

R. Rajasekar (✉) · K.S.K. Sasi Kumar
Department of Mechanical Engineering, Kongu Engineering College, Erode, Tamil Nadu, India
e-mail: rajasekar.cr@gmail.com; rajamech@kongu.ac.in

J.K. Kim et al. (eds.), *Multicomponent Polymeric Materials*,
Springer Series in Materials Science 223, DOI 10.1007/978-94-017-7324-9_10

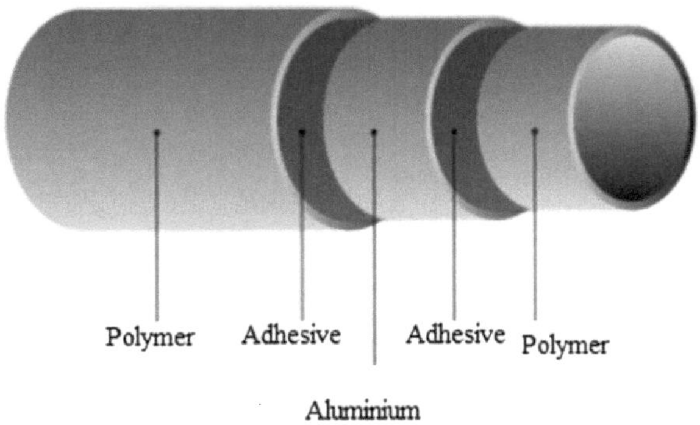

Fig. 10.1 Layers of multi-layer piping system

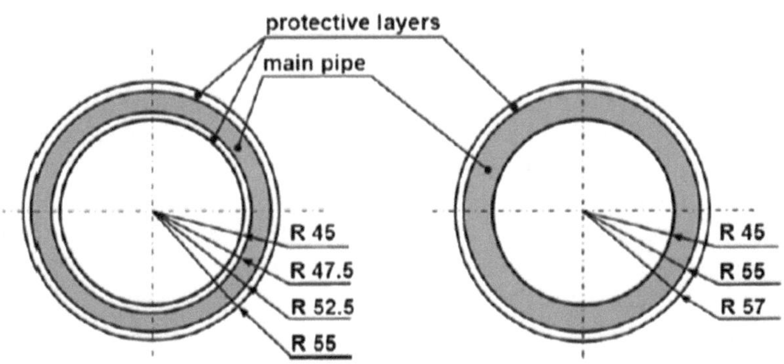

Fig. 10.2 Geometry of two layer and three layer pipe system [5]

material for shielding layers should be selected properly; in particular, it has to have greater resistance against surface scrabbling.

As far as polymeric materials selection is concerned, an amalgamation of HDPE (high-density polyethylene) is mostly used, for example; PE 100 with PE-Xd (peroxide cross-linked), resp. PE-Xb (silane crosslinked) and PP (block and random copolymers) [1]. The life period of a multi-layer pipeline system is usually restricted by slow crack growth (SCG) [2, 3] and rapid crack growth (RCP) [4] states, but there are other situations e.g. elastic failure and loss of adhesive strength that contribute to this limited lifetime.

10.2 Materials of Multi-layer Pipes

Depending on the application area and the respective requirement profile, the following polymer material may be selected.

PE-Xc (physically cross-linked polyethylene) based on PE-HD

PE-Xc based on PE-MD
PE-HD (high density polyethylene)
PE-MD (medium density polyethylene)
PE-RT (polyethylene with raised temperature resistance)

10.3 Production of Multi-layer Pipes

Two technology processes were available for production of PEX piping by moisture induced silane crosslinking (PEX-b) for nearly three decades. The first technology invented and patented by Dow Corning was the Sioplas™process. This process involves extruding pipe from a reactive compound composed of silane groups which have been attached onto polyethylene (PE) polymer chains by the adding of organic peroxide in an offline compounding process. The second technology was developed by Maillefer as a one-step-process where all of the components of PE base resin, additives, peroxide and silane are grafted in a dedicated compounding extruder which also extrudes the pipe in-line. Later the third method, the Vinyl Silane Copolymer or Visico™ process, was invented and patented by Borealis A.S. This process copolymerizes the silane directly into the PE polymer chain during the polymerization process. This guarantees an excellent homogeneous distribution of the reactive silane sites and evades the heavy use of organic peroxides required by the other PEX-b technologies to graft the silane component. The resulting product offers greater process stability and higher purity for superior taste and odor performance in PEX piping. All of these technologies need post treating the pipe with steam or hot water to persuade the crosslinking reaction and produce PEX-b pipe although the new Borealis Ambicat™ catalyst system for MDPE PEX-b radiant heating pipe allows crosslinking to take place during pipe storage at ambient temperatures by absorbing moisture from the surrounding air in the warehouse or yard.

10.4 Silane Crosslinking Technology

Wire & Cable industry invented in the PEX-b method of producing PEX piping, often referred to as silane crosslinking technology or the moisture crosslinking process, as a means of insulating metal conductors. Moisture-induced silane crosslinking technology was considered to substitute PE compounds that were waterlogged in peroxide and then crosslinked by extrusion through a steam or molten salt vulcanization tube. This vulcanization process is not entirely unlike some current PEX-a process technologies used to produce PEX tubing.

A standard thermoplastic pipe extrusion line combined with a post-extrusion hot water treatment apparatus to flush and crosslink the tubing allowed a pipe manufacturer to enter the fast growing PEX market using Sioplas™ process technology. Invented and patented by Dow Corning in 1968, the Sioplas process extrudes tubing pipe from a reactive compound composed of silane groups which have been grafted onto PE polymer chains [5]. This is accomplished by a custom compounder who grafts the silane onto an HDPE by adding organic peroxide in a high temperature compounding, extrusion and pelletizing process. This reactive compound, when combined with a tin catalyst master batch and extruded into tubing by the pipe manufacturer, the crosslinking in PEX tubing occurs after being flushed with hot water. An alternate technology for PEX-b tubing production is referred to as the Monosil™ process and was developed jointly by Maillefer and BICC in 1974 as a one-step-process, where all of the components of PE base resin, additives, peroxide and silane are grafted in a specialized compounding extruder which also extrudes the pipe in-line Ribarits [6]. Like the Sioplastechnology, the Monosil technology also needs a post-extrusion treatment that flushes hot water through the pipe to initiate the crosslinking reaction. The key to this moisture-induced crosslinking is the silane molecule that has been grafted onto the PE polymer chain.

PEX-a and PEX-c technologies forge carbon to carbon bonds during their crosslinking process, but PEX-b crosslinking links PE polymer chains together by forming a silane bridge. Grafting seats a trimethoxysilane side group [-SI-(0CH3)3] onto the PE chain, during crosslinking first experiences a hydrolysis reaction in which the three trimethyl groups are smote and transformed into methanol (3 CH3OH). Two PE chains containing the silane hydroxyl [-Si–$(OH)_3$] pendants next form the crosslink site by undergoing a condensation reaction [7].

10.5 Failure Analysis of Multi-layer Pipes

The temporal sequence of the failure of composite pipes may be categorized into four ways [8]:

1. failure of internal layers
2. failure of external layers
3. simultaneous failure of external and internal layers
4. failure of the whole pipe system.

As far as mechanical modes of fracture of pipes are concerned, there exists two ways of diminishing the probability of failure. The first includes a suitable choice of materials with greater resistance to slow crack initiation and propagation, since, from a practical point of view, the most significant damage to polymer pipes is determined by slow crack growth [9, 10].

In the case of a normal mode of loading, the relevant controlling fracture mechanics parameter is stress intensity factor K_I. A crack initiates at a critical point and its stable (slow) crack growth is controlled by the equation [9, 11]

$$\frac{da}{dt} = AK_I^m$$

The crack becomes unstable when certain conditions are reached

$$K_I(a_{crit}) = K_{IC}$$

Here A and m are corresponding material parameters, da/dt is crack propagation rate, a is a characteristic crack dimension, a_{crit} its critical value, and K_{IC} fracture toughness.

With the data taken from the literature, the majority of problems of homogeneous pipes can easily be solved. A two dimensional approximation is sufficient in most cases of the testing specimens [13]. The values of T-stress characterizing constraint level can be also seen in the literature [13]. Because of a relatively simple pipe geometry, (characterized by a cylinder with given ratio of inner and outer diameters) corresponding K-calibration curves for cracked pipes have been reliably set for both two and three-dimensional configurations. An accurate estimation of the stress intensity factor in the case of pressure pipe taking into account, change of the axial crack shape during propagation was presented in paper [12].

With respect to multilayer pipes, the situation is less straightforward and the fracture mechanics approach is cumbersome due to the existence of interfaces between single layers, where material parameters are altered by a step, see [14, 15]. Moreover, the corresponding data have not been reliably recognized in the literature. The K_I and T values relies on the material combination and on the full geometric specification of a multilayer pipe (i.e. on inner and outer radii and thicknesses of all layers). This rises the number of possible variations of multilayer pipe structures. Because of the production technologies and applicability, most specific multilayer pipe constructions utilized in practice are similar to each other, with relatively small deviations in geometric and material parameters. This fact enables generalization of the results gained from a specific multilayer structure.

Hutar et al. [12] considered a three layer plastic pipe representing a category of multilayer pipes currently used for water and gas supply. The pipe consisted of a main (functional) pipe and two (inner and outer) protective layers, see Fig. 10.3. The basic material for the pipes studied is a high density polyethylene composed of carbon and hydrogen atoms joined together forming high molecular weight products. The property characteristics of HDPE depend upon the arrangement of the molecular chains. Due to good adhesion between layers (these polyethylene pipes are produced by co-extrusion technology), the assumption of perfect adhesion is acknowledged. The existence of the interface complicates the calculations and has a negative influence on the results' accuracy. All calculations correspond to the assumptions of linear elastic fracture mechanics. In all cases, it is presumed that the cracked pipe is loaded by inner pressure P and the crack is subjected to a normal mode of loading, i.e. $K = K_I$.

Figure 10.4 shows the calculated values of the stress intensity factor K and T—stress are presented for a three-layer pipe with various crack-pipe configurations. It is supposed that the pipe is loaded by inner pressure P and cracks can be initiated either at a free surface of the inner protective layer or at an outer free surface of the three layer pipe (external crack). In all cases, the values of K and T are also

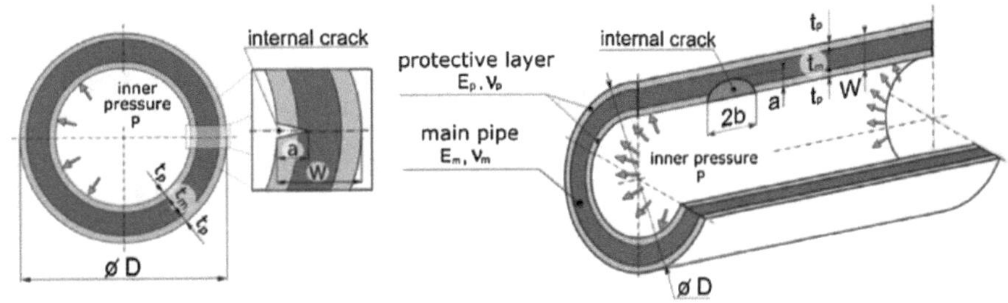

Fig. 10.3 Two-dimensional (2D) and three-dimensional (3D) model of a pipe with an internal crack [12]

Fig. 10.4 Normalized values of the stress intensity factor $K_I/P\sqrt{\pi a}$ for the studied composite pipe with the internal crack. P is the internal pressure, a crack length (the depth in the 3D case) and W is the pipe thickness. The K values in 3D case correspond to the deepest point of the crack. The value a/W = 0.25 represents the interface between inner protective layer and the main pipe [12]

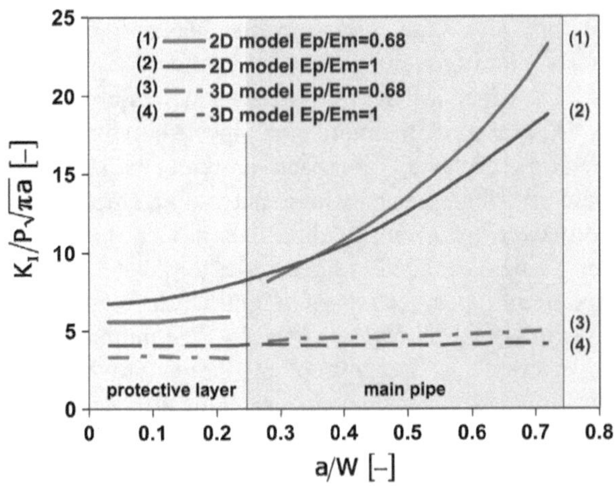

calculated for an equivalent homogeneous pipe. The geometry of the pipe resembles to the three layer commercial plastic pipe such as Wa-win TS with outer and inner layer from extremely durable PE material XSC 50 and with main pipe from PE 100. The geometry of the cracked pipe corresponds to those shown in Fig. 10.2. The values of K and T are presented as a function of normalized crack size a/W, where a is a crack length (or depth). Note that a/W = 0.25 (0.75) corresponds to the interface between inner (outer) layer and the main pipe.

10.6 Lifetime Estimation

In a composite pipe, a crack in the inner protective layer initially propagates in all directions. As soon as the crack front approaches the interface, it can stop or pierce into the main pipe. Because of the non-homogeneous structure of the pipe, the resulting values of stress intensity factor are influenced by an elastic mismatch of both materials used and can be articulated by the ratio of Young's modulus E_p/E_m.

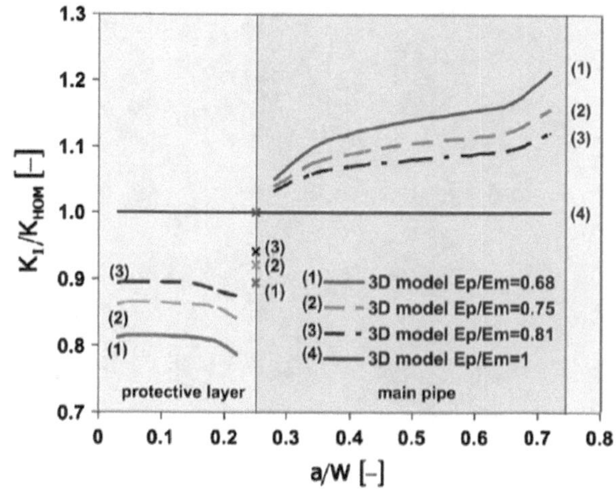

Fig. 10.5 Details of the dependence of stress intensity factor for a semi-elliptical inner crack in the three layer pipe considering the step change at the interface between the inner layer and the main pipe [12]

Figure 10.5 shows the differences between normalised values of the stress intensity factor for the homogeneous and the composite three-layer pipe are less than 20% for the currently used composite pipe materials. For cracks growing in the protective layer, the K_I values are smaller than those corresponding to the homogeneous pipe; see Fig. 10.5. This is associated with a lower degree of stiffness in the protective layer in comparison with the main pipe. Theoretically, the values of the stress intensity factor of the crack tip approaching the interface incline towards zero for cracks which reach the interface [16].

10.7 Fittings for MLP

Figure 10.6 shows the several fittings used in multi-layer pipe systems.

10.8 Environmental Impact of Multi-layer Pipes

The environmental impact of each pipe material is normally assessed against six different criteria across its full life cycle (Figs. 10.7 and 10.8) and its comparison to copper (Fig. 10.8) were also discussed.

'Abiotic-depletion' potential: the over-extraction of minerals, fossil fuels and other non-living, non-renewable materials which can lead to exhaustion of natural resources.

'Acidification' potential: emissions, such as sulphur dioxide and nitrogen oxides from manufacturing processes, result in acid rain which harms soil, water supplies, human and animal organisms, and the ecosystem.

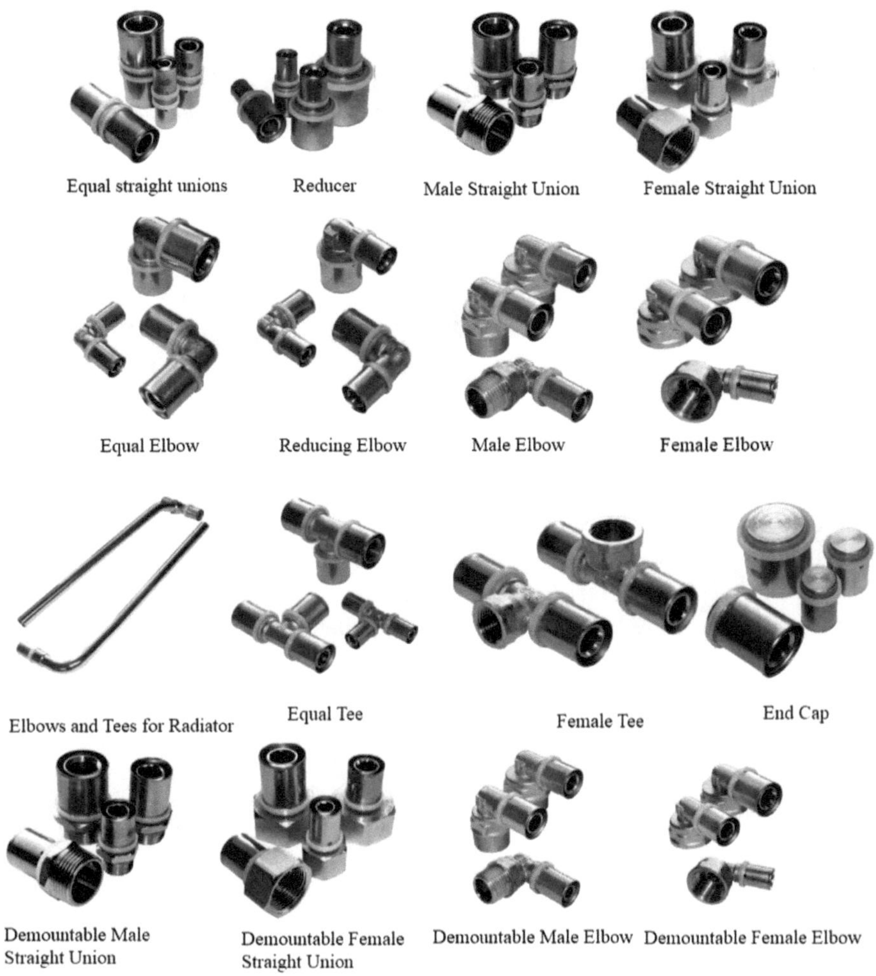

Fig. 10.6 Various fittings for multi-layer pipes (Courtesy RIFENG)

'Eutrophication' potential: which arises from the over-fertilisation of water and soil by nutrients (such as nitrogen and phosphorous). This speeds up plant growth and kills off animal life in lakes and waterways.

'Global warming' potential (its carbon footprint): the insulating effect of greenhouse gases—CO_2 and methane—in the atmosphere is a major contributor to global warming, affecting both human health and that of the ecosystem in which we live.

'Ozone-depletion' potential: depletion of the ozone layer in the atmosphere caused by the emission of chemical foaming and cleaning agents allows the passage of greater levels of UV from the sun, causing skin cancer and reducing crop yields.

IMPACT CATEGORY	Abiotic depletion	Acidification	Eutrophication	Global warming	Ozone layer depletion	Photochemical oxidation
Life cycle phases	kg Sb eq	kg SO2 eq	kg PO4--- eq	kg CO2 eq	kg CFC-11 eq	kg C2H4 eq
PRODUCT STAGE						
Production raw materials for polymer ML pipes	0,00401	0,00079	0,00007	0,23096	2,45E-10	0,000076
Transport of raw materials for polymer ML pipe to converter	0,00004	0,00002	0,00001	0,00589	9,69E-10	0,000001
Production of aluminium layer for the polymer ML pipe	0,00071	0,00057	0,00021	0,11981	7,30E-09	0,000047
Transport of aluminium layer to converter	0,00001	0,000005	0,000001	0,00124	2,04E-10	0,00000
Extrusion polymer ML pipes	0,00076	0,00038	0,00024	0,09853	5,84E-09	0,000016
Production of PPSU fittings	0,00103	0,00041	0,00040	0,10530	4,91E-08	0,000116
Production of brass fittings	0,00030	0,00148	0,00160	0,04336	2,90E-09	0,000056
Production of metal compression rings	0,00016	0,00022	0,00001	0,04074	0,00E+00	0,000010
CONSTRUCTION PROCESS STAGE						
Transport of complete polymer ML pipe system to the building site (apartment)	0,00023	0,00012	0,00003	0,03282	4,92E-09	0,000005
Installation of polymer ML pipe system (in apartment)	0,00066	0,00031	0,00018	0,09950	3,94E-09	0,000036
USE STAGE						
Operational use of polymer ML pipe system	0	0	0	0	0	0
Maintenance of polymer ML pipe system	0	0	0	0	0	0
END OF LIFE STAGE						
Transport of polymer ML pipe system to EoL (after 50 years of service life time)	0,00004	0,00002	0,00001	0,00630	9,55E-10	0,0000008
EoL of PEX pipe system (after 50 years of service life time)	-0,00026	-0,00012	-0,000099	0,05252	-1,38E-09	-0,0000062
Total	0,00768	0,00421	0,00265	0,83697	0,0000000750	0,000358
A: contribution > 50 %: most important, significant influence						
B: 25 % < contribution ≤ 50 %: very important, relevant influence						

Fig. 10.7 Environmental profile of the Polymer/Al/Polymer composite pipe system for hot and cold water (cradle-to-grave) in absolute figures per functional unit (www.teppfa.eu)

'Photochemical-oxidation' potential: where the photochemical reaction of sunlight with primary air pollutants such as volatile organic compounds and nitrogen oxides leads to chemical smogs that affect human health, food crops and the ecosystem in general.

10.9 Application of Multi-layer Pipes

Multi-layer pipes find its application in vast areas such as water connection, concrete core activation, compressed air installation and underfloor heating (Fig. 10.9).

10.10 Coating for Corrosion Resistance

Transporting gases and liquids over long distances from their sources to the ultimate consumers requires pipelines. Oil and gas industries faces the corrosion of pipelines' coatings as a main problems for which a large amount of money is spent

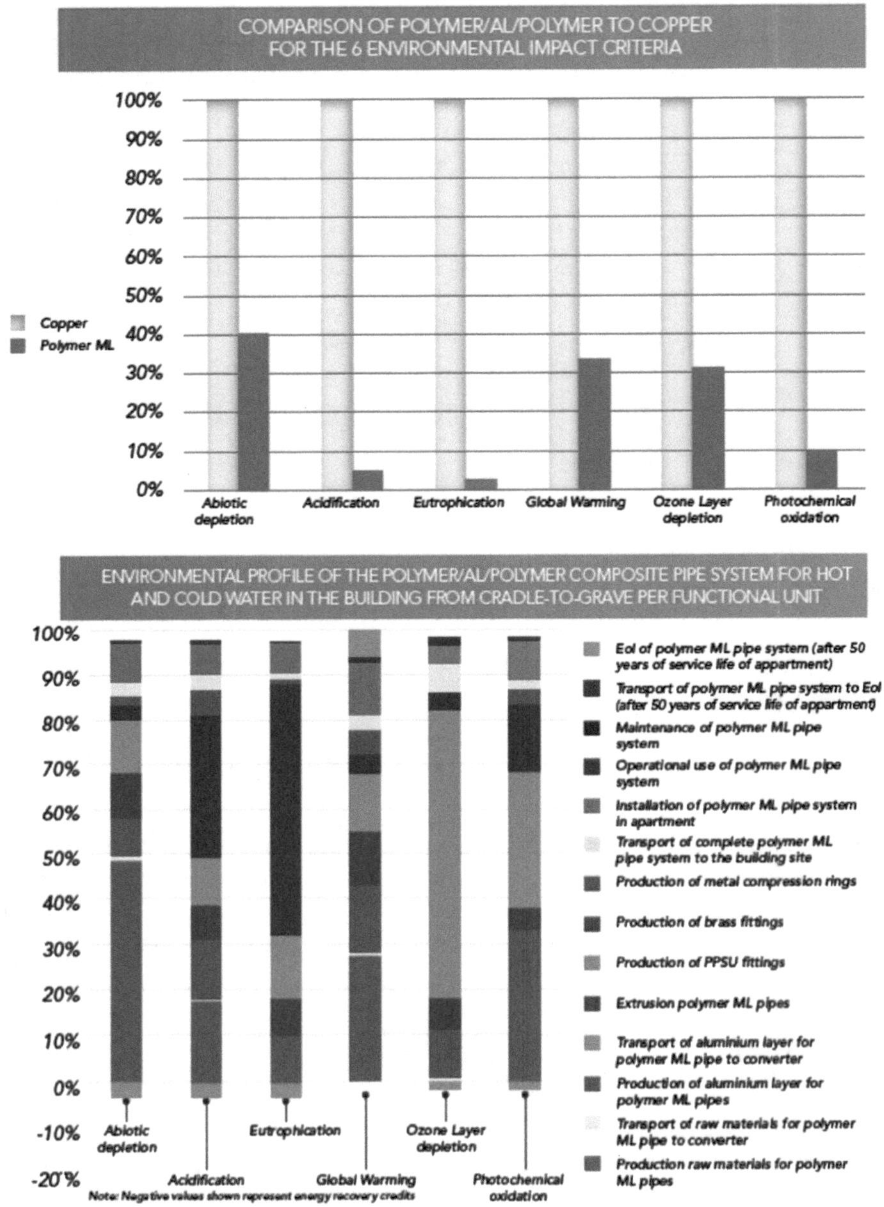

Fig. 10.8 Environmental profile of the polymer (www.teppfa.eu)

each year. Cessation of production makes a very huge loss in terms of hydrocarbon production or maintenance costs. The function of coatings is to control corrosion by isolating the external surface of the underground or submerged piping from the environment, to decrease corrosion protective requirements, and to improve

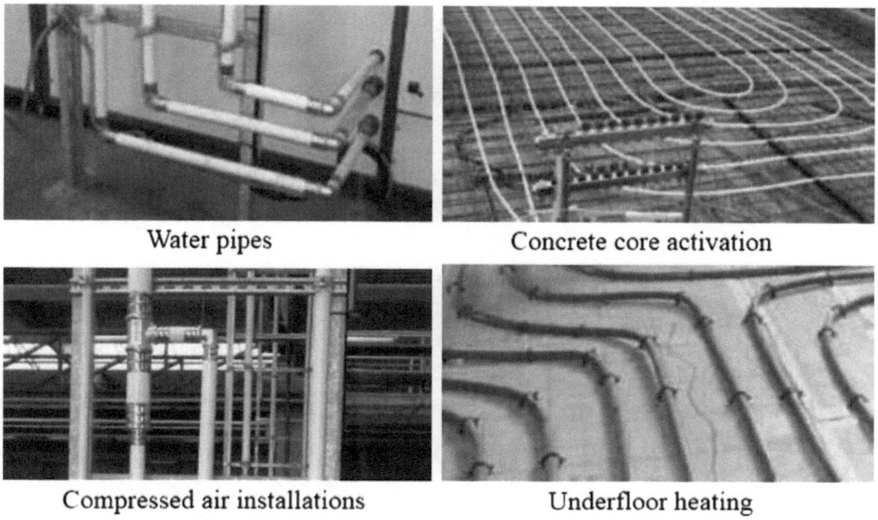

Fig. 10.9 Typical applications of multi-layer pipes

(protective) current distribution. Coatings should be properly selected and applied, and the coated piping should be carefully commissioned to fulfill these functions.

The desired characteristics of the coatings include:

- Effective electrical insulation
- Good adhesion to the pipe surface
- Ability to fight the development of holidays with time
- Ability to battle damage during handling, storage, and installation
- Ability to keep substantially constant resistivity with time
- Resistance to disbonding
- Resistance to chemical degradation
- Retention of physical characteristics
- Resistance to changes and deterioration during above-ground storage and long-distance transportation.

Common coatings are

- Bituminous enamels
  These are framed from coal-tar pitches or petroleum asphalts and have been widely utilized as protective coatings for many years. Coal-tar and asphalt enamels are offered in summer or winter grades. These enamels are the corrosion coating; they are mixed with various combinations of fiberglass and/or felt to obtain mechanical strength for handling.
- Asphalt mastic
  These are pipe coating with dense mixture of sand, crushed limestone, and fiber bound together with a select air-blown asphalt.
- Liquid Epoxies and Phenolics

These systems are primarily used on larger-diameter pipe when conventional systems may not be available or when they may offer better resistance to operation temperatures in the 95 °C range.

- Extruded plastic coatings
  These fall into two categories based on the method of extrusion, with additional variations resulting from the selection of adhesive. The first method of extrusion are the crosshead or circular die, and second is the side extrusion or T-shaped die. The four types of adhesives are asphalt-rubber blend, polyethylene copolymer, butyl rubber adhesive, and polyolefin rubber blend.
- Fusion-bonded epoxy (FBE)
  FBE coatings are heat-activated, chemically cured coating systems. The epoxy coating is supplied in powdered form and applied to preheated pipe, special sections, connections, and fittings using fluid-bed, air spray, or electrostatic spray methods.
- Tape
  For normal construction conditions, prefabricated cold-applied tapes are applied as a three-layer system consisting of a primer, corrosion-preventive tape (inner layer), and a mechanically protective tape (outer layer).
- Three-Layer Polyolefin
  These systems contain of an FEB primer, an intermediate copolymer layer, and a topcoat consisting of either polyethylene or polypropylene. The purpose of the intermediate copolymer is to bond the FBE primer with the polyolefin topcoat.
- Wax coatings
  Microcrystalline wax coatings are usually used as a protective overwrap. The wax helps to waterproof the pipe, and the wrapper defends the wax coating from contact with the soil and affords some mechanical protection.

10.10.1 Corrosion Control Methods

Corrosion in industries is controlled by one of the following methods.

1. Corrosion-resistant alloys
2. Corrosion inhibitors
3. Stabilization method

First layer
Every pipe one form of film of liquid or gum of epoxy is produced with Minimum dryer thickness between 20 and 60 micron. According to ISO 2808, epoxy powder has some materials that is used for three-layered poly ethylene coatings for steel pipes and must be specially formulated and designed (Fig. 10.10). Epoxy powders used in three-layered coatings can be classified in two different groups. The first group posses primer property and the second group possess coating quality. These two materials differs majorly in applying, temperature and thickness; there is a trend

Fig. 10.10 View of pipe with three layer [17]

in industries to use epoxies with coating quality. Epoxy layer must have such an enough thickness that prevent holiday formation. Experiences and experiments done in the field shows 40 holiday in 40 ft for a layer with the thickness of 150 μm. Time plays an important role in creating adhesive and poly ethylene layers as sensitive and critical factor. First the adhesive creates a very strong chemical bond with chemical groups in epoxy powder which is uncure, therefore into this stage the epoxy must not completely cured.

The following is Dennis Neal's opinion about this matter obtained from experts:

(i) The FBE layer is under cured because the application temperature is low to allow the adhesive to chemically bond to FBE.
(ii) There is no adhesion between the FBE and adhesive because the temperature is higher and the FBE (fusion bonded epoxy) is fully cured before the adhesive is applied.

Second layer
Second layer polymer makes adhesiveness between layers 1 and 3 and must be well-matched with both layers. Minimum thickness should be between 160 and 200 μm. Thickness may increase or reduce according to the requirement of customer but minimum thickness must be examined safely.

Third layer
Polyethylene coating must be shaped in this layer. Thickness must be uniform in all through the pipe and minimum general thickness must be acceptable.

10.10.2 Erosion Resistance of Fusion Bonded Epoxy Coating

Most of the fluid transporting components such as hydraulic turbines, slurry pumps, pipeline and agitators are worked under the conditions where erosion–corrosion happens. The resistance of slurry erosion–corrosion is a dominant factor for the lifetime of the above components. Previous researches presented that the combined action of erosion–corrosion consequences in a synergistic reinforcement of their effects, and the erosion–corrosion rate is often greatly bigger than the sum of pure erosion rate and pure corrosion rate [18, 19]. Slurry erosion of the organic coating is mainly caused from the impingement of solid particles suspended in a carried corrosive fluid. Some researchers have considered the influence of environmental factors and the performance of reinforcement filler on erosion resistance of the polymer and the solvent-based organic coatings [20–23].

The fusion-bonded epoxy powder coating, developed by the Institute of Corrosion and Protection of Metals, Chinese Academy of Sciences, was coated on the pipeline steel substrate [24]. The coating contained of 40–60% epoxy resin, 40–60% filler, 1–5% curing agent and 1–5% couple agent. The test medium was a simplified industrial medium which contained of 1 wt% NaCl, 1–20 wt% emery sand (Hv, 2200–2300) and tap water. The emery sand contained of about 98.5% Al_2O_3 and 0.8% Na_2O. All tests were performed at ambient temperature.

Although the test medium contained 1 wt% NaCl, the coating degradation produced by corrosion was not noticed by electrochemical impedance spectroscopy and infrared optical spectroscopy. Those results showed that the effects of corrosion on erosion could be neglected, therefore the mass loss rate, as shown in Figs. 10.11, 10.12, 10.13 and 10.14, was caused by the mechanical impact of abrasive particle. The relationship between erosion time and mass loss rate is demonstrated in Fig. 10.11. Except particle content of 1 wt%, the mass loss rate declined with time and stabilized in 10–15 h. Figure 10.12 displays the variation of mass loss rate with flow velocity while the particle content were 10, 15, 20 wt%, respectively, and the abrasive particle size was a constant (112_m). It was obvious that l gE was proportional to l gV at different particle content with the linear correlated coefficient higher than 0.99 obtained by using the linear least-squared method. Figure 10.13 depicts the variation of the mass loss rate of the coating with the particle content within the flow velocity range of 6–10 m/s.

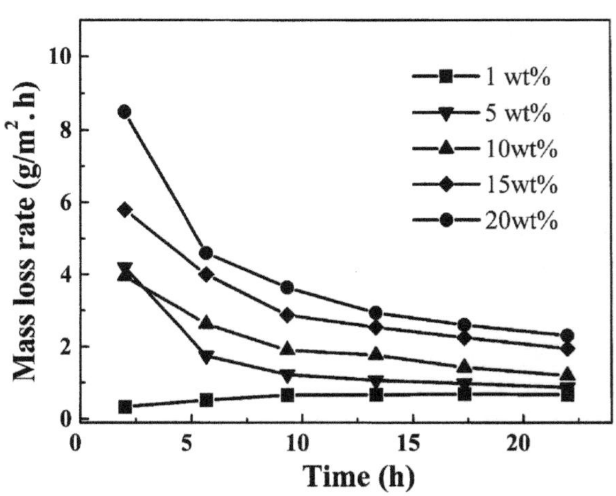

Fig. 10.11 Relationship between mass loss rate and erosion time [24]

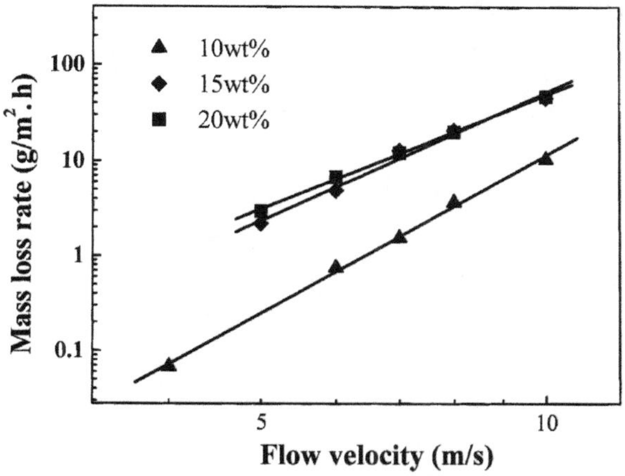

Fig. 10.12 Relationship between mass loss rate and flow velocity [24]

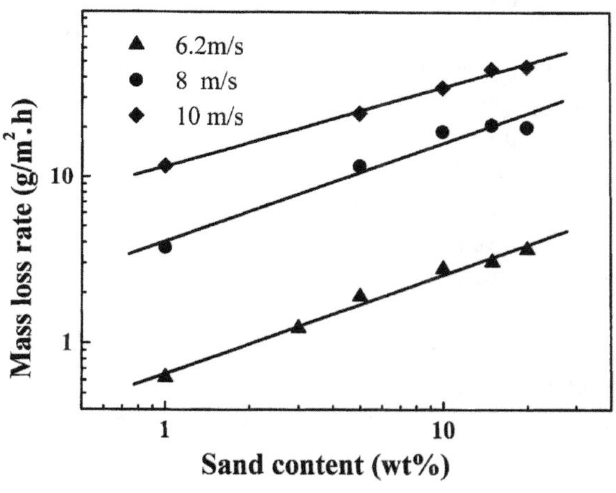

Fig. 10.13 Variation of mass loss rate with abrasive particle content [24]

The relationship between mass loss rate and particle size is exposed in Fig. 10.14. Obviously, there were two transition points between which the effect of particle size on mass loss rate was remarkable, in contrast, beyond which the effect was very small.

Furthermore, when the abrasive particle size was smaller than SCD, the mass loss rate also increased according to a power law with particle size (Fig. 10.15).

The relationship between mass loss rate and particle size could be described by the equation

$$E = K_3 D^{n3}$$

where D and n3 are the particle size and the exponent of particle size, respectively.

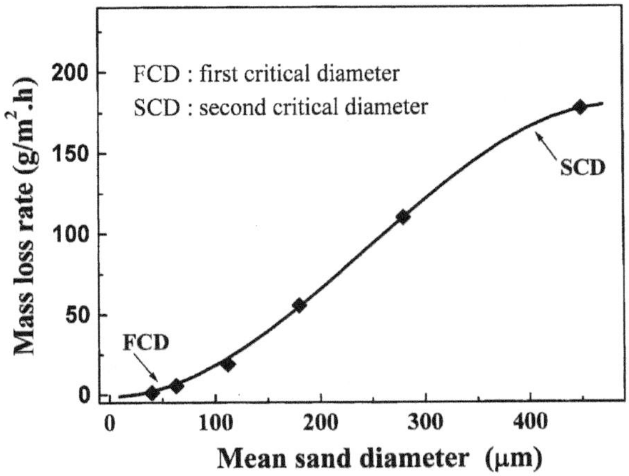

Fig. 10.14 Influence of mean diameter of emery particle on mass loss rate [24]

SEM images of the untested and eroded surface of the coatings under different test condition are offered in Fig. 10.16. The morphology of untested surface displays that there was a thin layer of resin matrix formed in the curing process and fillers in the coating did not expose after polished by the abrasive paper; only the scratched grooves could be witnessed. In Fig. 10.16b–d, the fillers protrude on the surfaces after the softer resin matrix was removed by the abrasive particles and the fillers were neither cracked nor detached from the resin matrix. There were similar morphologies among the specimens tested under the following conditions: flow velocity of 6.2 m/s and mean particle size of 112 μm within the sand content range of 1–20 wt%, sand content of 10 wt% and mean particle size of 112 μm within the flow velocity range of 6.2–10 m/s. In Fig. 10.16e and f, deeper impinging pits happened on the resin matrix and the fillers dropped from coating surface or were broken by the impact of abrasive particles. Figure 10.17 shows the SEM morphologies of the untested and eroded surface [25]. The morphology of untested

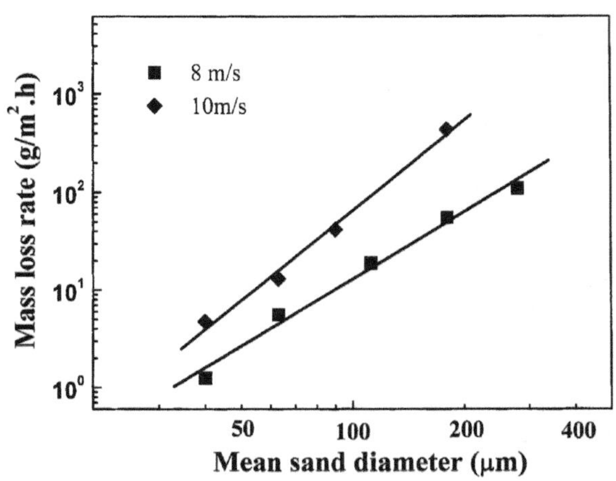

Fig. 10.15 Variation of mass loss rate with particle mean diameter [24]

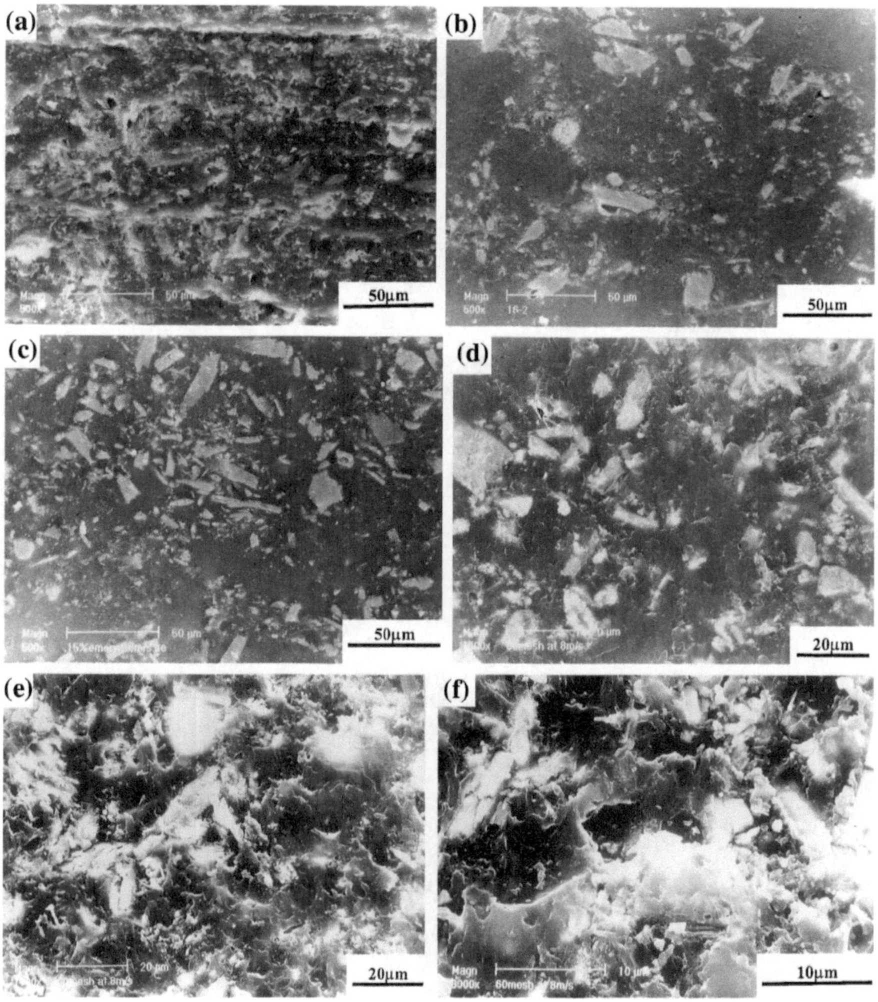

Fig. 10.16 Morphologies of the coating before and after erosion [25]

surface shows that there was a thin layer of resin matrix formed in the curing process. Fillers in the coating have not been exposed after being polished; only the scratched grooves could be observed. Figure 10.17b–e showed the eroded surface of the coatings with different curing degree after slurry erosion. A lot of protruding fillers could be observed on the surfaces after the softer resin matrix was removed by the eroding particles.

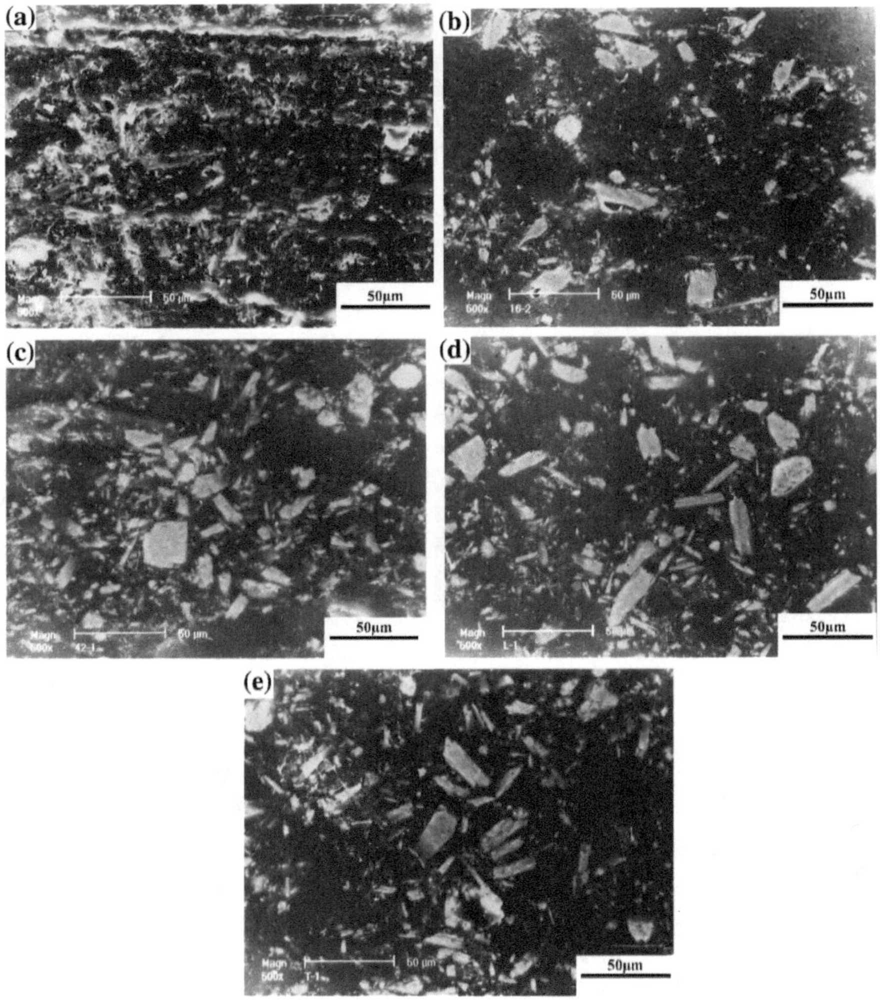

Fig. 10.17 Morphologies of coatings before and after erosion in a slurry containing 1 wt% emery particles (112 μm) at 6.2 m/s [25]

10.11 Analysis of Failure of Fusion Bonded Powder Epoxy Internal Coating

The failures of pipe line samples were examined both in the field and in the laboratory in order to draw reliable conclusions [26]. The complete coating delamination was observed in some pipelines in addition to curling and blistering in some other parts of the pipeline. The macroscopic examination was achieved on

Fig. 10.18 Backscatter images from SEM analysis

three epoxy samples using a stereomicroscope and with the naked eye. Before examination, the samples were cleaned for oil remnants in order to fully see the powder epoxy surfaces. The microscopic examination was completed on three thin fluorescent impregnated sections from three powder epoxy samples. Three slices of each sample were comprised in each thin section. Oil was cleaned off the epoxy samples before the preparation of the thin section.

The main goals of the metallographic analysis were to examine the sample for the following defects:

- Trapped vapor bubbles,
- brittleness and cracks,
- voids or inclusions, and
- Presence of corrosion products on the pipeline substrates.

Photo microscopic examinations of both samples contained orange coloured corrosion products as well as small, shiny yellowish crystals on the exterior face of the powder epoxy flakes (pipeline interface sample 2). Additional analysis such as SEM-EDX (Fig. 10.18) confirmed that this material was iron-hydroxide and iron.

10.11.1 Differential Scanning Calorimetry (DSC) Analyses

Differential scanning calorimetry (DSC) was carried out for the forwarded samples of powder epoxy, to identify the reason for the peeling off of the epoxy layer. A higher Tg of failed coating shows that it had cured and hardened further during the operation (Fig. 10.19) validating the earlier research results [27].

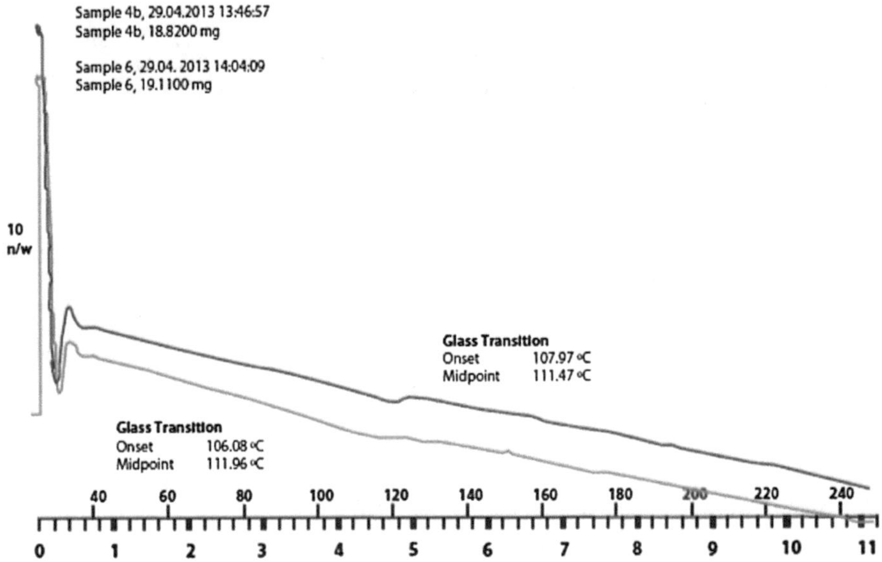

Fig. 10.19 DSC curves

10.12 Summary

This chapter discusses the multi-layer pipe materials, production methods and applications. It gives a glimpse of the fittings for the MLP. It also deals with development and analysis of coating performed on steel pipes. Finally, it provides the information about corrosion resistive coating on pipes used in commercial applications.

References

1. Janson LE (1999) Plastics pipes for water supply and sewage disposal. Borealis, Stockholm
2. Nezbedova E, Zahradnickova A, Salajka Z (2001) Brittle failure versus structure of HDPE pipe resins. J Macromol Sci Phys B 40:507
3. Lu X, Brown N (1992) A test for slow crack growth failure in polyethylene under a constant load. Polym Test 11:309–319
4. Šindelarˇ P, Nezbedova E, Šimkova P, Buranˇ Z, Bohaty′ P (2005) Effect of structural parameters on rapid crack propagation and slow crack growth in high density polyethylene pipeline materials. Plast Rubber Compos 34:329–333
5. Hutař P, Náhlík L, Šestáková L, Ševčík M, Knésl Z, Nezbedová E (2010) A fracture mechanics assessment of surface cracks existing in protective layers of multi-layer composite pipes. Compos Struct 92:1120–1125
6. Ribaritis E (1987) Visico crosslinkable polyethylene silane copolymer for cable insulation. Neste Polyeten AB
7. Plueddemann E (1982) Silane coupling agents. Plenum Press, pp 32–33

8. Farshad M (2005) Determination of the long-term hydrostatic strength of multilayer pipes. Polym Test 24:1041–1048
9. Frank A, Pinter G, Lang RW (2009) Prediction of the remaining lifetime of polyethylene pipes after up to 30 years in use. Polym Test 28:737–745
10. Andena L, Rink M, Frassine R, Corrieri R (2009) A fracture mechanics approach for the prediction of the failure time of polybutene pipes. Eng Fract Mech 18:2666–2677
11. Frank A, Freimann W, Pinter G, Lang RW (2009) A fracture mechanics concept for the accelerated characterization of creep crack growth in PE-HD pipe grades. Eng Fract Mech 76 (18):2780–2787
12. Hutar P, Zouhar M, Nahlik L, Ševčík M, Mašab B (2013) Multilayer polymer pipes failure assessment based on a fracture mechanics approach. Eng Fail Anal 33:151–162
13. Anderson TL (1991) Fracture mechanics—fundamentals and application. CRC Press Inc., Boca Raton
14. Hutař P, Nahlik L, Šestáková L, Ševčík M, Knesl Z, Nezbedova E (2010) A fracture mechanics assessment of surface cracks existing in protective layers of multilayer composite pipes. Compos Struct 92:1120–1125
15. Zouhar M, Vallet L, Hutař P, Nahlik L (2011) Life time estimation of the multilayer plastic pipes. Key Eng Mater 452–453:33–36
16. Romeo A, Ballarini R (1995) A crack very close to a bi-material interface. Trans ASME 32:614–619
17. Samimi A, Zarinabadi S (2011) An analysis of polyethylene coating corrosion in oil and gas pipelines. J Am Sci 7(1)
18. Wang P, Yang Y, Ding G, Qu J, Heshen S (1997) Laser adding coating against wear and its application to mining machine parts. Wear 209:96
19. Zheng Y, Yao Z, Wei X, Ke W (1995) The synergistic effects between erosion and corrosion in acidic slurry medium. Wear 186/187:555
20. Ma X, Lin F, Shao H (1991) Study on the wear-resistant adhesive coating in scouring. Lubr Eng. (4):11 (in Chinese)
21. Iwai Y, Nambu K (1997) Slurry wear properties of pump lining materials. Wear 210:211
22. Zhang JX, Richardson MOW, Wilcox GD et al (1996) Assessment of resistance of nonmetallic coatings to silt abrasion and cavitation erosion in a rotating disk test rig. Wear 194:149
23. Parslow GI, Stephenson DJ, Strutt JE et al (1997) Paint layer erosion resistance behavior for use in a multilayer paint erosion indication technique. Wear 212:103
24. Luo S, Zheng Y, Li J, Ke W (2001) Slurry erosion resistance of fusion-bonded epoxy powder coating. Wear 249:733–738
25. Luo S, Zheng Y, Li J, Ke W (2003) Effect of curing degree and fillers on slurry erosion behavior of fusion-bonded epoxy powder coatings. Wear 254:292–297
26. Husain A, Al-Bahar S, Chakkamalayath J, Vikraman A, Al Ghamdi A, Kamshad T, Siriki RS (2015) Differential scanning calorimetry and optical photo microscopy examination for the analysis of failure of fusion bonded powder epoxy internal coating. Eng Failure Anal 56:375–383
27. Weldon DG (2005) Failure analysis and degree of cure. J Protective Coat Linings (July)48–55

Chapter 11
Multilayer (Fuel) Storage Tank

R. Rajasekar, K.V. Mahesh Kumar, K. Krishnamurthy and P. Sathish Kumar

11.1 Introduction

In today's environment the storage of fuels in onboard vehicle, underground and above ground storage are of great challenge due to its surface temperature, evaporation loss and corrosiveness. In summer, due to solar radiation, which has an intensity of 1.373 kW/m^2 at the earth's atmosphere surface, causes the surface temperature of oil storage tanks to rise. Due to this temperature rise the light hydrocarbon components of oil vaporize in the space between the oil and the tank roof, called as evaporation loss. Oil evaporation loss which pollutes the environment and also the quality of the oil is reduced. It is generally known that the evaporation loss of oil storage tanks depends significantly on solar radiation and local climate conditions. In Thailand, the total evaporation loss of gasoline from storage sites and service stations were estimated to be 21,000 tons/year. In the Shanghai area of China, the evaporation loss of gasoline from a 5000 m^3 oil storage tank in the summer is 350 kg/day and diesel oil from a 5000 m^3 oil storage tank in the summer is 20 kg/day respectively, and the evaporation loss of other oil products is somewhere between these two figures. During the three summer months, the value of the annual evaporation loss from 40 crude oil storage tanks with a capacity of 1 million barrels in Khark Island, Iran, with the presence of coating (absorptivity $\alpha = 0.3$) is 8000 barrels. Now it is important to consider about reducing the evaporation loss of oil storage tanks is of particular significance for nonrenewable energy conservation and environmental protection [1].

R. Rajasekar (✉) · K.V. Mahesh Kumar · K. Krishnamurthy
School of Building and Mechanical Sciences, Kongu Engineering College, Erode, Tamilnadu, India
e-mail: rajasekae.cr@gmail.com; rajamech@kmgu.ac.in

P. Sathish kumar (✉)
Department of Mining Engineering, Indian Institute of Technology Kharagpur, Kharagpur, India

J.K. Kim et al. (eds.), *Multicomponent Polymeric Materials*,
Springer Series in Materials Science 223, DOI 10.1007/978-94-017-7324-9_11

A conventional method that is employed in reducing the oil evaporation loss is to cool storage tanks with water spray. For that a water spray pump (power: 37 kW; flow rate: 160 m^3/h) was needed to keep a 2000 m^3 oil storage tank cool in a hot climate. Assuming that if the pump works 6 h per day for 5 months, the average annual electricity and water consumption for an oil storage tank is 33,300 kWh and 14,400 m^3, respectively. It results that cooling oil storage tanks with a water spray result in a waste of water and electricity. In addition, this cooling method aggravates the corrosion of the tanks and significantly deteriorates the environment. An alternative method to reduce oil evaporation loss with a better cooling effect and fewer side effects is to paint storage tanks with cool coatings. There was only limited studies based on the coatings, also it focus mainly on the effects of the outer surface paint color of the oil storage tanks on oil evaporation loss. The evaporation loss of crude oil was found to increase by approximately 250–300% in summer months as the absorptivity of the outer surface paint increases from 0.1 to 0.9. Thus, oil evaporation loss may be considerably reduced simply by using cool coatings with high solar reflectance.

11.1.1 Need for Multilayer Storage Tanks

According to a study conducted in US, the cost estimate for the corrosion of the storage tanks are around 7 billion US dollars yearly with an approximate estimation of 4.5 billion dollar for aboveground storage tanks and 2.5 billion dollar for underground storage tanks [2]. Corrosion is the major parameter that accounts for about 65% the failure of these storage tanks, while the other 35% may be due to other reasons. As per the study the majority of UST fails due to the external corrosion as it is exposed to soils, while a minimum quantity is responsible for internal corrosion. This failure may also threaten to human life, sue to spills, overfill or leakage in tanks and pipes. Later the corrosion has been overcome by coating the tanks with external corrosion protection systems to underground tanks for flammable and combustible liquids, resulted in almost zero failures due to this coating.

11.1.2 Overview of Multilayer Assembly

The layer-by-layer (LbL) deposition technique which comes under the category of pattern assisted assembly. Here the pattern assisted assembly is much less time consuming process than that of self assembly/modification cycles whose outcome is very hard to predict. For LbL—deposition of materials, it can be customized to allow multimaterial assembly of numerous compounds without any chemical modifications, which provide access to multi layer films whose functionality falls into the following two categories, one is of surface interactions [3], where every object interacts with environment through its. Thus all surface interaction properties depends on its function such as Corrosion protection, antireflective coatings, stickness or non-stickness, antistatic coatings, surface induced nucleation, antifouling, biocompatibility, hydrophilicity or hydrophobicity, antibacterial properties, molecular recognition, microchannel flow control, chemical sensing or biosensing, etc., and the later one is the fabrication of surface based devices which is the sequence of deposition of different materials that defines the multilayer architecture and thus plays a role in the device properties.

However the multilayer materials act as challenging barrier for the permeation of fuels and gases. This Layer-by-layer (LbL) deposition is a technique which is used for precisely depositing nanostructured thin films that can exhibit exceptional properties. Alternately depositing materials with complementary functionalities, such as opposite electrostatic charges, allows for a layer by layer thin-film growth. Many types of materials can be used to construct these films, including polymers, nanosheets, nanotubes, quantum dots and biological molecules [4]. There are also various parameters which affects the interactions between these components such as molecular weight, pH, ionic strength, concentration and deposition time, which alter the thickness, morphology and properties of the multilayer thin film based on the above said parameters. These LbL developed material may be applied for delivering drugs, impart electrical conductivity, reduce gas permeability which is of great importance for the storage applications, resist abrasion, kill bacteria, separate gases and stop fire [5].

With the reduction of fossil fuels around the world and increasing pollution on the environment, hydrogen has emerged as a most important alternative fuel for use in transportation. Hydrogen can be stored in high pressure storage vessels, liquid hydrogen tanks, metal hydrides, chemical hydrogen storage. Hydrogen which may be an alternative to gasoline that offers many advantages such as no greenhouse gas emission. Although Hydrogen has many advantages there are some barriers which must be overcome before industrial commercialization. The storage tank for hydrogen should be of very important, i.e., they have to carry sufficient hydrogen on-board and to meet consumer requirements like cost and performance (refuelling time) without compromising passenger space. To meet industry's specifications, such as volumetric hydrogen density, internal pressure of this kind of tank has to be increased up to 700 bars.

11.2 Techniques Involved in Multilayer Coating

Though the LbL—deposition of materials can be customized to allow multimaterial assembly of numerous compounds without any chemical modifications, which provide access to multi layer films, the manufacturing of multilayer material is of more important in the research point of view and also in the industrial environment [6, 7]. In considering all these parameters many researchers and commercial peoples, they followed mainly three type of coating techniques for the development of multilayer materials, viz. spin coating, dip coating and spray coating. Also there are other techniques followed for the progress of generating multilayer materials, they are vapour deposition technique (Physical and Chemical) and using plasma coating technique [8]. All these techniques are explained in this chapter.

11.2.1 Spin Coating

Spin coating is one of the most common and important technique suggested for producing thin films of organic materials in uniform manner with thickness in the order of micrometers and nanometers. It is one of the prominent techniques employed for the production of layer-by-layer material in many applications [9]. The pioneering analysis of spin coating was performed more than fifty years ago where the coating of a thin axisymmetric film of Newtonian fluid on a substrate rotating with constant angular velocity [10]. Spin coating was first studied for coating of paints.

The process of spin coating can be divided into four stages as shown in Fig. 11.1, which is in the order as deposition, spin-up, spin-off and evaporation of solvents. The first three steps are mostly sequential, but spin-off and evaporation usually overlaps during coating. However flow controlled spin-off and evaporation control are the two stages that have the most impact on final coating thickness.

Spin coating can be carried out for producing minimum thickness of about less than 10 nm and a maximum of about 10 μm. Benefits of spin coating include fast process time (only a few seconds per coating) and high uniformity over the curved parts surface. Lenses with varying curvatures may be coated evenly and uniformly with minimal thickness variation or edge effects. However, only one part can be performed at a time in spin coating. Spin-coating's speed and low equipment cost may be most suited for developing multilayer materials and also used in small optical labs.

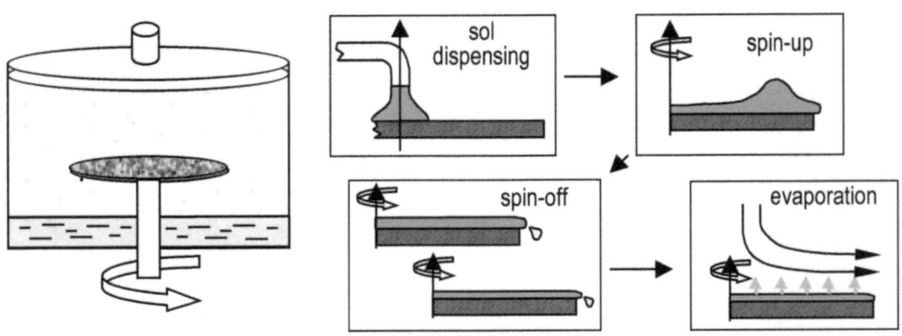

Fig. 11.1 Stages of spin coating process (*Source* http://homepages.rpi.edu/~plawsky/Research/Processing.htm)

11.2.2 Dip Coating

Dip coating refers to the immersing of a substrate into a tank that contains coating material and removing the piece from the tank, after that it is allowing it to drain. Then the coated piece can then be dried by force-drying or baking [11].

The dip coating process is carried out in five different stages viz,

1. Immersion: Here the substrate is immersed in the solution of the required coating material at a constant speed.
2. Start-up: The substrate has is allowed to remain inside the solution for a while and it is pulled out.
3. Deposition: The thin layer deposits on the substrate while it is pulled up. The substrate is withdrawn a constant speed to avoid any jitters. Here the speed determines the thickness of the coating when removed fast coating will be thick and vice versa.
4. Drainage: Excess liquid will drain from the surface.
5. Evaporation: The solvent evaporates from the liquid, which leads to formation of thin layer. For volatile solvents, such as alcohols, the evaporation starts earlier during the deposition & drainage steps.

As a continuous process, all the five steps are carried out after each other. The schematic sketch of the process is shown in Fig. 11.2. After the deposited layer is cured, another layer may be applied once again on top of it with another dip-coating process. In this way, a multi-layer material can be constructed. Dip coating processes were applied for optical coatings, solar energy control systems, anti-reflective coatings on windows, plate glass, etc.

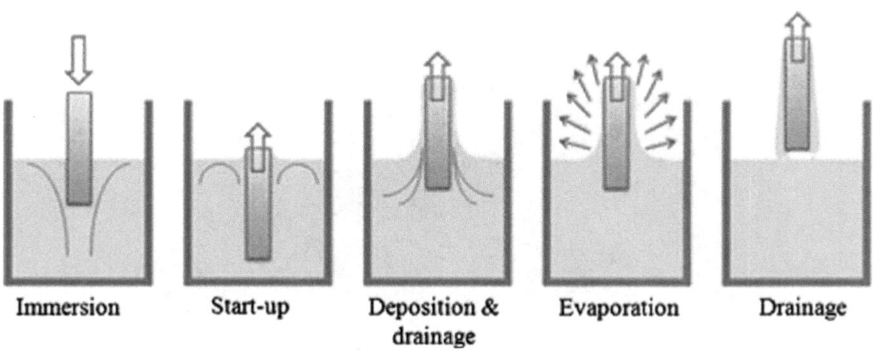

Fig. 11.2 Stages of the dip-coating (*Source* Raut et al. [12], anti-reflective coatings: a critical, in-depth review)

11.2.3 Spray Coating

Spray coating is a widely known method for car body painting and from artists by using spray cans. The fluid required to be coated is atomized at the nozzle of the spray head, which generates a continuous flow of material as shown in Fig. 11.3. Pneumatic-based systems use a stream of pressurized air or gas (e.g. nitrogen or argon) that breaks up the liquid into droplets at the nozzle. The quality of the coated layer is defined by the wetting behavior, surface properties, working distance, coating speed, droplet sizes, and the amount of sprayed layers. Surface temperature of the base material plays an important role as well. The advantage of spray coating is its high throughput and low material waste for large area depositions without patterning [13]. The combination of multiple spray head enables large area covering and/or multilayer coating.

11.2.4 Vapour Deposition Process

Vapor deposition is a process in which materials which is in vapor state are reduced through a stepwise procedure by condensation, chemical reaction or conversion in formation of solid material. Vapour deposition process are used for forming coatings to change the mechanical, electrical, thermal, optical, corrosion resistance, and

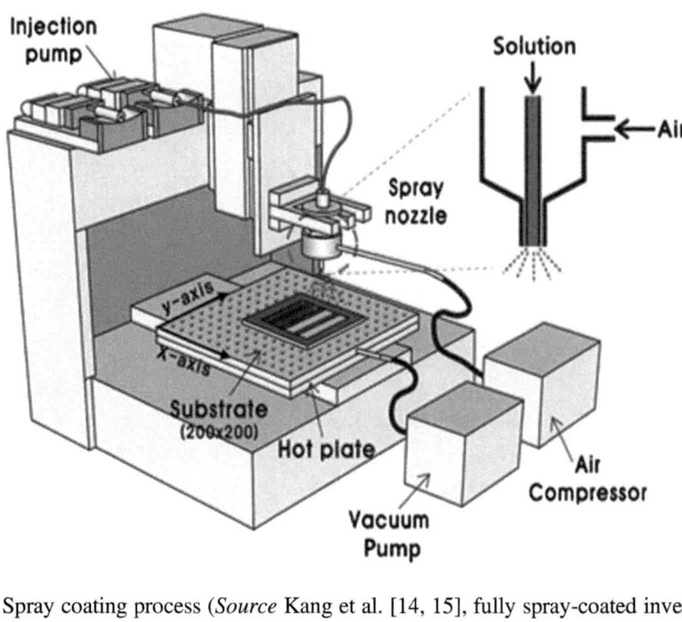

Fig. 11.3 Spray coating process (*Source* Kang et al. [14, 15], fully spray-coated inverted organic solar cells)

wear properties of the material. Also this process is used to form free-standing bodies, films and fibers for forming composite materials. This process takes place inside a vacuum chamber. Mainly there are two types of vapor deposition processes, they are Physical vapor deposition (PVD) and Chemical vapor deposition (CVD).

11.2.4.1 Physical Vapor Deposition (PVD)

Physical vapour deposition process is the process that deposits thin films by condensing the desired film material in vapourized form into various required surfaces. This coating method involves physical processes such as high-temperature vacuum evaporation with subsequent condensation, or plasma sputter bombardment. This process is recently being used for the property enhancement of various products such as automotive parts like wheels and pistons, surgical tools, barrier applications and guns. The process is shown in Fig. 11.4.

11.2.4.2 Chemical Vapor Deposition (CVD)

Chemical vapour deposition is a process used for producing high performance and high quality solid materials. The process is mainly used in the semiconductor industry for thinfilm production. In CVD process, the substrate is exposed to one or more volatile precursor, which reacts on the substrate surface in producing the required deposition. Here the byproducts produced during the process are removed

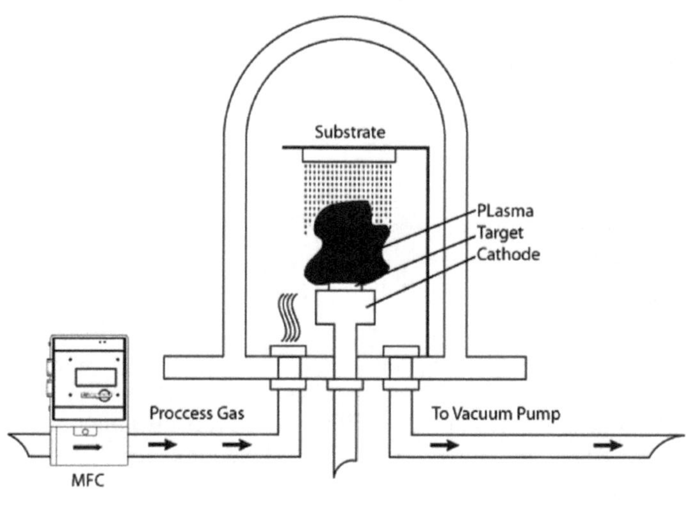

Fig. 11.4 Physical vapour deposition process (*source* http://www.sierrainstruments.com/products/pvd.html)

by gas flow in the reaction chamber. In the microfacrication environment CVD is widely used for depositing materials in numerous forms which includes monocrystalline, polycrystalline, amorphous, and epitaxial materials [16]. CVD is mainly used for the atomic layer deposition for depositing exceptionally thin layers of material.

11.2.5 Plasma Spray Coating

Plasma spraying is a technique extensively used for preparing protective coatings. In plasma spraying the hot, high-speed flame of a plasma gun can melt a powder of almost any ceramic or metal and sprays as a coating which acts as protection against corrosion, wear or high temperature. In plasma spray devices, an arc is formed between two electrodes in a plasma forming gas, which usually consists of either argon/hydrogen or argon/helium. As the plasma gas is heated by the arc, it expands and is accelerated through a shaped nozzle, creating high velocity. In this process the temperatures in the arc zone reaches up to 36,000 °F (20,000 K). Plasma spray coatings mainly find its application in the area of aerospace, automotive, medical devices, agriculture communication, etc. for producing coatings.

11.3 Materials and Particles Used for the Construction of LBL Assembled Storage Tanks

The design of fuel gas storage tanks is important in various engineering and industrial applications including hydrogen fuel cells, petroleum recovery and refining, production of biodiesel, food and chemical processing, aerospace, power generation, etc [17, 18]. Based on the need the researchers have done research for improving the barrier properties of many materials. Some of the materials that have been used are nano-fillers, such as Graphene oxide nanosheets (GONSs), layered clay minerals, nanoclays, layered double hydroxide, exfoliated graphite, graphene and metal nanoparticles, polyethylenimine (PEI), poly(acrylic acid) (PAA), montmorillonite (MMT) clay, polyethylene, poly(vinyl alcohol) [19], poly(lactic acid), polyurethane and regenerated cellulose, poly-p-phenylene oxide (sPPO) [20], polyurethane (PU) foam and polylactic acid (PLA), polyethylenimine (PEI), Cellulose, polyvinylpyrrolidone (PVP), PAMPAS [Poly (2-acrylamino-2-methyl-1-propanesulfonic acid-costyrene)], PVDF (PVDF refers to polyvinylidene fluoride) [21], Pt/Poly(diallyldimethylammonium chloride) (PDDA) [22, 23], Pd/PDDA, Au/PDDA, heteropolyacid/PDDA or hybrid polyelectrolytes.

For methanol barrier fuel cells the mainly used materials are Nafion [24], Aciplex, Flemion, XUS, poly(vinylalcohol)/poly(styrene sulfonic acid), polyvinyl alcohol, polybenzimidazol, Sulfonated polyimide and Sulfonated poly(ether ether ketone), bromomethylated poly(2,6-dimethyl-1,4-phenylene oxide) (BPPO).

11.4 Development of Multilayered Films for Fuel Storage Applications

Though a multilayer material improves the barrier property of base materials, it has been studied by many researchers on applying for various property improvement analysis. Some of the research findings are discussed in the following sections. Mainly the multilayer techniques are studied for barrier applications such as hydrogen, oxygen, methanol barriers and gas turbine applications [25, 26].

11.4.1 Multilayer Assembly for Transport Applications

Walther et al. [27] have made an attempt to make a natural multilayer material for transport and construction fields. They had achieved a mechanically excellent structure of nacre in laboratory scale models. Here, they introduced a simple and rapid methodology for large-area, lightweight and thick nacre-mimetic films and laminates with superior material properties. They have used MTM, polyvinyl alcohol (PVA) material for developing the new organic material using layer by layer deposition technique. Here they have followed three processes namely paper-making, doctor-blading, or painting for making the polymer-coated nanoclay particles (center) that gives rise to strong and thick films with tensile modulus of 45 GPa and strength of 250 MPa. The above said concept is environmentally friendly, energy-efficient and economic and can be developed by roll-to-roll processes. The developed material is in competence with metals and ceramics, thus suitable for many lightweight applications which also show low gas permeability and excellent fire resistance. Thus the developed material may be suitable for making multilayer tanks.

Ariga et al. [28] in the year 2014 have discussed about various methods for generation of layer by layer material and also about various application such as physical applications which may used for many storage applications, chemical and biochemical applications and biomedical applications. Also the rolling process for the industrial layer by layer deposition process were briefed by the authors.

Kim et al. [29] have discussed about grapheme which is a two-dimensional material that offers many opportunities for membrane applications because of ultimate thinness, flexibility, chemical stability and mechanical strength. They developed few- and multilayered graphene and graphene oxide (GO) sheets developed by using spin coating which exhibit desired gas separation characteristics. The required gas diffusion would be achieved by controlling gas flow channels and pores via different stacking methods. For multilayered (3- to 10-nanometer) GO membranes, the gas transport behavior strongly depends on the degree of interlocking within the GO structure. Here high carbon dioxide/nitrogen selectivity was achieved by well-interlocked GO membranes in high relative humidity.

11.4.2 Multilayer Assembly for Hydrogen Barrier Applications

Roh et al. [30] have designed a tank for storing type IV hydrogen. Figure 11.5 shows a schematic representation of a hydrogen storage tank which is of 1400 mm in length and 465 mm in diameter, can hold 5.6 kg of hydrogen at 70 MPa. To help wring out excessive resin during the winding process, the helical and hoop layers were wound alternately over the liner. They have done a finite element analysis to determine the lowest helical and hoop layer thicknesses which is needed for the assurance in structural integrity of the tank. They also analyzed the use of doilies to reinforce the dome and the boss sections of the tanks in reducing the number of helical layers which is wound around the cylindrical section. The obtained FE results showed that the doilies reduce the stresses generation near the dome end and the stresses at the shoulder of the tank were not affected. This study was mainly for the hydrogen storage tanks.

Subrahmanyam et al. [31] in the year 2011, discussed about Storage of solid hydrogen using few layer graphene which is of great interest and it has been suggested that carbon nanotubes may be useful for storing hydrogen, but only up to 3 wt%. It has been shown that single-layer graphene on reacting with hydrogen atoms leads to generation of sp^3 C–H bonds on the basal plane and the reduced material gets dehydrogenated on photothermal heating. Reaction of atomic hydrogen with graphene produces hydrogenated graphene giving rise to changes in

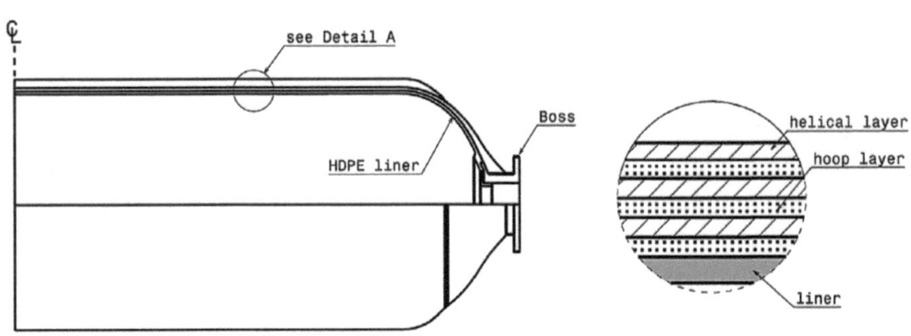

Fig. 11.5 Schematic of Type 4 compressed hydrogen storage tank [30]

electronic and phonon properties. On reaction of hydrogen a tunable band gap were induced in graphene. The nature of hydrogen in the hydrogenated samples or the amount of hydrogen uptake by graphene has not been outlined. It has been suggested that hydrogenation of graphene may be more possible with multilayer graphene than that with single-layer graphene. Also they analyzed that reduction of graphene give rise to the storage of up to 5 wt% of hydrogen. From the spectroscopic studies it is revealed that presence of sp^3 C–H bonds in the hydrogenated graphenes. When it is heated up to 500 °C or by irradiation with UV developed few layer graphene releases all the hydrogen, which confirms the possibility of few-layer graphene for the use chemical storage of hydrogen.

Rajasekar et al. [32] in the year 2013, developed a multilayered storage material for main hydrogen gas barrier application. They fabricated multi-layered films containing poly(diallyldimethylammonium) chloride (PDDA) and sulfonated polyvinylidene fluoride (SPVDF)-graphene oxide (GO) composites through layer by-layer (LBL) assembly for enhancing the hydrogen gas barrier properties. Polyethylene terephthalate (PET) substrate was made hydrophilic by treatment with aqueous sodium hydroxide solution before LBL construction. Positively-charged PDDA and negatively-charged SPVDF or SPVDF/GO composites were made by spin-coating and were tightly packed by electrostatic attraction. The assembled LBL material was characterized by Fourier transform infrared (FT-IR) spectroscopy and suggested a successful sulfonation of PVDF to SPVDF and also the layer by layer formation is clearly shown in scanning electron microscopy (SEM) analyses as in the Fig. 11.6.

Also the developed Electrostatic LBL assembled PDDA/SPVDF-GO films showed improved mechanical and gas barrier properties when compared with to PDDA/SPVDF LBL assemblies without GO. The hydrogen gas transmission rate (GTR) of a 16 bi-layer LBL assembly with 2 wt% GO was 11.7 cc/m^2 d atm, which was much lower when compared with that of PET substrate (329.1 cc/m^2 d atm) and a one bi-layer LBL assembly without GO (277.9 cc/m^2 – d atm) which is

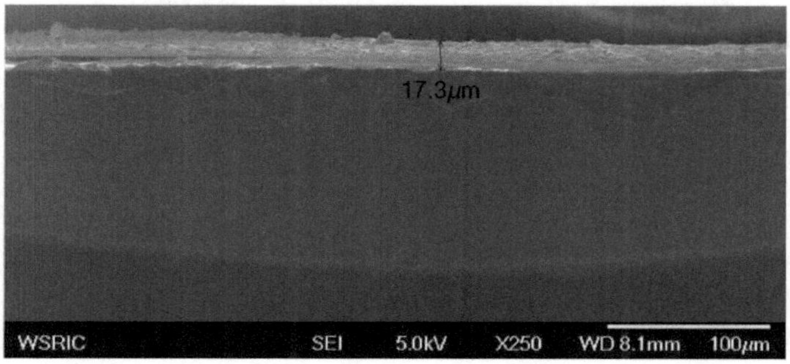

Fig. 11.6 SEM image of the developed material [32]

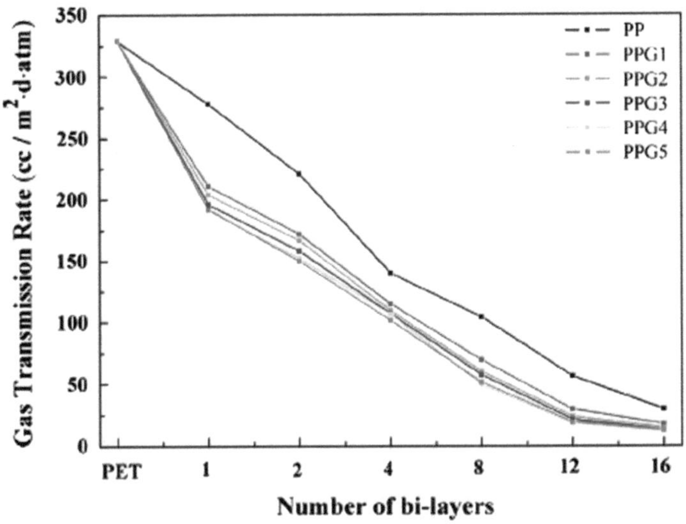

Fig. 11.7 Gas transmission rate [32]. Where; PP-Bilayers of PDDA/Sulfonated PVDF. PPG 1, 2, 3, 4, 5-Bilayers of PDDA/Sulfonated PVDF-Graphene oxide 0.5, 1, 1.5, 2, and 3 wt%

evident from the graph shown in the Fig. 11.7. This decrease in gas transmission rate indicates the suitability for use in high hydrogen barrier applications.

Nathan et al. [33] in 2010 discussed about layer-by-layer (LbL) films which is composed of poly(diallyl dimethyl ammonium chloride) (PDAC) and sulfonated poly(2,6-dimethyl 1,4-phenylene oxide) (sPPO) (PDAC/sPPO). The study was to find the variation of the ionic strength of assembly solutions and determine the nature of the exceptionally high ionic conductivity of the system. Both methods spray technique and dip coating technique was used for making the layer by layer material. Here the film growth is modulated from 6.91 nm/bilayer (BL) when made with no salt to 62.2 nm/BL when made with 0.5 M salt in all solutions. An optimized assembly conditions of 1.0 M salt in only the sPPO solution, fully humidified PDAC/sPPO films have ionic conductivity values of 7.00×10^{-2} S cm^{-1} at 25 °C, which was the highest value reported for any LbL assembled system. Selectively screening charges when salt is added to sPPO assembly solution shows decrease in ionic crosslink density of the films and increase in water uptake, which yields high ionic conductivity. Thickness measurements made at 0.5 BL increments indicate that the film composition can also be tuned by the ionic strength of the assembly baths. Additionally, PDAC/sPPO films fabricated using a LbL-Spray technique were allowed for preliminary characterization of the mechanical properties of free-standing membranes.

Paul A. Lessing with his coworkers [34] they have submitted a report in the year 2003, in which they have mentioned a suitable technique for developing multilayer hydrogen storage tank using spin coating and spray coating technique. They have started a project with the aim of developing hydrogen diffusion barrier which should posses less permeability to hydrogen, fix well to the polymer liner, stiffness

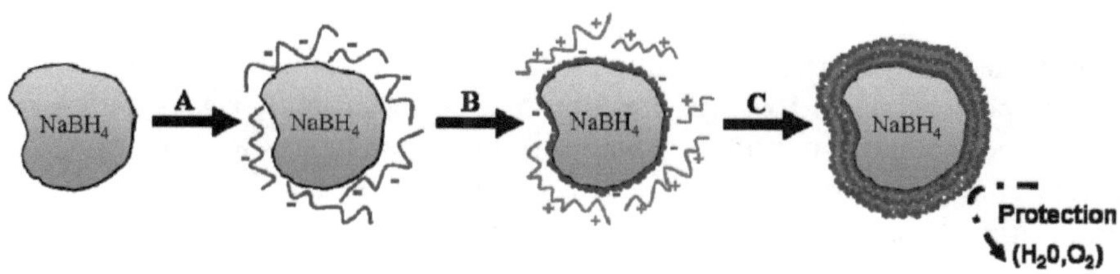

Fig. 11.8 Scheme of the polyelectrolyte coating of sodium borohydride particles by layer-by-layer adsorption: (step A) polyanion adsorption; (step B) polycation adsorption; (step C) multilayer nanofilm formation [35]

of the coating should match the underlying polymer to avoid cracking when the tank is under pressure and the cost of the material and coating should not be much more as of overall cost. Here the material used are H-PEDOT polymer (PEDOT + a small amount of alcohol) and PAMPAS [Poly (2-acrylamino-2-methyl-1-propanesulfonic acid-costyrene)]. Tri-layers of these materials were first made by spin-coating on glass. Later, the layers were spin-coated on polymer substrates. Significant reduction in permeation was shown with addition of the tri-layer. The same have been applied for patent.

Borodina et al. [35] in the year 2009 described their work which aims in making of hydrogen storage materials from organic solvents. Due to complex formation the shell provides stability and selective permeability. Here sodium borohydride particles were encapsulated within polymer films by using layer-by-layer technique of oppositely charged polyelectrolytes (polyethyleneimine and poly(acrylonitrile-co-butadieneco-acrylic acid)). The polymer nanofilm fabrication was performed using dichloromethane as a working media. The stepwise procedure for making layer by layer deposition is shown in Fig. 11.8. As a result in the X-ray diffraction (XRD) measurements the peaks observed on the pattern of untreated SBH after storage correspond to the crystallinity of NaB(OH), showing the hydrolysis of unprotected SBH upon reaction with environmental water and also IR spectroscopy was applied to investigate the chemical interaction between the polyelectrolytes. The multilayer film preparation was verified by Z-potential measurements, scanning electron microscopy and confocal laser microscopy. The explained approach for hydride protection can find its applications such as hydrogen storage systems and may also be used as hydrogen fuel cells.

11.4.3 Multilayer Assembly for Oxygen Barrier Applications

Hagen et al. [36] had taken an effort to reduce the deposition time of the depositing layers and also for achieving high gas barrier. Multilayer films were deposited using 1 s exposures for the first four bilayers (BLs) and followed 1 min for succeeding dips. Here the assemblies of polyethylenimine (PEI) and poly(acrylic acid)

(PAA) were deposited onto poly(ethylene terephthalate) [PET] using the layer-by-layer deposition process by spin coating. By varying the exposure time of PET to polyelectrolyte solutions (i.e., dip time) considerably changes the growth rate of the multilayer thin films. As a result from the findings they informed that PEI/PAA system grows linearly with 1 s dip times and exponentially with longer times. Also the oxygen transmission rate (<0.005 $cm^3/(m^2$ day)) was undetectable for the eight bilayers which is 650 nm thick using 1 min deposition steps, but this barrier was obtained with only 6 BLs (552 nm) using 1 s deposition of the initial layers, resulted in the reduction of deposition time by 73%. This time based concept made the layer-by-layer assembly faster and more commercially feasible and act as a good barrier for gases.

Huang et al. [37] demonstrated that cellulose is considered as a better material for biodegradable packaging films, but its main weakness is its poor gas barrier performance. Here they have used a simple, efficient, low cost, recyclable, non-toxic solvent an aqueous solution of NaOH/urea for fabricating graphene oxide nanosheet (GONS)/regenerated cellulose (RC) nanocomposite films with an ultra-low O_2 permeability and high mechanical performance. And the prepared nanosheet was studied using Transmission electron microscopy and two-dimensional wide-angle X-ray diffraction measurements which showed that the grapheme oxide nanosheets were fully exfoliated, equally dispersed and highly aligned along the surface of the cellulose nanocomposite films. Also the Rheological and Fourier transform infrared spectroscopy measurements were carried out for verifying the existence of strong H-bonding interactions between the GONSs and the cellulose matrix. Also a considerable improvement in the barrier properties for the prepared cellulose nanocomposite was achieved. The O_2 permeability coefficient was reduced by about 1000 times when compared with pure cellulose film at a low GONS loading of 1.64 vol.%. The tensile strength and Young's modulus of the cellulose nanocomposite films were also improved by about 67 and 68%, when compared with the RC film.

Tzeng et al. [38, 39] made a Layer-by-layer assembly using dip coating technique for creating quadlayers (QLs) of chitosan (CH), poly(acrylic acid) (PAA), CH, and graphene oxide (GO). The obtained material was studied using Electron microscopy which confirmed GO coverage on the film and a highly ordered nanobrick wall structure was present in the prepared specimen. By changing the pH deviation between CH and PAA, a thick and interdiffused polymer matrix ware generated because of the altered chain conformation. A 5 QL assembly (48 nm) exhibits very low oxygen permeability. They also made an effort to maintain barrier performance under high humidity, so GO was thermally reduced to increase hydrophobicity of the film. This distinctive water-based multilayer nanocoating is a distinctive material that may be used for gas purification and packaging applications.

Priolo et al. [40] in the year 2015 stated that, Layer-by-layer (LbL) assembly has emerged as the most important non-vacuum technique for the development of transparent, gas barrier films. The gas barrier presentation of LbL deposited films was discussed in abundant studies, with a variety of polyelectrolytes such as Polyethylene (PE), Polyamide (PA), Oriented polypropylene (OPP), Poly(ethylene terephthalate) (PET), which compete with the metal and metal oxide-based barrier films. This review paper main presents the LbL-based multilayer thin films with a nanobrick wall structure which comprises polymers and nanoplatelet that posses high gas barrier to permeable polymer substrates. They also mentioned that increased clay spacing influence the gas barrier performance, quadlayers (QL) of PEI/poly(acrylic acid) (PAA)/PEI/MMT were deposited onto PET to produce ultrathin, transparent gas barrier assemblies. Also nonclay particles such as grapheme based materials are reviewed for gas barrier applications and mentioned improved gas barrier property, also only polymer LBL material showed improvement in the gas barrier property.

Priolo et al. [41], have insisted that the development of flexible and transparent polymeric material for the barrier application is of due importance. Because of its need and high barrier property requirement, the development of Layer-by-layer assembly technique for the fabrication of layered, polymer-clay superbarrier thin films is very important for todays requirement. At about a thickness 51 nm the developed nanocomposite thin films is composed of 12 polymer and 4 clay layers, which exhibits a lower oxygen permeability.

Liu et al. [42], have prepared clay nanopaper hybrid composite montmorillonite platelets in a continuous matrix of nanofibrillated cellulose (NFC) with the aim of connecting the toughness of fibrillar networks. The clay was a sodium montmorillonite and Nanofibrillated cellulose (NFC) was used for making the matrix. For LbL self-assembly, films with multilayers are formed through sophisticated alternating deposition of negatively and positively charged layers. Hydrocolloid mixtures were used in a filtration approach similar to paper processing. The obtained multilayered structure of the nanopaper was studied by FE-SEM, FTIR, and XRD. Uniaxial stress-strain curves measured in tension and thermal analysis were carried out by DMTA and TGA. At 50% by weight of montmorillonite, the tensile strength was 124 MPa and the modulus 8.7 GPa, which shows improved interacting of NFC-MTM. Also, fire retardance and oxygen permeability characteristics were measured and showed good results for both. Hence the developed nanopaper may be used for oxygen barrier applications.

Aulin et al. [43] in 2013 have used layer-by-layer (LbL) deposition method for the build-up of alternating layers of nanofibrillated cellulose (NFC) or carboxymethyl cellulose (CMC) with a branched, cationic polyelectrolyte, polyethyleneimine (PEI) on flexible poly (lactic acid) (PLA) substrates. By using this procedure, optically transparent nanocellulosic films with tunable gas barrier properties were formed. The developed multilayer material posses good oxygen permeability. The adsorption of the various components to the PLA substrate was characterized using the quartz crystal microbalance with dissipation (QCM-D) technique. Atomic force microscopy imaging was used to characterize the surface morphology and surface roughness of the LbL assembly films. And as a result the height image of the film shows network structure with monodispersed fibrils and an average rms roughness of 7.0 nm, determined from $3 \times 3\ \mu m^2$ areas. Cross-sectional SEM images of LBL coated PLA films were done to measure the thickness of the LbL assemblies on the PLA substrates and the measured result of 30 bilayer PEI/PAA assembly prepared from PEI at pH 10 and PAA at pH 4, respectively, resulted in a multilayer thickness of 4.74 μm. Owing to its flexibility, transparency and high performance gas barrier properties, these thin film assemblies are promising candidates for several large-scale applications which indicates low oxygen permeance.

11.4.4 Multilayer Assembly for Hydrogen and Helium Barrier Applications

Tzeng et al. [38, 39] have made a polymer subtracted which acts as a light gas (i.e. H_2 and He) barrier by using thin films of polyethylenimine (PEI), poly(acrylic acid) (PAA), and montmorrilonite (MMT) clay which were deposited via layer-by-layer (LbL) assembly using dip coating method. A five quadlayer (122 nm) coating deposited on 51 μm polystyrene has shown less permeability of hydrogen and helium. As the number of QDs increased it showed one to three orders of magnitude when compared with commercial barriers, demonstrating better performance than thick-laminated ethylene vinyl-alcohol (EVOH) copolymer film and even metalized poly olefin/polyester film. The uniqueness of the developed polymer/clay barrier thin film was due to high clay loading and strong ionic interactions between polyelectrolytes. The impermeable clay platelets, with highly interdiffused PEI and PAA mortar, formed a nanobrick wall structure which acts as an important light gas barrier. This low permeability indicates the application of packaging and also for protective application.

Kang et al. [14, 15] have developed a layer for the protection of the surface of metals such as Fe and Cu from oxidizing which is of due importance because of its increased usage. Here, oxidation resistance of these Fe and Cu foils were achieved by coating them with reduced graphene oxide (rG-O) sheets. The rG-O-coated Fe and Cu foils were prepared by transferring rG-O multilayers from a SiO_2 substrate onto them by using spin coating technique at the speed of 3000 rpm for 60 s. The oxidation resistance of the rG-O-coated metal foils was studied by Raman spectroscopy, optical microscopy, and scanning electron microscopy after heat treatment at 200 °C in air for 2 h. From the study it was resulted that the bare metal surfaces were oxidized severely when compared with rG-O-coated metal surfaces which shows less oxidation.

11.4.5 Multilayer Assembly for Methanol Barrier Applications

Paula T. Hammond [44] in the year 2011 have explained about the commonly used membrane i.e., sulfonated perflouroether ionomer which is composed of Nafion, is highly effective in hydrogen fuel cells and it is moderately permeable to liquid methanol, because the methanol molecules diffuse through sulfonated pores within the Nafion matrix. Also Poly-p-phenylene oxide that has excellent mechanical and thermal stability is less permeable to liquid methanol. By developing highly sulfonated poly-p-phenylene oxide (sPPO) fully water-soluble polymer that can be used for fuel cell membrane. However, when layered with a polycation, such as polydiallyl dimethylammonium chloride (PDAC), sPPO can be used as a material for low-methanol permeability. Various methods such as spin coating, spray coating and dip coating, for the preparation of the multilayer material were clearly discussed. Also automated spray coating method is explained.

Li and Manthiram [45] in the year 2010 developed new poly (arylene ether sulfone) copolymers which contains different amount of pendant sulfonic acid groups using polymerization reaction. They suggested that the properties of the synthesized sulfonated poly(diphenylsulfone-diphenol) (SDPS-DP) copolymers clearly depend on the sulfonic acid group content present in the copolymers. Mostly the copolymers show superior thermal stability, less liquid uptake and high methanol barrier property, they reveal low proton conductivity than Nafion or sulfonated poly(ether ether ketone) (SPEEK) [46]. By taking this low methanol barrier as the opportunity, multilayer membranes consisting of the SDPS-DP copolymer as a methanol-barrier in the center layer and SPEEK as the proton-conducting outer layers were developed by the authors and the same were characterized using FTIR and found that the intensity of the absorption peaks corresponding to the aromatic sulfonic acid group (1030 cm^{-1}) increases with increasing sulfonated monomer content. Also it is revealed that methanol crossover is 66 and 42% of that of plain SPEEK for 5 and 15 μm thick methanol-barrier center layers. On increasing the center layer thickness furthermore the methanol crossover is reduced up to 37%. They also concluded that the multilayer membranes show decreased methanol crossover. The multilayer membranes with an optimized center-layer thickness of 15 μm show much improved fuel cell performance compared to plain SPEEK membrane.

Hasani-Sadrabadi et al. [47] have made an attempt for making a tri-layer proton exchange membrane which is comprised of two thin layers of chitosan that was structurally modified, that acts as methanol barrier layers, by coating Nafion on both the sides and the prepared sample were tested for high performance fuel cell applications. Also it was identified that there were strong bonding between the newly designed membrane which was confirmed by using SEM and EDX data.

As per the permeability test data it showed improved transport properties for the multi-layer membrane when compared to pure Nafion of same thickness. Also direct methanol fuel cell test revealed high open circuit voltage, output of power density and overall efficiency of fuel cell for the triple-layer membrane when compared with Nafion, particularly for concentrated methanol solution. And 72% more power output were obtained when compared the multilayer membrane with nafion. The fuel cell efficiency for multi-layer membrane was measured about 19.55 and 18.45% for 1 and 5 M methanol concentrations. Due to its significant reduced methanol crossover, easiness in preparation and little cost, the triple-layer membrane under study were considered as an important polyelectrolyte for methanol fuel cell applications.

Zhang et al. [48] in the year 2011 prepared methanol-blocking polymer composite membranes by placing purple membrane (PM) in Nafion resin using dip coating. By analyzing the prepared composited by scanning electron microscopy (SEM) resulted that the lateral dimension of 5 μm and achieved thickness of 5 nm. confocal laser scanning microscopy (CLSM) analyses show that PM was dispersed well in Nafion. Also the methanol barrier property has reached a maximum of 31.6% for the prepared multilayer composite when compared with pure nafion. Also they have indicated that many nanoparticles such as Pt/Poly(diallyldimethylammonium chloride) (PDDA), Pd/PDDA, Au/PDDA, heteropolyacid/PDDA or hybrid polyelectrolytes can be used with Nafion for the preparation of surface by layer-by-layer (LbL) assembly for reducing the methanol permeation. Purple membrane (PM) has found importance in preventing proton leakage and assists proton transfer in Nafion membranes by approximately 8% [38].

Zhao et al. [49] have made an attempt to modify Sulfonated poly(arylene ether ketone) bearing carboxyl groups (SPAEK-C) membranes by deposition of oppositely charged polyaniline (PANI) and phosphotungstic acid (PWA) using the layer-by-layer method via dip coating technique. This attempt was tried for improving the crossover of methanol in a methanol fuel cell. The FTIR and UV spectra result indicate that the PANI and PWA were assembled in multilayer. Compared to the SPAEK-C and Nafion membranes, the methanol permeability of SPAEK-C-(PANI/PWA)n was much lesser and shows a decreasing trend by increasing the number of bilayers. The SEM images indicate the occurrence of thin PANI/PWA layers coated on the SPAEK-C membrane.

Zhang et al. [50], in their findings they have fabricated Nafion composite membranes of about 1 of about 1 μm with improved methanol barrier properties from chitosan (CS) and silicotungstic acid (SiWA) using Layer-by-layer (LbL) self-assembly technique by dip coating technique which is shown in Fig. 11.9. From the Fourier transform infrared attenuated total reflection (FTIR-ATR) spectra and scanning electron microscope (SEM) study the successful

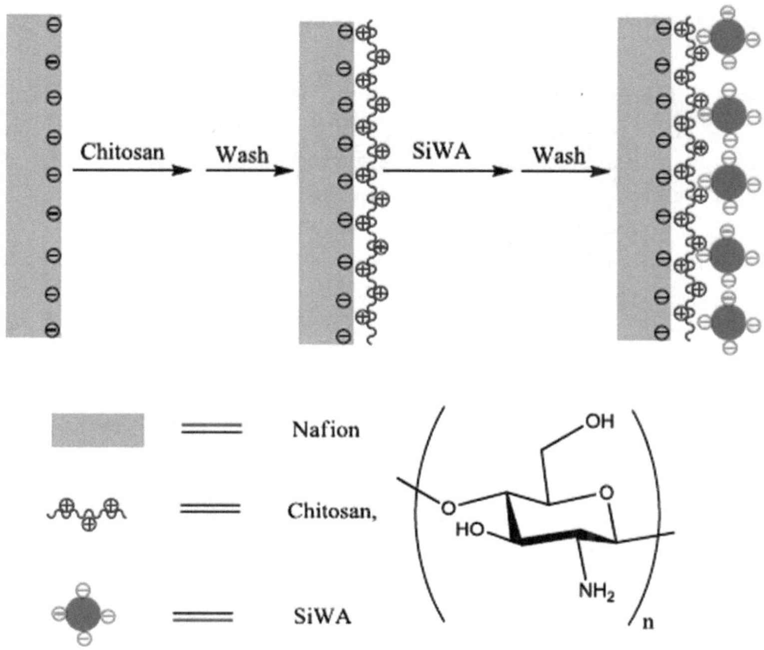

Fig. 11.9 Self-assembly mechanism of chitosan and SiWA on Nafion membrane [50]

formation of methanol blocking layer were confirmed. As the number of CS/SiWA bi-layer number is increased both proton conductivity and methanol permeability of Nafion composite membranes showed decreasing trend. The prepared Nafion multilayer composite membrane showed a reduced methanol permeability of about 47% which was determined using voltammetric method.

Mannarino et al. [51] in the year 2013 discussed about the preparation of composite membranes which is composed of highly conductive and layer-by-layer (LbL) films and fiber mats. The spray layer-by-layer deposition method was used to fabricate the multilayer rapidly for the fuel blocking character. The prepared material was characterized for mechanical strength and electrochemical selectivity. These layer by layer membranes consist of highly selective poly(diallyl dimethyl ammonium chloride)/sulfonated poly(2,6-dimethyl-1,4-phenylene oxide) (PDAC/sPPO) films. The uncovered mats were annealed for the improvement in the mechanical properties, which is to be maintained in the prepared composite membranes. The mat that is coated was as thin as 15 μm. The permeability study revealed that the prepared composite membrane has methanol permeability 20 times lower than that of the pure Nafion.

11.4.6 Multilayer Assembly for Gas Turbine Applications

Padture et al. [52] in their review paper they informed that thermal barrier coatings (TBCs) which is carried out using low-thermal conductivity ceramics are now becoming more important due to its thermal insulation character with metallic components as application for hot gas stream in gas-turbine engines in aircraft propulsion, power generation, and marine propulsion. They also insisted that the structure of a thermal barrier-coated section consists of four layers, two metallic and two ceramic layers, with each of its four layer having different physical, thermal, and mechanical properties. There are four layers in the TBC system is made of different materials with specific properties and functions. These layers are (i) the substrate, which is generally nickel or cobalt based super alloy that act as a temperature gradient across the component wall, (ii) the bond-coat, is a metallic layer of about 75–150 μm thickness which is made of NiCrAlY or NiCoCrAlY alloy and is deposited by using the plasma-spray or the electron-beam physical-vapor deposition methods. Other types of bond-coats are made of aluminides of Ni and Pt and are deposited by electroplating in conjunction with diffusion-aluminizing or chemical-vapor deposition. (iii) the thermally grown oxide (TGO), which is of 1 to 10 μm in thickness is coated between the boand coat and ceramic top coat and (iv) the ceramic top-coat which is made of Y_2O_3-stabilized ZrO_2 (YSZ), that provides a thermal insulation. They also mentioned that this YSZ has a high melting point of about 2700 °C which may be suitable for high temperature applications.

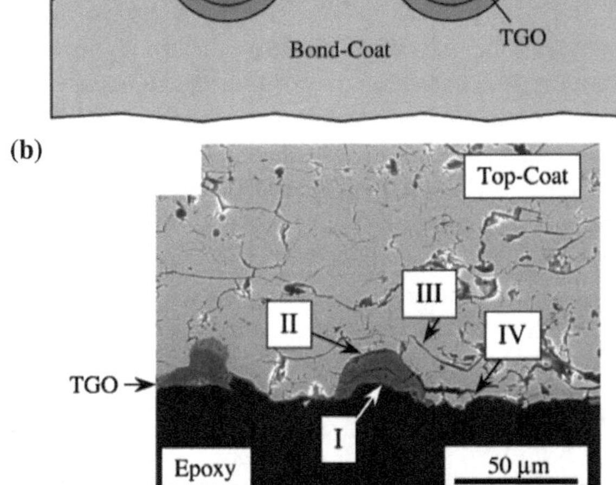

Fig. 11.10 Schematic diagram and SEM image showing the four different cracking mechanisms in APS TBC [52]

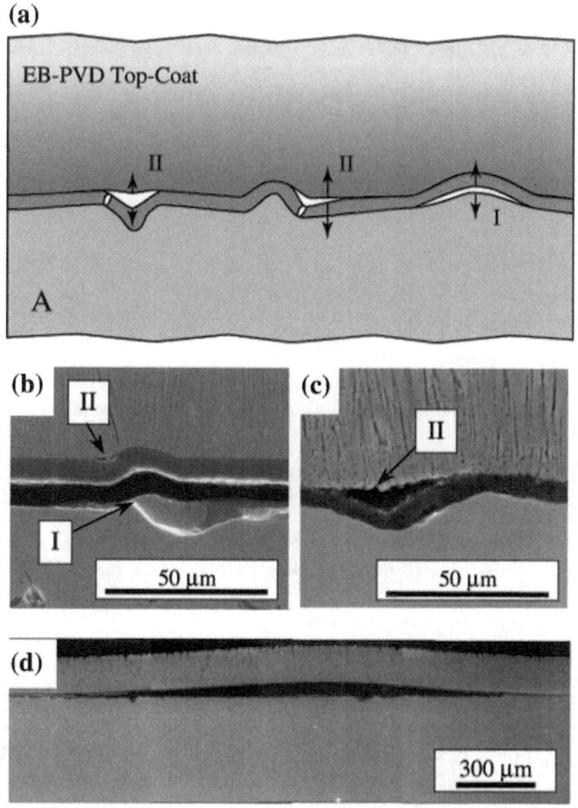

Fig. 11.11 Schematic diagram and SEM image showing two of the three different cracking mechanisms in EB-PVD TBC [52]

The ceramic top coat is deposited by air-plasma-spray (APS) deposition and electron beam physical-vapor deposition (EB-PVD) which is as shown in Fig. 11.10 and the SEM image of the coating is shown in Fig. 11.11.

As a result they informed a growth of the TGO during engine operation is the most important phenomenon which is dependable for failure of TBCs. This TGO sometimes results in a constrained volume expansion that leads to compressive stresses which continue at all temperatures. The strain energy in the TGO shows linearity with the TGO thickness and quadratically with the TGO stress, and drives fracture. The stresses developed in bond coat are mainly accountable for TBC failure. Also the failure mechanisms in the two classes of TBCs (APS and EB-PVD) were different.

11.5 Summary

The importance and advances of multilayer materials for fuel storage application has been highlighted. Resplendent research works had been carried out on multi-layer deposition using the techniques like spin coating, dip coating, spray coating, vapor deposition, plasma spray coating for the construction of multilayer assembly. Depositions of layer through the latter techniques are extensively analyzed through morphological studies. The development of multilayer materials using layer-by-layer assembly technique displayed promising fuel barrier properties, which can be employed for fuel storage applications.

References

1. Chikh L, Delhorbe V, Fichet O (2011) (Semi-) interpenetrating polymer networks as fuel cell membranes. J Membr Sci 368(1–2):1–17
2. Guan SW (2002) External corrosion protection systems for flammable and combustible liquids underground storage tanks. The 2nd China international corrosion control conference, Beijing
3. Lin HD, Zhao CJ, Ma WJ, Li HT, Na H (2009) Low water swelling and high methanol resistant proton exchange membrane fabricated by cross-linking of multilayered polyelectrolyte complexes. J Membr Sci 345:242–248
4. Krogman KC, Zacharia NS, Schroeder S, Hammond PT (2007) Automated process for improved uniformity and versatility of layer-by-layer deposition. Langmuir 23(6):3137–3141
5. Deligoz H, Yilmazturk S, Karaca T, Ozdemir H, Koc SN, Oksuzomer F (2009) Self-assembled polyelectrolyte multilayered films on Nafion with lowered methanol cross-over for DMFC applications. J Membr Sci 326:643–649
6. Krogman KC, Lowery JL, Zacharia NS, Rutledge GC, Hammond PT (2009) Spraying asymmetry into functional membranes layer-by-layer. Nat Mater 8:512–518
7. Krogman KC, Zacharia NS, Schroeder S, Hammond PT (2007) Automated process for improved uniformity and versatility of layer-by-layer deposition. Langmuir 23:3137–3141
8. Polak PL, Mousinho AP, Ordonez N, da Silva ZL, Mansano RD (2007) Deposition of polymeric perfluored thin films in proton ionic membranes by plasma processes. Appl Surf Sci 254:173–176
9. Cho J, Char K, Hong JD, Lee KB (2001) Fabrication of highly ordered multilayer films using a spin self-assembly method. Adv Mater 13(14):1076
10. Sahu N, Parija B, Panigrahi S (2009) Fundamental understanding and modeling of spin coating, process: a review. Indian J Phys 83(4):493–502
11. Scriven LE (1988) Physics and applications of DIP coating and spin coating. MRS Proceedings 121:717. doi:10.1557/PROC-121-717
12. Raut HK, Anand Ganesh V, Nairb AS, Ramakrishna S (2011) Anti-reflective coatings: a critical, in-depth review. Energy Environ Sci 4:3779
13. Izquierdo A, Ono SS, Voegel J-C, Schaaf P, Decher G (2005) Dipping versus spraying: exploring the deposition conditions for speeding up layer-by-layer assembly. Langmuir ACS J Surf Colloids 21(16):7558–7567
14. Kang J-W, Kang YJ, Jung S, Song M, Kim DG, Kim CS, Kim S (2012) Fully spray-coated inverted organic solar cells. Solar Energy Mater Solar Cells 103:76–79
15. Kang D, Kwon JY, Cho H, Sim J-H, Hwang HS, Kim CS, Kim YJ, Ruoff RS, Shin HS (2012) Oxidation resistance of iron and copper foils coated with reduced graphene oxide multilayers. ACS Nano 6(9):7763–7769
16. Maruyama T, Kanagawa T (1994) Electrochromic properties of niobium oxide thin films prepared by chemical vapor deposition. J Electrochem Soc 141(10):2868–2871
17. Costamagna P, Srinivasan S (2001) Quantum jumps in the PEMFC science and technology from the 1960s to the year 2000: part II. Engineering, technology development and application aspects. J Power Sources 102:253–269
18. Hwang JJ, Chang WR, Weng FB, Su A, Chen CK (2008) Development of a small vehicular PEM fuel cell system. Int J Hydrogen Energy 33:3801–3807

19. Chang Y-W, Wang E, Shin G, Han J-E, Mather PT (2007) Poly(vinyl alcohol) (PVA)/sulfonated polyhedral oligosilsesquioxane (sPOSS) hybrid membranes for direct methanol fuel cell applications. Polym Adv Technol 18(7):535–543
20. Yun S-H, Woo J-J, Seo S-J, Liang W, Dan W, Tongwen X, Moon S-H (2011) Sulfonated poly (2,6-dimethyl-1,4-phenylene oxide) (SPPO) electrolyte membranes reinforced by electrospun nanofiber porous substrates for fuel cells. J Membr Sci 367(1–2):296–305
21. Piboonsatsanasakul P, Wootthikanokkhan J, Thanawan S (2008) Preparation and characterizations of direct methanol fuel cell membrane from sulfonated polystyrene/poly (vinylidene fluoride) blend compatibilized with poly(styrene)-*b*-poly(methyl methacrytlate) block copolymer. J Appl Polym Sci 107(2):1325–1336
22. Zhang DF (2009) Surface modification for the perfluorosulfonate proton-conducting electrolyte membrane. Appl Surf Sci 255:4119–4122
23. Schlenoff J, Dubas S, Farhat T (2000) Sprayed polyelectrolyte multilayers. Langmuir 16 (26):9968–9969
24. Wang JT, Xiao LL, Zhao YN, Wu H, Jiang ZY, Hou WQ (2009) A facile surface modification of Nafion membrane by the formation of self-polymerized dopamine nano-layer to enhance the methanol barrier property. J Power Sources 192:336–343
25. Ho D, Chu B, Lee H, Montemagno CD (2004) Protein-driven energy transduction across polymeric biomembranes. Nanotechnology 15:1084–1094
26. Alberti G, Casciola M (2003) Composite membranes for medium temperature PEM fuel cells. Annu Rev Mater Res 33:129–154
27. Walther A, Bjurhager I, Malho J-M, Pere J, Ruokolainen J, Berglund LA, Ikkala O (2010) Large-area Lightweight and thick biomimetic composites with superior material properties via fast, economic, and green pathways. Nano Lett 10:2742–2748
28. Ariga K, Yamauchi Y, Rydzek G, Ji Q, Yonamine Y, Wu KC-W, Hill JP (2014) Layer-by-layer nanoarchitectonics: invention, innovation, and evolution. Chem Lett 43:36–68
29. Kim HW, Yoon HW, Yoon S-M, Yoo BM, Ahn BK, Cho YH, Shin HJ, Yang H, Paik U, Kwon S, Choi J-Y, Park HB (2013) Selective gas transport through few-layered graphene and graphene oxide membranes. Science 342:91–95
30. Roh HS, Hua TQ, Ahluwalia RK (2013) Optimization of carbon fiber usage in Type 4 hydrogen storage tanks for fuel cell automobiles. Int J Hydrogen Energy 1–8
31. Subrahmanyam KS, Kumar P, Maitra U, Govindaraj A, Hembram KPSS, Waghmare UV, Rao CNR (2011) Chemical storage of hydrogen in few-layer graphene. PNAS 108(7):2674–2677
32. Rajasekar R, Kim NH, Jung D, Kuila T, Lim JK, Park MJ, Lee JH (2013) Electrostatically assembled layer-by-layer composites containing graphene oxide for enhanced hydrogen gas barrier application. Compos Sci Technol 89:167–174
33. Ashcraft JN, Argun AA, Hammond PT (2010) Structure-property studies of highly conductive layer-by-layer assembled membranes for fuel cell PEM applications. J Mater Chem 20:6250–6257
34. Lessing PA, Yang Y, Ma LP, Chen FC, Shrotriya V, Sirosh N, Warner MJ (2003) Low permeation liner for H2 gas storage tanks, hydrogen, fuel cells & infrastructure technologies program, 2003 merit review and peer evaluation meeting, 19–22 May 2003. Berkeley, California
35. Borodina TN, Grigoriev DO, Andreeva DV, Mohwald H, Shchukin DG (2009) Polyelectrolyte multilayered nanofilms as a novel approach for the protection of hydrogen storage materials. Appl Mater Interfaces 1(5):996–1001
36. Hagen DA, Foster B, Stevens B, Grunlan JC (2014) Shift-time polyelectrolyte multilayer assembly: fast film growth and high gas barrier with fewer layers by adjusting deposition time. ACS Macro Lett 3:663–666
37. Huang H-D, Liu C-Y, Li Dan, Chen Yan-Hui, Zhong Gan-Ji, Li Zhong-Ming (2014) Ultra-low gas permeability and efficient reinforcement of cellulose nanocomposite films by well-aligned graphene oxide nanosheets. J Mater Chem A 2:15853–15863

38. Tzeng P, Lugo EL, Mai GD, Wilhite BA, Grunlan JC (2015) Super hydrogen and helium barrier with polyelectolyte nanobrick wall thin film. Macromol Rapid Commun 36:96–101
39. Tzeng P, Stevens B, Devlaming I, Grunlan JC (2015) Polymer−graphene oxide quadlayer thin-film assemblies with improved gas barrier. Langmuir 31:5919–5927
40. Priolo MA, Holder KM, Guin T, Grunlan JC (2015) Recent advances in gas barrier thin films via layer-by-layer assembly of polymers and platelets. Rapid Commun, Macromol. doi:10.1002/marc.201500055
41. Priolo MA, Gamboa D, Holder KM, Grunlan JC (2010) Super gas barrier of transparent polymer-clay multilayer ultrathin films. Nano Lett 10:4970–4974
42. Liu A, Walther A, Ikkala O, Belova L, Berglund LA (2011) Clay nanopaper with tough cellulose nanofiber matrix for fire retardancy and gas barrier functions. Biomacromolecules 12:633–641
43. Aulin C, Karabulut E, Tran A, Wagberg L, Lindstrom T (2013) Transparent nanocellulosic multilayer thin films on polylactic acid with tunable gas barrier properties. ACS Appl Mater Interfaces. dx.doi.org/10.1021/am401700n
44. Hammond PT (2011) Engineering materials layer-by-layer: challenges and opportunities in multilayer assembly. AIChE J 57(11):2928–2940
45. Li W, Manthiram A (2010) Sulfonated poly(arylene ether sulfone) as a methanol-barrier layer in multilayer membranes for direct methanol fuel cells. J Power Sources 195:962–968
46. Cai H, Shao Ke, Zhong S, Zhao C, Zhang G, Li X, Na H (2007) Properties of composite membranes based on sulfonated poly(ether ether ketone)s (SPEEK)/phenoxy resin (PHR) for direct methanol fuel cells usages. J Membr Sci 297(1–2):162–173
47. Hasani-Sadrabadi MM, Dashtimoghadam E, Mokarram N, Majedi FS, Jacob KI (2012) Triple-layer proton exchange membranes based on chitosan biopolymer with reduced methanol crossover for high-performance direct methanol fuel cells application. Polymer 53:2643–2651
48. Zhang J, Lan F, Liang D, Xiao Y, Lu S, Xiang Y (2011) Bulk modification of Nafion® with purple membrane for direct methanol fuel cell applications. J Memb Sci 382:350–354
49. Zhao C, Lin H, Zhang Q, Na H (2010) Layer-by-layer self-assembly of polyaniline on sulfonated poly(arylene ether ketone) membrane with high proton conductivity and low methanol crossover. Int J Hydrogen Energy 35:10482–10488
50. Zhang H, Huang H, Shen Pei Kang (2012) Methanol-blocking Nafion composite membranes fabricated by layer-by-layer self-assembly for direct methanol fuel cells. Int J Hydrogen Energy 37:6875–6879
51. Mannarino MM, Liu DS, Hammond PT, Rutledge GC (2013) Mechanical and transport properties of layer-by-layer electrospun composite proton exchange membranes for fuel cell applications. ACS Appl Mater Interfaces. dx.doi.org/10.1021/am402204v
52. Padture NP, Gell M, Jordan EH (2002) Thermal barrier coatings for gas-turbine engine applications. Science 296:280–284

Chapter 12
Multilayer Bottles

Rane Ajay Vasudeo, A.R. Ajitha, Shivendra S. Patil, Krishnan Kanny and Sabu Thomas

12.1 Why Do We Need Multilayer Polymer Films?

The last 50 years have seen major changes in the way that food is produced and retailed. Self service supermarkets have grown to dominate the market with corresponding growth in pre packed food. More food is prepared before retail scale, even to the extent of ready to eat dishes. Such food production has become concentrated into fewer sites, with more products being distributed and marketed on a planned basis. Changes have driven an increasing sophistication in food packaging self service demands products that stands out on the supermarket shelf. Consumers place increasing value in being able to see the food and freshness of the food at the time of purchase. Processed and partially processed food requires new packaging techniques to achieve an adequate shelf life. The concentration of food production facilities results in larger distribution lines and thus creates a need for packaging that can extend the shelf life especially for fresh and chilled products. These changing requirements for packaging have become more and more difficult to achieve with single materials [1]. To achieve the required performance it has become necessary to

A.R. Ajitha · S. Thomas (✉)
International and Inter University Centre for Nanoscience and Nanotechnology, Mahatma Gandhi University, Kottayam 686 560, Kerala, India
e-mail: sabuchathukulam@yahoo.co.uk

S.S. Patil
Institute of Chemical Technology, Nathalal Parekh Marg, Matunga 400019, Mumbai, India

S. Thomas
School of Chemical Sciences, Mahatma Gandhi University, Kottayam 686 560, Kerala, India

R.A. Vasudeo · K. Kanny
Composite Research Group, Department of Mechanical Engineering, Durban University of Technology, Durban, South Africa

J.K. Kim et al. (eds.), *Multicomponent Polymeric Materials*,
Springer Series in Materials Science 223, DOI 10.1007/978-94-017-7324-9_12

combine the properties of different materials. The chapter deals with the way in which multilayer achieves this. The use of multilayer materials is driven by the need to design a pack that combines different functional requirements [2]. A package must collate and contain the food, to do this it may need strength, sealing. Very importantly the packaging must help preserve the food, protecting it from external influences like moisture, oxygen, light, other flavors, odors and chemicals and could cause a deterioration in quality. The appearance of a pack is important to attract the consumer and to present a quality image that is consistent with that of the product it contains, the rigidity of materials can be important to allow it to stand up on the shelf and prevent it looking creased and tired. Food is packed on high speed automated machinery. Efficient operations require the use of packaging materials with precisely defined characteristics. Different machines make different demands on the material. Required characteristics include tensile strength to prevent stretching or snapping, softening or melting when heated to provide a sealing medium, conversely heat resistance to prevent stretching or other damage during the sealing operation slip, example to allow materials to pass over static machine parts or lack of it, e.g. to allow materials to be gripped firmly or to be stacked. Ideally the package will include design features that make it easier for the consumer to handle, to open easily and in some cases to reclose [3]. There is a continuous pressure from food manufacturers and retailers to reduce packaging cost, both of the materials themselves and of the packaging operation. Increasingly the Industry is not just thinking of the monetary cost of packaging but also of its environmental impact.

In many cases it is not possible to meet all the packaging design requirements using a single monolayer material. The whole rationale for the use of multilayer is to create a single packing structure that will combine the different properties of different base materials in order to meet these design requirements, A barrier to moisture and gases such as oxygen is needed [4]. The actual permeability will depend on the precise grade, thickness and the final structure in which the material is used, it is seen that moisture barrier and oxygen barrier do not go hand in hand. For polymers transmission rates are inversely proportional to layer thickness. For multiple layers the total transmission rates of individual layers i.e.

$1/T_{R\ \text{multilayer}} = 1/T_{R\ \text{layer1}} + 1/T_{R\ \text{layer2}} + \ldots\ldots\ldots\ldots\ldots$, these consideration don't apply to thin coatings of polymers or to non polymers.

When the design requirement for a pack demand two or more properties that cannot be found in a single material, there are two ways of proceeding. Firstly the pack can be made out of two components, each one contributing at least one of the desired properties. Secondly whereby two or more components are combined to make a multilayer material. The following examples illustrate how this is done: Heat sealable coextruded oriented, oriented polyester/polyethylene laminate. The use of pack for liquids such as milk and fruit juices is well established. Extended shelf life of six or more months can be achieved by the aseptic packing. The key requirement for such pack are: high quality leak proof seals, barrier to gases to prevent oxygen ingress as well as to prevent flavor loss, light barrier to prevent light induced of flavors as well as accelerated oxidation is needed for some products, rigidity so that the pack can stand on shelf and be easily handled by the consumer, easy opening and reclose in the pack, the outer layer of polyethylene protects the board from moisture [5]. In some cases PE to PE seal occurs where the outside of the pack to seal to the inside. The board is primarily used to give structural rigidity for display and handling. It is the cheapest way of achieving this. Not all requirements can be met using multi layers. At the moment the most effective easy open and easy reclose system still rely on adding separate components such as caps and spouts to the main body of the pack.

Glass bottles have been used to pour the correct amount of viscous sauces like tomato ketchup or mayonnaise, a squeeze plastic bottle offers increased convenience. However many such sauce products are susceptible to oxidation, causing discoloration and off flavors, so the design requirements may be summarized as: a level rigidity that allows the bottle to be squeezed, resistance to flex cracking potentially caused by repeated squeezing of the bottle in the presence of fats and acids in the food product, barrier to moisture to prevent product drying out, barrier to oxygen to maintain product quality, reclosure system to allow multiple use in the home, transparency to allow the contents to be seen as with glass [6]. Polyethylene terephthalate (PET) provides the necessary balance of rigidity and flexibility with the required chemical resistance and adequate moisture barrier. However its oxygen barrier is inadequate to prevent spoilage of the product, this is provided by a layer of Ethylene vinyl alcohol (EVOH). Polyamide or oxygen scavengers could also be used. The bottle shape is made by co extrusion blow molding or co injection stretching molding. A separate "snap open" closure is often used to give consumer easy access. This could be a point for oxygen to enter the pack. To prevent this, at least within the retail distribution chain, an impermeable membrane is sealed across the mouth of the bottle, to be removed by the consumer before use. Such a membrane could be made of polyester for strength and heat resistance, aluminium foil for barrier, peelable polymer blend to achieve a hermetic seal to the PET, preventing loss of product and ingress of oxygen yet with sufficiently weak peel strength to allow easy removal by the consumer.

Migration of substances from multilayer bottles is also an important measure, the components of multilayer may contain chemical substances that have a potential to migrate into the packed food, and such substances could include: residue of starting substances used to make a material examples monomers, additives used to achieve specific technical effects examples antioxidants and slip agents, non intentionally added substances examples impurities in raw material, reaction by products [7]. To ensure the safety of consumers of packed food it is essential to control the amount of such migration. Starting with the model of a monolayer plastic containing a potential migrant that is placed in contact with food, the following factors must be considered: the initial concentration of the migrant in polymers, the ratio of the surface area of the plastic to the mass of the food; the greater this is the quicker the substance will be transferred, the diffusion coefficient which describes the rate at which the substance can move through plastic to be available for transfer into the food, the diffusion coefficient is itself depended on temperature, the higher temperature the quicker the substance will move, the time the plastic is in contact with the food, the partition coefficient is defined as the ratio of the migrant concentration in the packaging to its concentration in the food at equilibrium, partition coefficient measures the relative affinity of the polymer and the food for the substance, for example a polar substance will partition between a polar and a non polar medium so that there is a higher concentration in the former, the thickness of film of plastic combined with the ratio of surface area to weight of food gives the mass ratio of plastic to food; the higher the mass the higher the concentration of the substances in the food at equilibrium [8].

The basic principles governing the migration from single webs into foods also apply to all plastic multi layer materials but the complexity increases, the rate and amount of migration from one layer to another which starts as soon as the multi layer is manufactured must also be considered. In consequence a migrant in the food contact layer might move not only into the food, but also into other outside layers, thereby reducing the concentration in the layer and hence the eventual equilibrium concentration in the food. Conversely a migrant in an outer layer might move into the food contact layer and from there into the food [9]. There will be a time lag in such behaviour which might be significant depending on the delay between the manufacture of the multilayer and its use to pack the food. In multi layer structure of the type layer A/layer B/layer C and where layer C is in contact with food, it is possible that layer B forms a barrier to migrants from layer A entering layer C, hence layer B is known as functional barrier as it reduces the level of migration from layer A to layer C. Further we shall discuss on properties, preparation techniques, quality test and applications of multilayered plastic bottles.

Plastic bottles are widely used in many applications includes the food packaging industry, medical application, cosmetic applications etc. Poly Ethylene Terephthalate has large market application, it is widely used for the carbonated soft drink, fruit juice etc. The permeability properties of PET are good only for products having shelf life of 8–12 weeks. So it is very necessary to improve the gas permeability properties of plastic bottle materials. The better technique adopted for this purpose is the multilayer bottle construction. In this method thin layers of polymers which have low gas permeability properties are incorporated into a sandwich like structure. The barrier layers with different barrier materials are placed in the centre of the bottle wall encapsulated by the polymer on either side, in majority cases PET is the polymer. In many cases plastics such as cyclo-olefins and amorphous nylons are incorporated as core layers to reduce the permeation of moisture and gases to achieve shelf life requirements for the bottle. Another method includes the coating of some materials such as graphite carbon, silica and epoxy resins to the external or internal walls of the bottles to lower the overall ingress of gas into the food.

12.1.1 Materials Used for Multilayer Bottles

Most of the plastic bottles are made of PET. It provides a clear, lightweight, recyclable plastic that can be processed economically. PET has a viscosity between 0.75 and 0.85 cm^3/g is used for bottles and sheet. PET has high strength, good transparency, excellent dimensional stability, lightweight, and good air tightness, hence it's bottles can be used as a substitute for glass bottles. Also the PET is a recyclable plastic and can prepare economically. PVC is mainly used for film and pipe. But it is also used for rigid plastic box and bottles. PVC have good transparency, high strength and have good heat sealing property. The other polymers include Low density polyethylene (LDPE), High density polyethylene (HDPE) and polypropylene (PP).

In many multilayer bottles nanocomposites are used as barrier layers. This materials hinders the permeability of gases through the polymer matrix, there by increases the shelf life of bottles. Mainly nylon nanocomposites are used as barrier layer. In which a combination of nylon and nanoclay is widely used as a barrier for multilayer bottles. In the early stages ethylene vinyl alcohol copolymer (EVOH), had a major role as a barrier for oxygen and carbon dioxide in food and nonfood packaging applications. Polyamide compositions are used as barrier layer in multilayer packaging in some composites. Nanoclay containing composites have a major role as barrier, since the Nanoclays create a tortuous path for gas molecules to go through the matrix resin and also they do not degrade in the matrix. Additions of nanoclay increases the Young's Modulus and reduces the oxygen transfer rate (OTR) of polymer matrix. The different polymers used for clay–polymer nanocomposites are PA (polyamides), nylons, polyolefins, polystyrene (PS), ethylene-vinylacetate (EVA) copolymer, epoxy resins, polyurethane, polyimides and polyethylene terephthalate (PET). And the nano clay used in nanocomposites is bentonite.

12.1.2 Processing of Multilayer Bottles

Nathaniel C. Wyeth, a son of well-known painter N.C Wyeth and an engineer for the Du. Pont Corporation, was the first to develop a usable bottle. The first mass production of plastic bottles was done in America in 1939. Various parameters need to be considered while manufacturing bottle. Wall-thickness needs to be consistent, necks of bottle should be in regular shape, lower extensional viscosity so that it can stretch easily, excess material formed during molding needs to be trimmed off which happens to be difficult, melt strength should be considered so that we get uniform wall thickness and no holes are formed, crystallization rate suggested to be slow, etc. Polymer which is used to manufacture bottles must be shatter proof under normal conditions so that bottles on falling won't get damage [10]. Also bottles should not deform under pressure. They should have good creep resistance. Other factors include-A level of rigidity that allows the bottle to be squeezed, resistance to "flex cracking", potentially caused by repeated squeezing of the bottle in the presence of fats and acids in the food product, barrier to moisture to prevent product drying out, barrier to oxygen to maintain product quality, reclosure system to allow multiple use in the home [11]. Transparency to allow the contents to be seen, as with glass. Also during polymerization reaction various impurities are produced which are difficult to eliminate from finished product. For example, during polymerization of polyethylene terephthalate from terephthalic acid and ethylene glycol, diethylene glycol and acetaldehyde were the impurities in finished product which happens to give funny taste to the soft drink because of excess of acetaldehyde. It was difficult to take all these factors into account which resulted into very low production of bottles. Different polymers used for making a bottle include Low density polyethylene (LDPE), High density polyethylene (HDPE), polypropylene (PP), polyvinylchloride(PVC), polyethylene terephthalate (PET). Multilayer bottles are composed of layers of selected plastic materials which are coextruded to retain and utilise the unique characteristics of each material. Most commercial containers consist of outer layer of polypropylene (for resistance to heat) or polyethylene and an inner layer such as EVOH (ethylene vinyl alcohol) or PVDC (polyvinylidene chloride copolymer). The plastic gasoline tanks are produced from a coextruded parison that contains six layers of different resins. For example, HDPE/adhesive/EVOH/adhesive/regrind/HDPE [12].

Layering two or more materials can economically overcome individual materials shortcomings. Multilayer blow molding can combine the strengths of two or more thermoplastic materials to economically manufacture bottles containing far better properties arising due to individual layer. For example, Polypropylene having relatively low cost, good clarity, excellent water vapor barrier, good heat and impact resistance but poor oxygen barrier can be combined with Ethylene vinyl alcohol (EVOH) copolymer which has high cost, excellent oxygen gas barrier but sensitive to water. Multilayer bottles are composed of layers of selected plastic materials which are coextruded to retain and utilize the unique characteristics of each material [13]. Production of multi-layered bottles involves combination of the separate skin and core materials in a precise and controlled manner to create a desired multi-layered structure. Coextrusion is the extrusion of multiple layers of material simultaneously. This type of extrusion utilizes two or more extruders to melt and deliver a steady volumetric throughput of different viscous plastics to a single extrusion head (die) which will extrude the materials in the desired form. The layer thicknesses are controlled by the relative speeds and sizes of the individual extruders delivering the materials. In many real-world scenarios, a single polymer cannot meet all demands for an application. Coextrusion allows for mediation of this problem, optimizing a wide range of properties such as oxygen permeability, strength, stiffness, and wear resistance [14] (Figs. 12.1 and 12.2).

Multilayered bottle structure can be achieved by several ways. Some of them are:

(a) *Extrusion molding of multilayer parison*
(b) *Injection molding of multilayer preform*
(c) *Stretch blow molding of multilayer parison*
(d) *Interior or exterior coating of injection molded preform or blow molded bottle*
(e) *bottle treatment*

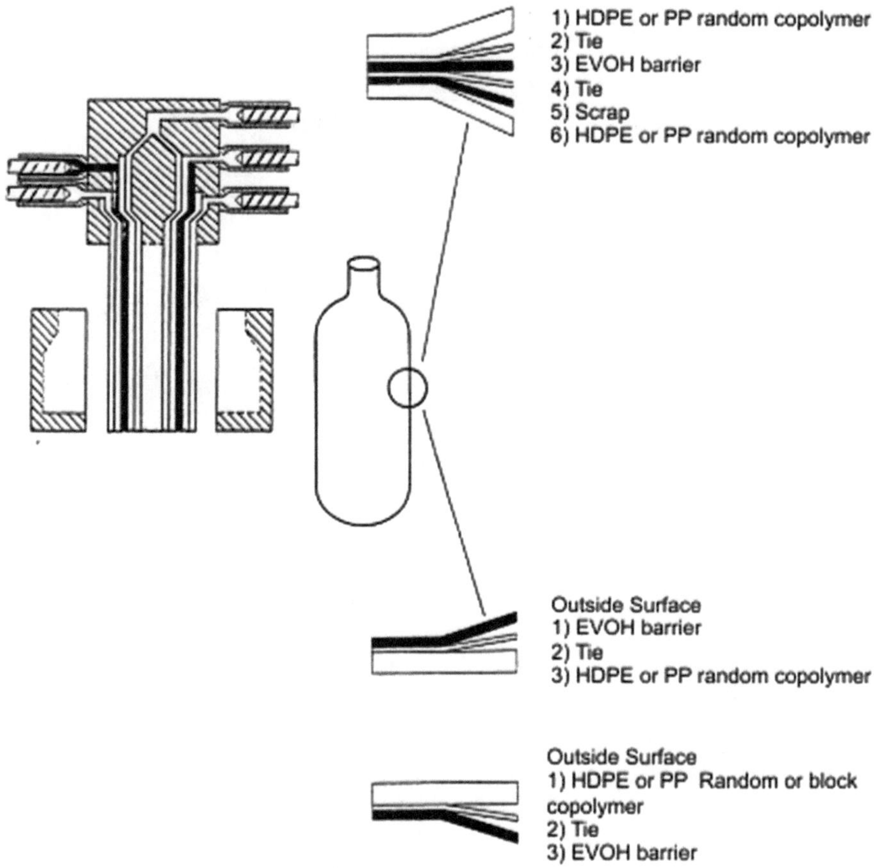

Fig. 12.1 Coextrusion of multilayered bottles by blow-molding process

12.1.2.1 Extrusion Multilayer

Blow molding machinery form the multilayer parison by simultaneously extruding two or more thermoplastic materials in continuous layers within the parison's wall. Creating the multilayer structure is relatively straightforward because melted thermoplastic materials flow in a laminar fashion and the high-molecular weight materials will not mix while passing weight materials will not mix while passing through the extrusion head [15]. As the extrusion head establishes the tubular cross-sectional form of the parison from a flow of pigmented thermoplastic material, before the parison exists the head, the pigmented material flow splits at one point with a secondary flow of unpigmented material from a separate plasticizer, thus creating a pigmented parison with an unpigmented translucent stripe. The stripe is in a position usually corresponding with a side of the blow-molded bottle.

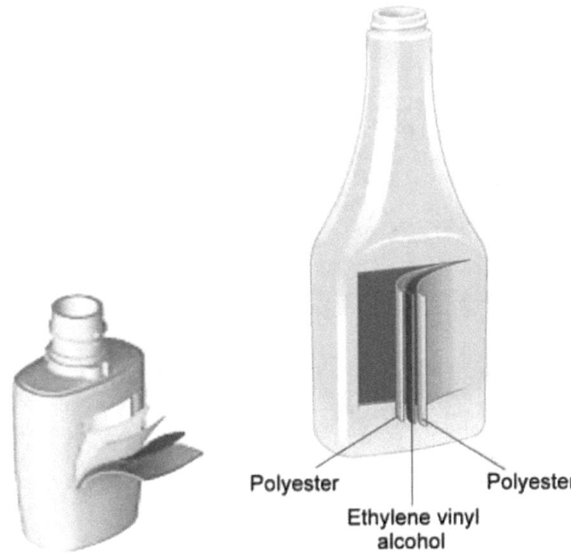

Fig. 12.2 Multilayered bottle

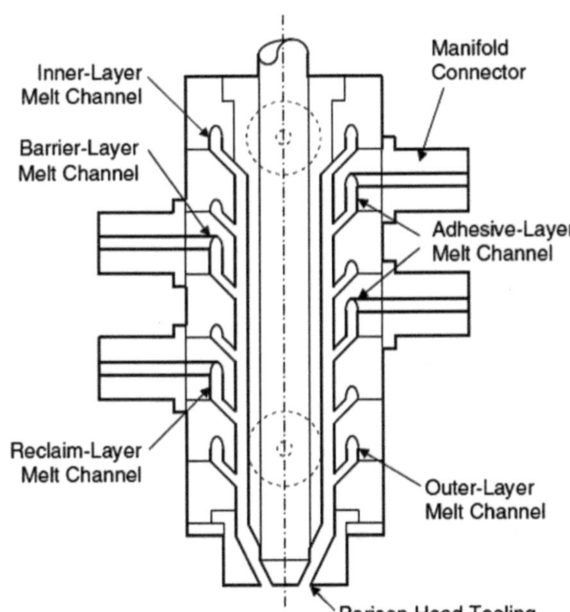

Fig. 12.3 Cut-away schematic of a typical multilayered parison extrusion head

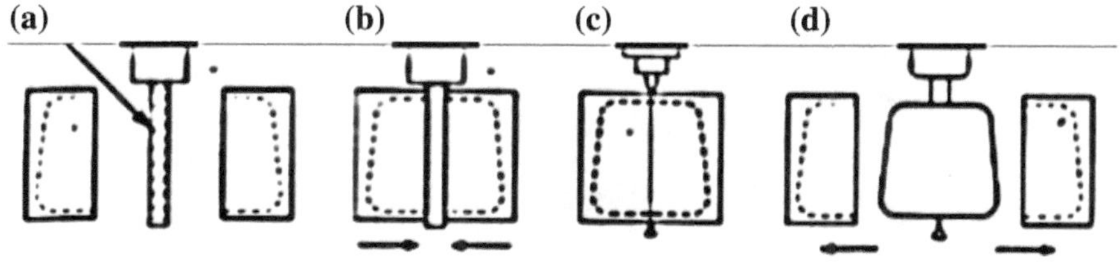

Fig. 12.4 Extrusion blow molding. **a** Parison extrusion (Parison denoted by *arrow*). **b** Mold halves close onto Parison. **c** Parison inflated against internal mold walls. **d** Mold halves open, blow molded part ready for ejection

Graduation marks on the bottle in a position along the view stripe allow consumers to easily judge content volume [16] (Figs. 12.3 and 12.4).

12.1.2.2 Injection Multilayer

As with the extrusion multilayer, injection multilayer also relies on laminar flow of thermoplastic materials entering the preform cavity to create individual layers Although there are subtle process variations between several techniques, two or three individual plasticizers inject material into the preform cavity through concentric injection nozzles, in generally either a simultaneous or a sequential manner, to establish sequentially three or five-layer structures [17]. Creating an injection based two or four layer structure requires an insert or an insert equivalent approach; that is, an apparatus usually places the insert made of at least one layer of material over the insert to complete the multilayer preform. Process technicians often refer to this approach as "over-molding". In general, the over-molding approach is not as

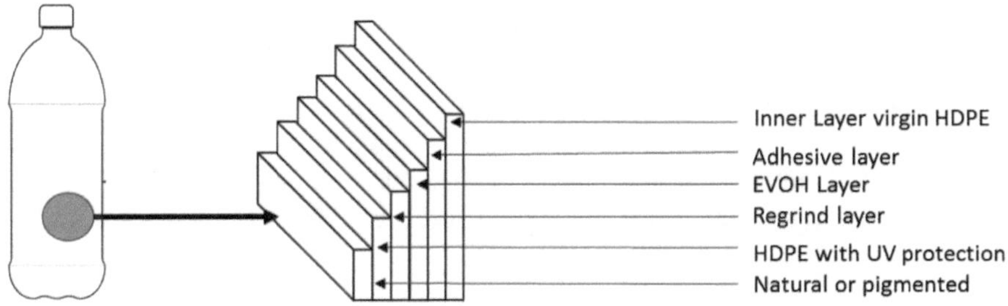

Fig. 12.5 Common multilayer bottle wall structure

common or as efficient, but it does have an interesting advantage in that the insert can be thermoformed, injection molded, or even extrusion blow-molded [18] (Figs. 12.5 and 12.6).

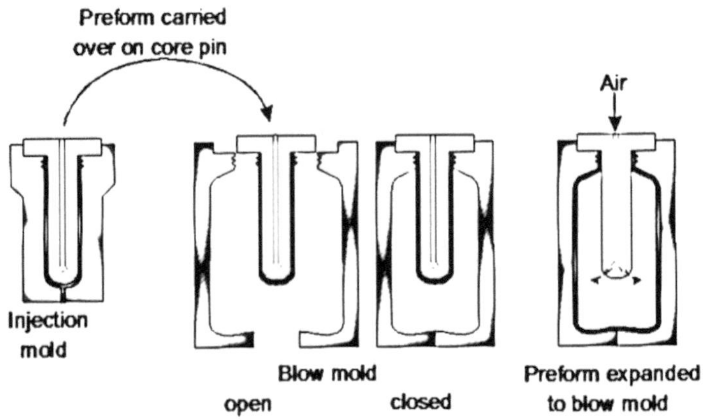

Fig. 12.6 Injection blow molding

12.1.2.3 Stretch Blow Molding

PET beverage bottles are made using a process known as stretch blow molding (also called orientation blow molding). First, PET pellets are injection molded—heated and put into a mold—into a thin walled tube of plastic, called a parison. The parison is then cooled and cut to the proper length. Next, the parison tube is re-heated and placed into another mold, which is shaped like a soda bottle, complete with screwtop [19]. A steel rod (a mandrel) is slid into the parison. Highly pressurized air then shoots through the mandrel and fills the parison, pressing it against the inside walls of the mold. The pressure of the air stretches the plastic both radially ("out") and axially ("down"). The combination of high temperature and stretching in the desired direction causes the molecules to polarize, line up and essentially crystallize to produce a bottle of superior strength. The entire procedure must be done quickly, and the plastic must be pressed firmly against the wall, or the bottle will come out misshapen. In order to give the bottom of the bottle its proper concave shape—so that it can stand upright—a separate bottom piece is attached to the mold during the blowing process. The mold must then be cooled [20]. Different cooling methods are used. Water in pipes may flow around the mold, or liquid carbon dioxide, highly pressurized moist air, or room air is shot into the bottle to cool it more directly. The procedure is preferably done quickly, to set the bottle before creep (flow) occurs. The bottle is then removed from the mold (Fig. 12.7).

12.1.2.4 Preform or Bottle Coating

Multilayered structures can also be created by applying coating. A water based flow material and process have been developed for covering preform exterior surfaces for improved gas barrier in the blow molded bottle. Stand- alone apparatus applies and then cures the coating on the preform before the blow molding machine reheats

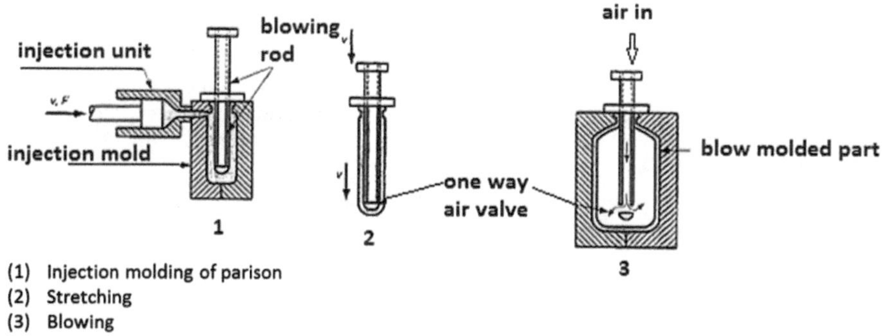

Fig. 12.7 Stretch blow moulding

and then expands that preform into that bottle [21]. Coatings tend to be thin, so instead of applying on preform it can directly be applied on the bottle. A more recent organic coating development uses epoxy amine chemistry to create a tough, transparent, glossy thermoset polymer film on bottle exterior surface. Stand-alone apparatus typically inline with the blow molding machine sprays or dips the bottle and then cures the coating material [22].

12.1.2.5 Bottle Treatment

Bottle treatment approaches seek to improve gas barrier of volatile hydrocarbon chemicals in polyolefin bottles by allowing polymer molecules at or near the container's surface to react with certain gases, thereby creating a thin integral gas barrier layer on the surface. There are two approaches: fluorination and sulfonation [23].

The fluorination approach is either downstream or part of the blow molding process. When part of the blow molding process, the molding machine uses pure nitrogen gas to initially inflate the parison in the blow mold cavity, quickly followed by a mixture of small amount of fluorine gas with nitrogen gas to complete the blow molding process [24]. The fluorine gas quickly and easily reacts with polyolefin material still hot from parison extrusion. Chemically the fluorine replaces hydrogen atoms within the polymer molecules that are on the interior surface. To promote reaction, the downstream approach requires reheating the bottle before fluorine exposure. Fluorine is a dangerous gas requiring special handling by trained technicians.

Fluorination makes high-density polyethylene and polypropylene bottles suitable for packaging nonpolar chemicals, such as those in cleaners, mineral spirits, turpentine, solvents such as carbon tetrachloride, and gasoline additives, insecticides, herbicides, and the like [25]. Although treatment barrier is not as effective, the sulfonation approach introduces sulfur trioxide to the inside surface of the bottle or container downstream of the blow molding machine. The sulfur trioxide reacts with the HDPE, causing sulfonic groups to form on the container's surfaces. The

Table 12.1 Description of different processing techniques

Extrusion blow molding	Injection blow molding	Stretch blow molding
Parison is produced by conventional extrusion method using a die similar to that used for extrusion pipes	Parison is produced by injecting a polymer into a hot injection mold around a blow tube or core rod	Stretch blow molding is similar to injection plow molding
The parison is extruded vertically in downward direction between two mold halves When the parison reaches the required length the two mold halves close resulting in pinching the top of parison end and sealing the blow pin in the bottom of the parison end Parison is inflated by air blown through the blow pin, taking a shape conforming that of the mold cavity. The parison is then cut on the top. The mold cools down, its halves open, and the final part is removed	Then the blow tube together with the parison is removed from the injection mold and transferred to a blow mold [26] Following operations are similar to those in the extrusion blowing molding Injection blow molding is more accurate and controllable process as compared to the extrusion blow molding It allows producing more complicated products from a wider range of polymer materials However production rate of injection blow molding method is lower than that of extrusion blow molding	Stretch blow molding involves injection molding of a parison, which is then stretched in the downward direction by means of the blow tube The extended parison is then inflated in a blow mold In this method biaxial molecular orientation is produced. The specific molecular orientation provides higher mechanical strength, rigidity and transparency of the material

approach then introduces ammonium hydroxide to neutralize any acidic byproducts followed by a water rinse (Table 12.1).

12.1.3 Properties of Multilayered Bottles

Imagine buying juice or soft drink bottle in a supermarket. Once you pick a bottle, things that we first check are the seal of the bottle, expiry date of the product as well as look for any physical damage to it. All these aspects are the basis of the properties that we thrive to obtain in multilayer bottle. Let's relate this aspect to properties [27].

Seal of product—Mechanical properties
Expiry date of product—Barrier Properties
Foul Smell—Chemical properties

12.1.3.1 Impact Strength

Impact strength is the capability of plastic to withstand sudden load and it is expressed in terms of energy. Impact strength determines the load at which the bottles will break. The multi layered bottles have greater impact strength than the conventional bottles. This makes them ideal for use of juices, oils containers. It is important that during transport of these juice bottles, no damage is done to the containers as it will lead to leakage of juices and make it useless [28].

12.1.3.2 Chemical Resistance

While selecting the plastic it is also important to see whether the polymer selected reacts with inner polymer layers. As this could lead to serious lead health concerns as well as fouling the products. Chemical resistance is also affected by temperature, relative humidity, concentration and driving force.

12.1.3.3 Barrier Properties

Food products are packed under certain atmospheric condition in the multilayered bottle to maintain its freshness. Thus, protecting the gases filled to leak or enter the bottle is important. Here barrier property of multilayer bottle plays an important role. Barrier property tells us about the permeability of the polymer material for gases, water vapour, liquid and organic substances. Permeability of an object is different for different materials. It also depends on the conditions under which it take place [29].

The permeability coefficient, or simply permeability, characterizes the steady state rate of mass transport of penetrant molecules through polymers. In a dense polymer film, the permeability P is defined as the molar flux of penetrant through the polymer relative to a fixed coordinate system N_A normalized by the film thickness L and the difference between the upstream (p_2) and downstream (p_1) partial pressures:

$$P = \frac{N_A}{(p_2 - p_1)}$$

In recent studies, it has been found that recycled PET can also be used as one of the barrier layer. Although, as compared to virgin PET, the properties will be less effective but sufficient enough for prevention. In terms of gas permeability, EVOH has high gas barrier property at low humidity. As humidity increases, the property will significantly decrease. Poly(m-xylene adipamide)–Kaolinite and Poly (m-xylene adipamide)–Montmorillonite nanocomposites show improved gas barrier properties when used as a neat resin [30].

Table 12.2 MVTR and OTR for Polymeric materials used in bottle manufacturing

Material	Clarity	MVTR*	O_2^{**}	CO_2^{**}	Impact strength	Recycle code
PET (oriented or stretch blown polyethylene terephthalate)	Excellent	2.0	75	540	Good	1
HDPE (high density polyethylene)	Poor	0.5	4,000	18,000	Good	2
PVC (polyvinyl chloride)	Good	3.0	150	380	Fair	3
PP (polypropylene)	Poor	0.5	3,500	7,000	Fair	5
PS (Polystyrene)	Excellent	10.0	6,000	18,700	Poor	6
PLA (polylactide—oriented/stretch blown bottles)	Very good	18–22	38–42	201	Good	7

*MVTR stands for Moisture Vapor Transmission Rate in g-mil/100 in. 2/24 h. MVTR is a measure of the passage of gaseous H_2O through a barrier. The lower the rate, the longer the package protects its contents from moisture and ensures the moisture content of the product remains the same
**O_2 and CO_2 stand for Oxygen Transmission Rate (OTR) and Carbon Dioxide Transmission Rate (COTR) in cm^3-mil/m^2/24 h. OTR and COTR are measures of the amount of gas that passes through a substance over a given period. The lower the readings, the more resistant the plastic is to letting gasses through

Preservation of food products is important factor that rules the products of packaging industry.

Moisture: In fruit juices, moisture loss is major concern as it decreases the quality of juice. Moisture loss is calculated in terms of water vapour transmission rates (WVTR) with units g/m^2/day. Transmission rates vary with temperature and relative humidity (RH) [31] (Tables 12.2 and 12.3).
Oxygen: Oxygen can oxidize the juices which lead to vitamin C from the juices. Also, it reduces shelf life (Tables 12.2 and 12.3). Oxygen transmission rates are generally used to measure permeability and are expressed as cc/m^2/day/bar
Light: U.V or any light may cause oxidation of products/juices. Thus, multilayer bottles must at least have one layer that will act as a physical barrier for light.
Chemicals: barrier needs to be provided in multilayer bottles to protect the products from absorbing unwanted odour. It plays an important role in preventing the flavours of the product.
Seal Quality: The seal must match the barrier properties of the multilayered bottles. A hermetic seal will be adequate for such a bottle [32].

12.1.3.4 Physical Properties

This group covers basically the physical strength and performance properties of packages on converting or packaging equipment. The different properties are thickness, tensile and elongation, heat seal strength, bond strength, hot tack, shrinkage, flexes resistance co-efficient of friction, pin holes, de-lamination,

Table 12.3 Properties of Polymers used for manufacturing bottles

Comparison of bottle polymer materials								
Material	Clarity	O_2	CO_2	Flexural modulus × 10^3	Impact strength	Maximum hot fill (°F)	Minimum tolerance (brittleness) °C	Density g/cc
LDPE	P	9,500	42,000	50	G	150°	−100°	0.92
HDPE	P	4000	18,000	150	G	190°	−100°	0.96
PP	P	3500	7000	200	F	200°	0°	0.91
PS(Styrene)	E	5000	16,700	400	P	150°		1.05
PVC	G	150	380	300	F	140°	30°	1.35
PC	E	4500	8225	350	E	240°	−125°	1.20
Unoriented Pet	E	100	720	300	F	140°		1.33
Oriented PET	E	75	540	450	G	120°	−40°	1.36
PETG	E	400	1200	300	G	140°	−40°	1.27
K-Resin	E	250		205	G			1.01

identification, leakage test, dart impact, seam strength, environmental stress crack resistance, closure leakage test, adhesion test, torque test.

- Physico-chemical Properties

These properties include transmission properties like permeability to gases, water vapours, odours, and extractability test.

- Optical Properties

These properties include haze and gloss.

Besides the above, tests which are carried out for ascertaining the quality of the packaging material or its conformity to laid down specifications, there are some other studies which are of great significance while developing packaging systems or selecting packaging materials for food products. These are:

- *Extractability/migration test*
- *Shelf-life Determination and Compatibility test*

Some of the important tests and test equipment of plastic packages are discussed Table 12.4.

12.1.4 Tests on Multi Layer Plastic Bottles

For a new material or a new application or a new product we have to prove the performance before introducing into the market. The consumer needs to be satisfied on the product to be used. In order to get good quality product we have to carry out some of the quality control tests that has been standardized. Quality Control means the control of "goodness" or the "excellence" of a product. In this ever-changing pattern of producing and making things, if "quality" of a product is not maintained then it is difficult for the product to survive. Selection of raw materials to the assessment of finished product; this entire chain of functions is effectively linked through quality control. Unless a job is thoroughly and continuously checked for quality, at the strategic stages, it is never possible to always get a good quality finished product. Quality control increases output and reduces breakdown. It seeks to ensure that the finished products conform to the specified standards of performance, utility and reliability. Hence, in any field today, quality control has become an indispensable tool of modern management.

Total quality control can be divided into two components as follows:

- *Product quality control*
- *Packaging quality control*
- Packaging quality control involves three main stages namely:
- *Quality control on incoming raw materials*
- *Quality control on in-process materials and*
- *Quality control on finished products*

Table 12.4 The important tests and test equipment of plastic packages

1. Caliper/thickness	
Standard No	IS 2508-1984/1S 1060
Equipment used	Dial Gauge micrometer
Units for results	mm/m/gauge
Significance	To check conformity of thickness to desired specification. It is an important property affecting mechanical properties, sealability, barrier properties, performance, etc.
2. Tensile strength and % elongation	
Standard No	IS 2508-1984/IS 1060
Equipment used	Universal testing machine (INSTRON)
Units used for results	kg/cm^2
Significance	It is a mechanical property significant in high—speed operations. By knowing the amount of deformation (strain) introduced by a given load (stress), it is possible to predict the response of the material under end-use conditions. Tensile strength is the maximum tensile stress, which a material can sustain and is taken to be the maximum load exerted in the film specimen during the test divided by the original cross section of the specimen. Elongation is usually measured at the point at which the film breaks and is expressed as the percentage of change of the original length of the specimen between the grips of the testing machine. Its importance is a measure of the film's ability to stretch. During the unwinding operation, elongation is usually measured at the point at which the film breaks and is expressed as the percentage of change of the original length of the specimen between the grips of the testing machine. Its importance is a measure of the film's ability to stretch. During the unwinding operation, elongation is dangerous as any sudden imbalance in the unwinding operation could lead to breaking of the film. A certain amount of tension is necessary during the unwinding operation so that films could lead to breaking of the film. A certain amount of tension is necessary during the unwinding operation so that films with low yield strength are in danger of being stressed beyond their yield point
3. Heat seal strength	
Standard No	ASTM F 88-68 (1973)
Equipment used	Universal testing machine (INSTRON)
Units used for results	g/15 mm width
Significance	It is the force required to pull open a seal. It is of relevance to the integrity of a package. Heat sealability of a packaging film is one of the most important properties when considering its use on wrapping or bag making equipment. It is directly related to dwell time, temperature, pressure, seal contamination, thickness variation, MFI, type of sealant layer, type of sealing process (impulse/high frequency/ultrasonic)

(continued)

Table 12.4 (continued)

4. Bond strength	
Standard No	ASTM D 903/F904
Equipment used	Universal testing machine (INSTRON)
Units used for results	g/15 mm width
Significance	The performance of multi-layer structure depends upon the ability of the laminate to function as a single unit. Hence this test is done to check the quality of lamination. The test indicates: • Reliability of continuing adhesion during packaging, sealing and storage • The influence of oils/solvents • The adhesive used • The quantity of adhesive • Affinity to moisture • Excess solvent retention • Climatic conditions
5. Hot tack strength	
Equipment used	Universal testing machine (INSTRON)
Units used	g/15 mm width
Significance	It is the critical factor in selecting sealant layers in applications such as vertical form/fill/seal lines where liquids and other substances may contaminate the seal area. Testing is done while the seal is still hot
6. Shrinkage	
Standard No	ASTM D 1204-1978
Equipment used	Instron hot air oven
Units used	%
Significance	As a result of manufacturing process, internal stresses are locked into the film and this can be released by heating. For any given type of film the temperature at which shrinkage will begin are related to processing techniques
7. Flex resistance	
Equipment used	Gelbo flex tester
Units used	No of cycles
Significance	This test determines the resistance of flexible packaging materials to flex formed pin holes. The resistance to repeated flexure or creasing is important. Some films are highly resistant whereas others will fail by pinhole or total fracture after bending only a few times. In essence, the resistance to bending is measured by repeatedly folding the film backwards and forwards at a given rate. The number of cycles to failure is recorded as the flex resistance

(continued)

Table 12.4 (continued)

8. Coefficient of friction	
Standard No	ASTM D 1894
Equipment used	Buchel Vander–Korpt (Pendulum method)
	The co-efficient friction is related to the slip properties of plastic film. This empirical data can be used for control of film production. It is a measure of the ease with which the surface of one material will slide over another. Thus films which are slippery over various surfaces have a low COF. COF or slip properties of film are important in determining how that film will perform on conversion equipment and in final form such as in openability or stacking
9. Identification plastics	
Equipment used	DSC/TMA
Significance	DSC is a fast method to identify the substrates in co—extruded films from melting temperatures of individual substrates. Multi-layer films can be easily checked by DSC to find out the individual and number of components along with their concentration
10. Leakage in heat sealed packages	
Standard No	ASTM D 3078 (1977)
Equipment used	Dissector, vacuum pump
Units used	Pass/fail
Significance	To evaluate seal performance of the pouch
11. Dart impact test	
Standard No	IS 2508-1984
Equipment used	Dart of different weights with stand
Units used	gf for 50% failure
Significance	This test is carried out to evaluate the impact resistance of flexible plastic film. The impact strength of a film is a measure of its ability to withstand shock loading
12. Tear strength	
Standard No	ASTM D 1922
Equipment used	Elmendort Tearing strength tester
Units used	Gf
Significance	Tear strength is an important property of packaging films and knowledge of both resistance to tear
13. Melt flow index	
Standard No	ASTM D 1238

(continued)

Table 12.4 (continued)

Equipment used	MFI tester
Units used	g/10 min @190 °C for PE and 280 °C for PP
Significance	Indicates the flow characteristics of the material at different processing conditions
14. Seam strength	
Standard No	IS 3790
Equipment used	Universal testing machine (INSTRON)
Units used	kg/cm^2
Significance	Determines the strength of the fabric used for the manufacture of sacks
15. Environmental stress crack resistance	
Standard No	IS 6312-1980
Equipment used	Oven, soap solutions etc.
Units used	Pass/fail
Significance	Indicates the stress crack resistance of plastic material
16. Closure leakage test	
Standard No	IS 6312-1980
Equipment used	Vibration table
Units used	Pass/fail
Significance	Indicates defects in closure system
17. Adhesion strength of pressure sensitive tape	
Standard No	IS 2880-1978
Equipment used	Universal Testing Machine (INSTRON)
Units used	kg/cm
Significance	Evaluate the quality of adhesive used
18. Torque test	
Equipment used	Torque tester
Units used	kg/lbs
Significance	Determines the torque required for proper closure

(continued)

Table 12.4 (continued)

19. Water vapour transmission rate	
Standard No	1060/ASTM E-96
Equipment used	Lassy/MOCON WVTR Tester
Units used	g/m^2/24 h
Significance	Most important property for moisture sensitive products. It decides shelf-life of the products and is directly proportional to thickness. This property varies from polymer to depending up on the thickness of the film
20. Oxygen transmission rate	
Standard No	ASTM F 3985
Equipment used	Mocon/Lysssi OTR TESTR
Units used	cc/m^2/24 h at 27 C under 1 atmospheric pressure
Significance	Most important property for gas sensitive products/vacuum packaging/gas packaging materials
21. Odour pick up test	
Standard No	4006-1972
Significance	Indicates odour transfer from packaging material as products
22. Haze	
Standard No	ASTM D 1003
Equipment used	Hazemeter
Significance	In certain applications, high clarity and minimal haze or frostiness is desirable. This is the case in many packaging applications where good clarity enhances and the polymer structure diffuses light as it passes through film and cause hazy appearance. The hazemeter is setup to transmit a beam of light, which is diffused or scattered from its original path. The results are reported in terms of percentage haze. The lower haze, the better the clarity of the film
23. Gloss	
Standard No	ASTM D 2457
Equipment used	Glossmeter
Units used	Percentage
	Specular gloss correlates to the shine or sparkle of film. This trait can influence desirability of consumers to purchase the product packed in it. Gloss in film can be optimized by adjustment of extrusion parameters. Once processing conditions are perfect, changing resins to a higher melt index and higher density at constant molecular weight distribution will yield in better gloss

Due to increasing consumer protection legislation and higher safety standards, more testing is needed. Very often investigation testing is needed for:

- *A new product—Established packaging for a new product formulation*
- *A new application—Over-wrap film for shrink packaging*
- *A new material—New plastic film for existing product*

There is a wide range of properties to be considered while selecting plastic materials for a particular purpose. The tests for plastic packages can be classified into three broad groups.

12.1.5 Applications of Multilayered Bottles

Potential applications for multi-layer bottles include:

- Edible Oil
- Juice
- Beverages
- Transparent food containers requiring increased barrier properties

Multi-layer PET bottles of both multi-layer and coated PET beer bottles are restricted to low volume breweries.

Pechiney Plastic Packaging has developed a family of PP barrier containers that are claimed to be as clear as multi layer PET bottles. The three layer PP/EVOH/PP barrier containers are made using modified reheat stretch blow mold machines.

Multi layer structures are far more prevalent than coatings and account for about 70% barrier PET bottles. The technology has succeeded for a decade in the PET Ketchup bottles. It is claimed that multi layer PET food bottle is the optimal solution in barrier performance functionality and cost. Today PET and PP are competing to determine which of these two base polymers, with other barriers will dominate the barrier plastic bottle market [33].

In a multi-layer structure a core layer or layers containing higher priced barrier materials are sandwiched by PET structural layers. There are several five layer PET bottles designed for beverages. A five layer PET bottle is used for Real life line of new age beverages, non carbonated flavors. The bottle incorporates virgin PET, EVOH provide necessary barrier performance and post consumer recycled (PCR) resin within the five-layer structures (Virgin PET/EVOH/PCP-Pt/VIRGIN PET) structure no adhesive or tie layers. Two thin layers of EVOH provide necessary barrier. Instead of EVOH, MXD 6 is used in a five layer structure (Virgin PET/MXD6/Virgin or PCR–PET/MXD6/Virgin PET) for Coca-Cola bottles [34].

Continental PET technologies supplies a five layer structure (PET/O_2Scavenger/PET/O_2 Scavenger/PET) single–serve PET bottle for Miller beer. Krones developed a five layer structure (PET/nylon 6/PET/nylon 6/PET) PET beer bottle for a swiss bravery, while Bass developed a multi-layer (PET/EVOH/PET)PET bottle for carls berg Black Label beer. Tetra Pak's patented two stage process Sealica was developed for molding a multi-layer PET perform using a new thermoplastic epoxy barrier resin called Blox from Dow plastics.

12.1.6 Conclusion

Multi layer bottles have great importance in the field of food packaging, cosmetics, agriculture and in the medical field. The contribution of multilayer bottles to our daily life is unpredictable. This chapter discusses the major areas in the field of multilayer bottles. In this chapter the preparation methods of multilayer bottles are discussed very clearly. The current trend in multilayer bottles and their applications in different areas are also discussed. The chapter also discusses the important parameters that must be needed for the multilayer bottles to be taken for any application. The chapter includes the important requirements for bottles for its best usage. Today, several researches are going on multilayer materials, so there is a chance of development in the field of multilayer bottles.

References

1. ACE (The Alliance for Beverage Cartons and the Environment) (2009) Beverage carton maintains growth in recycling rates. Press release of 6.7.2009. ACE, Brussels
2. BfR (German Federal Institute for Risk Assessment; Bundes instituts fur Risikobewertung) (2009) Recommendation XXXVI of the Federal Institute for Risk Assessment on plastics intended to come into contact with food. Paper and board for food contact. BfR, Berlin
3. BfR (German Federal Institute for Risk Assessment; Bundes instituts fur Risikobewertung) (2010) Recommendation XXVIII of the Federal Institute for Risk Assessment on plastics intended to come into contact with food. Cross-linked polyurethanes as adhesive layers for food packaging materials. BfR, Berlin
4. Busser S, Jungbluth N (2009) The role of flexible packaging in the life cycle of coffee and butter. Int J Life Cycle Assess 14(Suppl. 1):80–91
5. CEPE (European Council of producers and importers of paints, printing inks and artists' colours) (2009) Code of Practice for coated articles where the food contact layer is a coating
6. Commission of the European Communities (1993) Commission Directive 93/8/EEC of 15 March 1993 amending Council Directive 82/711/EEC laying down the basic rules necessary for testing migration of constituents of plastic materials and articles intended to come into contact with foodstuffs. Off J L090:22
7. Commission of the European Communities (1997) Commission Directive 97/48/EC of 29 July 1997 amending for second time Council Directive 82/711/EEC laying down the basic rules necessary for testing migration of the constituents of plastics materials and articles intended to come into contact with foodstuffs. Official Journal L222:0010–0015
8. Commission of the European Communities (2002) Commission Directive 2002/72/EC of 6 August 2002 relating to plastics materials and articles intended to come into contact with foodstuffs. Off J L220:18

9. Franz R, Stoermer A (2008) Migration of plastic constituents. In: Piringer O-G, Baner AL (eds) Chapter 11: plastic packaging—interactions with foods and pharmaceuticals. Wiley-VCH, Weinheim, pp 349–416
10. Guinee JB et al (2002) Handbook on life cycle assessment. Operational guide to the ISO standards. Kluwer Academic, Dordrecht
11. de Gennes P-G (1979) scaling concepts in polymer physics. Cornell University, Ithaca
12. Helfand E (1982) In polymer compatibility and incompatibility. In: Sole K (ed) Midland Macromolecular Institute, New York
13. Kovach GP (1959) In: Bernhardt EC (ed) Processing of thermoplastic materials. R. E. Krieger Publishing Co., Inc., Huntington, N.Y., pp 511–522
14. Presswood KJ (1981) In Oriented PVC bottles; process description and influence of biaxial orientation on selected properties, 39th annual technical conference, May 1981, Society of Plastics Engineers, Inc., Brookfield Center, Conn., pp 718–721
15. Date S (1981) In Co-Pak Multilayer Plastic Containers, 5th annual international conference on oriented plastic containers, March 1981, Ryder Associates, Inc., pp 37–48
16. Giacoletto GM in Tappi 1999 polymers laminations & coatings conference, p 79
17. Edenbaum J (1992) Plastic additives and modifiers handbook. Van Nostrand Reinhold Co., Inc., New York
18. Moore EM (1996) Polypropylene handbook: polymerization, Characterization, properties, processing, applications. Hanser Publishers, Munich
19. Van Der Hoeven JC, Wimberger-Friedl R, Meijer HEH (2001) Polym Eng Sci 41(1):32
20. Rauwendaal C, Osswald T, Gramann P, Davis B, in SPE ANTEC, 1998, pp 277–283
21. Hensen F, Knappe W, Potente H (eds) (1989) Handbuch der Kunststoff-Extrusions-technik I Grundlagen. Carl Hanser Verlag, Munich
22. Osswald T, Menges G (1966) Materials science of polymers for engineers, Carl Hanser Verlag, Munich
23. Rosato DV, Rosato DV (2000) concise encyclopedia of plastics: fabrication & industry. Kluwer Academic Publishers, Boston
24. Beall G (1982–2000) Plastic part design for economical injection molding, seminar manual
25. Sepe M (1998) Dynamic mechanical analysis for plastics engineering. Plastics Design Library, Norwich, N.Y, p 189
26. Kennedy P (1995) Flow analysis of injection molds. Hanser Gardner Publications, Cincinnati, p 237
27. Rieke JK (1981) In: Evans RE (ed) Physical testing of plastics, correlations with end-use performance (ASTM STP 736), American Society For Testing and Materials, Philadelphia, Pa., 1981, pp 59–76
28. Kinloch AJ, Young RJ (1983) Fracture behavior of polymers. Applied Science Publishers, London, p 107
29. Turner S (1983) Mechanical testing of plastics, 2nd edn. The Plastics and Rubber Institute, London, p 199ff
30. Bowden PB (1973) In Haward RN (ed) Physics of glassy polymers, Halsted Press, New York, Chap. "5", p 287ff
31. Koros W J, Hellums MW (1990) In Kroschwitz JI (ed) Encyclopedia of polymer science and engineering, Wiley, New York, Supplement volume, p 724
32. Weinkauf DH, Paul DR (1990) In: Koros WJ (ed) Barrier polymers and structures (ACS symposium series, vol. 423). American Chemical Society, Washington, D.C., p 60
33. Sina Ebnesajjad, Plastics Film in Food Packaging, (2012) Elsevier
34. Cirillo G, Spizzirri UG, Francesca I (2015) Functional polymers in food science and technology. Scrivener Publishing, Wiley

Chapter 13
Multiphase Materials for Tire Applications

R. Rajasekar, C. Sivasenapathy, C. Moganapriya and A. Sivakumar

13.1 Introduction

13.1.1 Multiphase Materials

In general for any material, the microstructure (internal structure) determines the mechanical property such as ductility, brittleness, toughness and resistance to creep at different geometric scales. This microstructure of any material is comprised of many different phases with different properties. The behaviour of all these phases (multiphase) determines the end property of the material. This study about the multiphase material helps in designing durable, high performing reliable products for this sustainable society. This multiphase materials have wide applications in the field of mechanical, civil and especially in tire engineering. As for as the mechanical field is concerned, the multiphase material is currently focussed on developing high temperature withstanding alloys, hard metals and pearlite steels.

The primary aim of materials science is to establish a correlation between microstructure and properties. For single phase alloys and metals, these correlationships are established well using a number of theories. However, for two-ductile-phase alloys and metal matrix composites (MMCs) these theories are not established well due to the need of an efficient method for quantitative characterisation of multiphase materials. Designing a theoretical concept for quantitative characterisation of multiphase materials and predicting it with experimental work will bring measurable advantage over these materials.

R. Rajasekar (✉) · C. Sivasenapathy · C. Moganapriya · A. Sivakumar
Department of Mechanical Engineering, Kongu Engineering College,
Erode, Tamil Nadu, India
e-mail: rajasekar.cr@gmail.com; rajamech@kongu.ac.in

J.K. Kim et al. (eds.), *Multicomponent Polymeric Materials*,
Springer Series in Materials Science 223, DOI 10.1007/978-94-017-7324-9_13

The main difference between single-phase and multiphase materials is the existence of different grain and phase boundaries. To improve the geometrical and the topological properties of multi-phase microstructures a systematic method should be developed which allows a two-phase microstructure with any grain shape, grain size and phase distribution to be transformed into a three-microstructural-element body (3-E body). Hall and Petch have discovered many experimental works related to single phase metals and later it was extended to multiphase materials known as Hall-Petch relation.

At the end of a tire's life cycle the compositions such as natural rubber, carbon black, butadiene rubber (BR) are present with BR in a major phase. This mixture of different rubbers complicates the devulcanization of tire. Proper devulcanization of the rubber using different mechanical, thermomechanical and thermo chemical processes improves the internal microstructure of the Ground tire rubber (GTR) material and especially thermo chemical process is the preferred one for better devulcanization process. Poor diffusion of the rubber particles during devulcanization is the major problem results in inhomogeneous devulcanization which affects the property of a devulcanized GTR. This problem is minimized by swelling of a GTR before devulcanization in an oil containing the chemicals so that it improves the diffusion resulting in improving the property of the GTR material.

13.1.2 Need for a Multiphase Material

Elastomeric compounds are often consist of more than one type of rubber as polymer matrix. Blending of two or more elastomers are done for several reasons, such as improving the mechanical and physical properties of the first elastomer, obtaining good characteristics of rubber compound or to decreasing the compound cost. For example, a tire compound has to be elastic and soft in order to adapt to the road surface. Also it has to be stiff and strong to bear load, and resistant to abrasion to deliver long service life. Likewise various rubber based goods such as seals and 'O' rings, have to be soft and meets the contour of the equipment; but also, they have to undergo minimum compression set to perform under high stress for along period of time. Since different elastomers provide different responses to external effects, blending of various rubbers has been practiced to meet all the needed properties. However, the mechanical, Rheological, thermal and other properties of a rubber blend depends mainly on its state of compatibility. Most of rubber blends have cure rate differences because of discrepancies in polarity and unsaturation. Also, various dispersion quality of fillers and other compound additives in two or more rubbers causes deterioration of mechanical properties of the blend.

Blends of rubber are often used as a basis for compound formulations. For instance, tire tread may contain the following rubbers Natural Rubber (NR), Butadiene Rubber (BR) and Styrene-Butadiene Rubber (SBR) to provide costeffective products with good dynamic properties, mechanical properties and abrasion resistance. Blends of Ethylene Propylene Diene Monomer Rubber (EPDM) grades are frequently used when formulating EPDM compounds. Scanning electron microscopy (SEM) can be employed to study the surface morphology of the blends. Consequently, these studies can also be reasoned to characterize compatibility of various rubbers such as EPDM and NBR rubber by many other analytical techniques like FTIR, atomic force microscopy (AFM), ultrasonic interferometry, and thermal techniques etc.

13.2 Fillers Used for the Preparation of Rubber Compounds for Tyre Application

13.2.1 Styrene Butadiene Rubber (SBR)

SBR is one of the synthetic rubber that is widely used for manufacturing various products such as in car tires, cable insulation material, automotive industry, in production of technical parts, shoe sole, etc. SBR is a random copolymer, does not crystallize on stretching, because it is totally an amorphous rubber. Therefore, it becomes a necessity of mixing SBR with filler materials. It is known as a cross-linkable polymer on irradiation. SBR in its raw form represents a homogenous and amorphous polymeric material and hence the bulk mechanical properties of radiation—vulcanized product depends on the basis of its chemical structure and irradiation conditions. Hence, self-reinforcement does not take place at high extension, as in the case for example of NR and there is no contribution to its tensile strength would be anticipated depends on its chemical structure. The tertiary substituted carbon atom on the macromolecules chain denotes the most susceptible carbon atom (Woods et al. [1]) and on irradiation breakage of this bond occurs favourably leading to the formation of the hydrogen atom and benzyl-type radical. The formed benzyl radical undertakes stabilization with phenyl ring of styrene monomer. Such resonance stabilization could be expected to reduce the efficiency of benzyl radical but it also increase its half-life. When such radicals collide favourably to each other, they may react among each other and results in the formation of a covalent. The formation of such bond would also be predicted to be a content of styrene in the copolymer, which makes 20–40% by wt. Under such circumstances, a less crosslinking density would be expected which leads to the low value of TS attained by vulcanized raw SBR.

13.2.2 Natural Rubber (NR)

NR is an elastic hydrocarbon polymer that occurs naturally as a milky colloidal suspension, or as latex, in the sap of some plants. The scientific name for the rubber tree is Heava brasiliensis. Natural rubber is a radiation crosslinking type of polymer because it contains a double bond in its basic polyisoprene units. Also, it is characterized by its stretching ability to crystalline and the absence of phenyl group in its structure as like SBR. Because of these properties, it would be predicted that NR attains higher TS than SBR. Elastomer blends are used for many reasons like lowering the compound cost and the complex shaped product can be easily fabricated during production. Natural rubber and styrene butadiene rubber have been blended from the beginning for these reasons. The mechanical properties of (NR/SBR) blends can be considerably improved. SBR/NR are a special performing blend which is applicable for lot of applications from race car tires to footwear applications. The SBR component adds the toughness whereas the natural rubber gives superior resilience and energy return.

13.2.3 Carbon Nanomaterials Based Elastomers

Araby et al. [2] studied the electrical, mechanical, thermal, and flame retardant properties of elastomers blended with carbon nanomaterials. In this article, the fabrication processes of elastomer composites, the importance of keeping fillers at nanoscale in matrices, and the importance of incorporating CB, CNTs, and graphene in the elastomeric composites have been explained briefly.

Of all materials, carbon nanomaterials occupies a unique position in materials research because of their amazing mechanical and physical properties. Since about 80 years, CB has being used as the leading filler for elastomers because of its significant reinforcement. It has also been used to increase the electrical conductivity of polymers. Since its discovery in 1991, CNTs have concerned widespread research. Since they are expensive, graphite sheets have been studied as an alternate, though they are lack in CNTs' superior properties. Since 2004 graphene has risen as a new carbon allotrope with nano plate-like structure which is considered as the strongest material with superior thermal conductivity and the electron mobility.

13.2.3.1 Carbon Black

CB is the oldest allotrope of carbon which has been used by ancient Egyptians and Chinese as a pigment for writing. It is produced by partial combustion or thermal cracking of heavy petroleum products or natural gas and it is in a colloidal form. A production reactor consists of two zones: high temperature zone and quenching zone, first being to crack the hydrocarbons produced by fuel like natural gas, and the latter to stop the reaction by water. Reaction time is limited by quenching, which decides the product morphology (i.e.) microstructure and surface area. Small-size product is attained by short reaction time at high temperature. The fine particles of CB always coalesce into aggregates which are in an irreversible and an isometric form. These aggregates attracts each other by van der Waal bonds, developing loosely-bounded agglomerates ranges from 100 to 500 nm in diameter. The degree and size of aggregates are important aspects which determines the mechanical performance of final composites. Because of increase in the size of aggregates, the inter-aggregate distance would get increase taking the same volume fraction. As a result, a high filler fraction (>30 vol%) is often required to attain required mechanical performance for elastomer composites.

13.2.3.2 Carbon Nanotubes

CNTs possess intrinsic electron mobility at room temperature with remarkable thermal and mechanical properties. CNTs have lattice-like tubular structure comprising of periodic honeycomb network of covalently bonded carbon atoms and it can be produced by laser ablation, chemical vapour deposition (CVD) and arc discharge. They exist in two types as single walled and multi-walled nanotubes. A single walled CNT (SWCNT) comprises a single graphene layer rolled up into a cylinder of 0.2–5 nm in diameter and a few centimetres in length. A multi-walled carbon nanotube (MWCNT) comprises of two or more concentric cylindrical shells of multi-graphene layers coaxially united by van der Waal forces with layer spacing 0.34 nm, an outer diameter 2–50 nm and a length of up to 10 μm. These exceptional intrinsic characteristics of CNTs have stimulated great and sustained benefits in materials research over the past few decades with various applications from one-dimensional quantum nanowires to functional nanocomposites.

13.2.3.3 Graphite Nanosheets, Graphene and GnO to GnPs

Graphite, an allotrope of carbon, has layered structure and it conducts heat and electricity. It is the most stable form of carbon under normal conditions. Graphite is the parental material for the plate-like carbon materials, including graphite intercalation compound, graphite nanosheets, expanded graphite, graphene and GnPs. Inserting different chemical species, such as alkali metals, alkaline earth metal, halogens and acids into graphite layers, improves the basal space and forms graphite intercalation compounds. Expanded graphite is regularly fabricated from graphite intercalation compounds by microwave, thermal shock, plasma and chemical treatments. During these processes, the fast evaporation of the intercalants produces immense gaseous pressure that can overcome the van der Waal forces between the layers producing worm like structure with an expansion volume ratio about over one hundred.

13.3 Researches Towards Rubber and Filler Based Composites

13.3.1 Silica Based Natural Rubber Composites

Synthetic silicon dioxide (silica) can be produced either by precipitation or by pyrogenic (thermal) process and, therefore, this filler can be classified into two groups; namely precipitated and pyrogenic (fumed) silica. It is the most important filler competing CB in the area of rubber technology. However, in the early stages of silica usage, their uses as reinforcing filler instead of CB were limited due to a number of problems:

- Silica-filled compounds sight higher viscosities
- They are more difficult to mix and process
- There is a concomitant increase in vulcanization time
- Silica-filled compounds often show lower crosslinking density.

Surface silanol groups of silica have received keen interest in recent years, because of their most important role of interfacial interactions in several applications, such as the reinforcement of synthetic or natural elastomers. The silanol groups show strong filler–filler interactions and cause the absorption of the polar materials such as curatives on the surface. Absorption of curatives results in a reduction in the crosslinking density and delay of the scorch time of the silica filled rubber compounds. The main concern in using silica as a reinforcing filler for elastomeric matrices is the difficulty of obtaining a very efficient dispersion of silica particles, because of the filler's surface behaviour, imposed by the chemistry of the last end silanol groups, with isolated. Silanols made the surface of hydroxylated silicas inducing strong interactions between silica particles and leading to a poor dispersion of the filler in the conventional natural or synthetic non-polar elastomer. Dispersal of silica in the matrix can be enhanced by reducing the polarity of the filler particles and sterically avoiding their aggregation. For example, silane coupling agents containing sulfur are widely used in the tire industry to increase dispersibility of silica into elastomeric matrices, thus promoting the formation of covalent bonds in between the phases. The strength and toughness of chemical interactions between silica of different chemical characteristics and NBR, was studied by infrared spectroscopy measuring the increase of C-N fundamental vibration frequency and by the density functional theory. The result shows that functional groups on the silica's surface plays a most important role on the compatibility of immiscible blends of NBR copolymers, as they strongly interact with the rubber phase.

Al-Hartomy et al. [3], studied the impact of carbon black CB/silica in the physical and mechanical properties of Natural rubber based composites. The aim of this research is to study how carbon black and silica ratio affects the curing characteristics, dynamic and mechanical properties of composites based on epoxidized natural rubber. The data achieved showed that the composites containing carbon black and silica at a ratio of 30/40, 20/50, and 10/60 ratio gives the finest complex of parameters. It is because of the intermingling of carbon black and silica particles which leads to hydrogen bond formation in between the silanol groups of silica particles. As a result, filler disperses well in the rubber matrix and silica particles aggregate very less.

13.3.1.1 NBR/EPDM Silica Reinforced Composites

Jovanovic et al. [4] studied the mechanical properties, cure kinetics, rheometric characteristics of the silica reinforced nanocomposite based acrylonitrile-butadiene rubber (NBR) and ethylene–propylene–diene monomer (EPDM) blends. Thermal aging has been done and mechanical properties were determined before and after the thermal aging. Using FTIR and SEM, The surface morphology of the blended rubber were studied. For EPDM/NBR blend in the ratio of 20/80 an increase in tensile strength were found. It was also determined silica-reinforced EPDM/NBR rubber blends were immiscible from the differential scanning calorimetry (DSC) test. The optimal curing time was found decreased with increasing NBR rubber content in the EPDM/NBR rubber blend during the rheometric characteristics test using rheometer.

13.3.2 Clay/NR Rubber Composites

Rubber-clay nanocomposites exhibit outstanding properties, at low loading levels of clay, as compared with unfilled rubber compounds or conventional filled composites. In general, the preparation methods for rubber-clay nanocomposite can be divided into four main groups, rendering to the handling technique:

- In situ polymerization
- Intercalation of rubber via solution blending
- Direct melt intercalation method
- Intercalation of rubber through latex compounding

The main objective of preparing organoclay nanocomposites is to achieve high degree of dispersion of organoclay aggregates within the polymer matrix that can produce very huge surface area. The efficient dispersion of organoclay platelets in the polymer matrix leads to improvement of the overall properties of the polymer. Clay minerals have hydrophilic character and hence a nanometric dispersion into a hydrophobic polymer is hard to achieve, particularly through solution and melt blending methods. To obtain a similar spreading in a hydrocarbon medium, the alkaline-earth cations of primeval clay crystals are replaced by hydrophobic organo modifiers, typically ammonium cations bearing long alkyl chains a compensating cations. The role of the alkyl ammonium cations is to lower the host's surface energy and thereby improve wetting-out by the polymers. The obvious outcome of the polymer-organoclay interactions is improved strength characteristics because of the strong interfacial interactions between silicate layers and the polymer chains. By appropriate choice of the alkyl ammonium cation, the interactions between the host layers and the intercalated polymer chains can be tailored to synthesize new organic–inorganic hybrids.

Magaraphan et al. [5], mentioned that MMT clays treated with long primary amines led to much more improved mechanical properties when incorporated into Natural Rubber matrices, than preserved with quaternary amines of the same number of carbon atoms. The length of hydrocarbon in the alkylamines had no effect on the curing time, whereas the nanocomposites prepared with long quaternary amine showed comparatively faster cure time.

Varghese et al. [6], arranged Natural Rubber, polyurethane rubber (PUR) and NR/PUR based nanocomposites from the related lattices by adding 10 phr pristine synthetic sodium fluorohectorite. It was observed that, in blends composed of polar PUR and apolar NR, the silicate layers were preferentially embedded in the polar PUR phase in well a intercalated/exfoliated stage. The properties of the PUR/NR-based nanocomposites were similar to those containing plain PUR. This result was claimed to be of great economic significance as NR latex is cheaper than PUR latex.

Epoxidized natural rubber (ENR), obtained by epoxidation of 1,4-polyisoprene, has a higher glass transition temperature, polarity and superior compatibility with Silicone Rubber. Rajasekar et al. [7], deliberated the effect of dual fillers on the properties of SBR and EPR compounds prepared using ENR as a compatibilizer. ENR-organically modified nanoclay (Cloisite 20A) composites were prepared by solution mixing and the obtained composite was incorporated in SBR and EPR matrices along with carbon black. The morphological studies proved the intercalation of nanoclay platelets in ENR and further incorporation of EC in SBR and EPR matrices leads to partial exfoliation of nanoclay platelets. A curing study validated quicker scorch time, cure time and improved supreme torque for the compatibilized SBR and EPR nanocomposites containing a dual filler system compared to control. Dynamic mechanical thermal analysis showed an increase of storage modulus for SBR and EPR compounds containing dual fillers as compared to compounds containing pure and single filler. The same compounds showed substantial improvement of mechanical properties.

13.3.2.1 Clay/EPDM Rubber Nanocomposites

EPDM rubber is a widely used engineering elastomer with outstanding resistance to ozone degradation, combined with high thermal stability and EPDM/clay nanocomposite materials have concerned special attention. However, this elastomer does not have any polar groups in the support and strong interfaces between EPDM chains and clay layers can hardly be achieved, unless a suitable compatibilizer is used.

Gatos et al. [8], studied the characteristics of EPDM/organoclay nanocomposites, resulting from dissimilar treating conditions. The modification in the rubber polarity was achieved by replacing half of EPDM rubber by maleic anhydride (MA) —and glycidyl methacrylate (GMA)-grafted EPDM. It was shown that incorporation of grafted EPDMs strongly improves the strength and stiffness of nanocomposites containing 10 phrorganoclay.

EPDM-g-MA was used as compatibilizer in EPDM/OMMT nanocomposites prepared by melt compounding. XRD analysis indicated that matrix chains were intercalated into the gallery space of OMMT effectively. An almost complete dispersion was obtained when the compatibilizer-filler ratio continued three. In the free sample holding a reduced amount of EPDM-g-MA, strong flocculation of the stacked dispersed clay particles was evidenced. Better dispersion of silicate layers was obtained in EPDM matrices with a higher Mooney viscosity, as they exert higher shearing during mixing. Also, the linear polymer chains can be more intercalated into OMMT galleries than long-branch EPDM chains.

13.3.2.2 Clay/NBR Nanocomposites

Rubber-clay nanocomposites have been prepared by modification of the naturally occurring Na montmorillonite clay followed by mixing with rubber. Intercalation and exfoliation of the clay layers has been perceived in SBR, Natural butadiene rubber, polyisobutylene-co-paramethylstyrene (BIMS), EOC and thermoplastic elastomers. The resulting structure and orientation of clay particles in the matrix depend on the nature of both rubber and clays. Usually nonpolar rubbers are better compatible with the organoclays where long aliphatic amine chains are existing, while some polar rubbers have better desirability towards unmodified clay. Once rubber and clay are companionable, extra uniform dispersions of clay are obtained when the width of the clay layers is thinner than in the case of lower compatibility.

The effect of clay modification on organo-MMT/NBR nanocomposites was further studied by Kim et al. The alkylamines used for modification were octylamine (C8), dodecylamine DDA (C12), and ODA (C18). It was found that the mechanical properties increased in the order C8-MMT\C12-MMT\C18-MMT, depending on the length of the alkyl chain in the alkyl ammonium. With increasing organo-MMT content, it showed substantial improvement of mechanical properties compared with unfilled NBR. This high reinforcement effect implied a strong interaction between the matrix and the clay interface, which was attributed to the nanoscale and uniform dispersion of the silicate layers in the NBR matrix. Study of the vulcanization reaction showed that as the chain length of the modifier increased the rheometric torque decreased and the optimum curing time were shortened. They also observed the extent of swelling in methyl ethyl ketone (MEK) clearly decreased with increase in organo-MMT loading. Thus the organo-MMT/NBR nanocomposites have excellent barrier properties compared with vulcanized NBR.

13.3.3 Carbon Nanotubes/Natural Rubber/Carbon Black Composites

Intrinsic dissimilarities in the mechanical properties, electrical conductivity and thermal property between CNTs and elastomers led to more number of studies on elastomer/CNT composites. As the elastomers are limited by their low mechanical properties, more studies are committed to improve the interface between the filler and the matrix through functionalization of CNTs by using various preparation methods. A wide range of increments have been observed even though the same fabrication was used, and some results seemed inconsistent. For example, when functionalized MWCNT was mixed NR, its tensile strength improved by 100%; but the strength improvement was stated 800% by using pristine MWCNT irrespective to the small difference in filler loading. The difference should be caused by the use of prevulcanized NR raw material in the first and a standard Malaysian NR in the latter.

Pongdhornsaeoui [9] in his investigation compared the properties of natural rubber (NR) filled with various fillers, like furnace black (N330), conductive carbon black (XE2-B), and carbon nanotube (CNT). Both sonicated and untreated carbon nanotubes were used and designated as S-CNT and U-CNT respectively. The filler constitution was varied from 0 to 8 phr. For whatever the filler type, the increase in the filler not only results in reduced cure time, increased compound viscosity and enhanced cross-link density but also increases the modulus and hardness of the vulcanizates. For N330 and XE2-B, with increase in filler content, the tensile strength increases constantly. However, for both S-CNT and U-CNT, the tensile strength increases with increasing filler content and decreases afterwards. For any given filler content, the CNTs give the highest values for thermal and electrical conductivities, tan δ and storage modulus followed by XE2-B and N330 respectively. The increase in the filler content, whatever the filler type may be results in greater compound viscosity, hardness modulus, tan δ, state of cure, as well as electrical and thermal conductivity. Elongation at break found to decrease with increasing filler content. Amongst all the fillers studied, CNTs has affected the above-mentioned properties to a larger extent than XE2-B and N330 respectively. The tensile strength increases continuously with increasing N330 and XE2-B content from 0 to 8 phr but for CNTs, the tensile strength increases slightly with increasing filler content up to 2 phr and decreases afterwards.

Zhan et al. [10], by using the ultrasonic assisted latex mixing process studied the microstructure and properties of NR/CB/CNT based composites. CNTs are made to disperse into NR using ultrasonic irradiation and finally the mixed latex were co-coagulated to obtain the NR/CNT composite. The results showed that CNT mixed well uniformly than the conventional mixing methods. At the weight ratio of CB/CNT of about 20:5 phr, the mechanical property is at its higher value. Dynamical mechanical analysis showed that with increasing CNT contents, the elastic modulus increases and the maximum loss tangent decreases with increase in CNT content.

Carbon nanotubes (CNT) can be visualized as graphene layers rolled into cylinders consisting of planar hexagonal arrangement of carbon–carbon bonds.

During their growth, depending on the synthesis methods, they can assemble either as concentric tubes (multiwall nanotubes, MWNTs) or as individual cylinders (single wall nanotubes, SWNTs). They attracted enormous attention for their fundamental behavior and for their use in a wide variety of applications in nanoelectronic devices, probe tips for scanning probe microscopes or in the automotive and aerospace industries for the dissipation of electronic charges. The extensive interest in CNTs arises from their unique structural and physical properties: their small size in the nanometer scale, their unique electronic performance, their incomparable properties of ballistic transport, their remarkably excessive thermal conductivity and high optical polarizability, as well as their unparalleled mechanical properties such as high elastic modulus and tensile strength. One of the biggest challenges in the preparation of polymer composites, is to obtain a homogeneous dispersion of carbon nanotubes in a polymer matrix because van der Waals interactions between individual tubes often lead to significant aggregation, thus falling the expected property improvement of the resulting composite.

In another work, it was found that the addition of 1 phr MWCNT to SBR leads to 45% increase of modulus and 70% increase of tensile strength. On the other hand, TEM analysis of composites has revealed reduced dispersal. The great reduction of equilibrium swelling in toluene with MWNT content was not ascribed to filler-matrix interfacial interactions but to the occlusion of rubber into the aggregates.

13.3.4 Polymer/Carbon Black Composites

CB was initially used as a class of fillers for elastomers to reduce the costs and to increase their mechanical properties. Some types of electrically conductive CB have been incorporated into polymers in the last 30 years. The improved electrical conductivity of this composite is mainly explained by a percolation threshold theory, i.e. a high level of conductivity in a polymer/CB composite can be achieved when a continuous conductive network is formed by breaking down the insulative nature of the matrix. It was studied when CB was added into a blend comprising of polystyrene and a styrene-butadiene block copolymer, the copolymer was found to improve the attraction of CB to polystyrene and also decreases the percolation threshold from 6 to 4 wt%. Many studies have focused on improving the mechanical properties of polymers by adding CB, but these did not yield favourable results. For example, tensile strength and Young's modulus of PP/SBR (89:11) composites were somewhat improved (50 and 4%, respectively) at 3 wt% (1.8 vol %) of CB. Most studies on polymers/CB composites focused on increasing electrical conductivity rather than mechanical property. An electrical percolation threshold depends generally on the CB structure and its dispersion inside the matrices. Adding CB into thermoplastics to attain electrical conductivity normally requires high concentration such as 15 vol% that may jeopardize the processing viscosity and the mechanical properties.

Wang et al. [11] studied the electrical, mechanical, and dynamic mechanical properties of conductive piezoelectric ceramic/carbon black/chlorobutyl rubber composites. The composites based on conductive carbon black (CCB), chlorobutyl rubber, and piezoelectric ceramic particles (CCB/CIIR/PMN) were prepared by vulcanizing and compounding process. Microstructure, dynamic mechanical, mechanical and electrical properties of CCB/PMN/CIIR with various PMN and CCB loading were measured by mechanical tests, SEM and electrical analysis. The results revealed that tensile properties decrease with increase in PMN loading, whereas it shows a higher value with CCB loading. This multiphase composites with appropriate composition can be used as good damping polymer materials. Compared to pure CIIR, both CCB/PMN/CIIR and PMN/CIIR exhibits greater temperature range. Dynamic mechanical testing revealed that the piezoelectric damping effect contributes to the damping properties of CCB/PMN/CIIR composites.

Liu et al. [12], studied the natural rubber based nanocomposites incorporating carbon black (CB) and organoclay through melt compounding. The hybrid filler displayed more important reinforcing effect over the same loading of CB on NR matrix, as discovered by dynamic mechanical analysis and mechanical tests. Transmission electron microscopy, wide angle X-ray diffraction, and scanning electron microscopy were used to study the microstructure of this hybrid NR nanocomposites. Dynamic rheological tests were done to study the mechanism of reinforcement to explain the better performance of NR containing the hybrid filler. The storage modulus of uncured rubber composites shows a three order increase in magnitude for rubber filled with 15 parts per hundred parts of rubber CB and 10 parts per hundred parts of rubber organoclay when compared with gum rubber, leads to solid-like low frequency behaviour. The results revealed that the reinforcement is due to a more developed formation of filler network in hybrid filler system compared to a single phase filler.

Alberdi-Muniain et al. [13], studied the dynamic properties of magneto sensitive (MS) natural rubber composites with the influence of carbon black, iron particles and plasticisers. The investigation were conducted at different amplitudes of shear strain within the lowest frequency range. The iron particles inserted in natural rubber are shaped irregularly and are distributed arbitrarily. The plasticisers helps in improving the iron particle blending process, whereas carbon black decreases the production costs and increases the mechanical properties. The end result exposes the effect of MS on shear modulus magnitude and it increases with increased plasticiser and iron particle content and decreases with increased carbon black content. Hence, the study gives a path for optimising the composition of MS natural rubber to meet different requirements, including vibration isolation, an important application area for MS materials.

13.3.5 Graphene/Natural Rubber Composites

Graphene possesses similar mechanical properties as CNTs, but has superior electrical and thermal properties and larger surface area because of its

2-dimensional crystal group. This material in addition to its derivatives as fillers for polymer matrix composites, have shown a great potential for various important applications. The interface between the dispersion layers and the matrix plays a crucial role in determining the structure and properties of nanocomposites. Surface functionalization, including non-covalent or covalent modification of nanoparticles, is an effective method for improving the interfacial interactions between nanoparticles and polymer matrix.

Graphene oxide (GO) is oxygenated graphene possessing abundant oxygen containing functional groups as well as hydroxyl, epoxide and carboxyl groups on the boundaries and surfaces. It has attracted increasing interest as a filler for polymer nanocomposites due to its high long coherence length, dispersive capacity and the barrier property. The improvement of mechanical properties of GO/polymer composites was attributed to its high elastic modulus (*650 GPa) and the efficient stress transfer through the polymer matrices to GO. This enhancement of mechanical properties mainly depends on two factors: filler dispersion and interfacial adhesion. It also should be noted that the oxygen-containing groups of GO assist its dispersion into polymer matrices and form strong hydrogen-bonding interactions especially with polar polymer matrices.

Bai et al. [14], effectively equipped, by a modified Hummers method, chemical exfoliated GO nanosheets with high aspect ratios. Such nanosheets were well mixed with carboxylated NBR (HXNBR) by a simple solution-mixing method and produced GO/HXNBR composites. The addition of 0.44 vol% of GO nanosheets enhanced the tensile strength and modulus at 200% elongation of HXNBR by more than 50 and 100%. It was claimed to be strong interfacial connections between the oxygen-containing functional groups of nanosheets and the carboxyl groups present in HXNBR. This speculation is further supported by the increase of the glass transition temperature of HXNBR from −23.2 to −21.6 °C, at a GO content of 1.3 vol%. The results indicated that GO efficiently reinforced HXNBR due to the good dispersion and strong interfacial interactions.

13.3.6 Polypropylene Blended with Scrap Rubber Tyres (SRT) and EPDM

da Costa et al. [15] studied the mechanical properties of EPDM and SRT blended with polypropylene (PP). Response surface methodology (RSM) is a collection of statistical and mathematical techniques useful for improving, developing and optimizing processes. In this work, RSM was applied to the examination of ternary mixtures of polypropylene (PP), ethylene–propylene–diene monomer (EPDM) and scrap rubber tires (SRT). After appropriate processing in a rotating twin extruder and injection moulding mechanical properties like impact strength and tensile strength were determined and used as response variables. Scanning electron microscopy (SEM) was also used for examination of the morphology of the different blends and results were interpreted. Mechanical properties shows sharp deterioration once SRT particles content improves in the ternary mixtures due to the poor adhesion between SRT and PP matrix. EPDM appears to span the interfaces between regions of SRT and PP, thus enhancing adhesion and compatibilization of the compound. Some tensile properties particularly impact strength together with SEM micrographs proved this observation. Highest impact strength can be reached for the experimental conditions used, only with a mixture of PP/EPDM/SRT when EPDM and SRT blends are retained around 25%.

13.3.7 Effect of Water Absorption on Mechanical Properties of Multiphase Material Rubber Blend

Hameed [16] studied the effect of water absorption on mechanical properties of Binary polymer blend, unsaturated polyester resin (UPE) and natural rubber (NR) with a weight ratio of (95/5)%. Standard dimensions of specimens were prepared from the polymer blend. Compression, impact and bending tests were carried out on the prepared specimens. The research was conducted with three types of water (distilled, tap and rain water) for different periods of time before and after the immersion of the specimens. The stress strain curves were obtained from compression test; whereas the Young's modulus values (E) were obtained from three-point bending test. The impact strengths were found from the impact test before and after immersion into water. The results have revealed that the mechanical behavior of the prepared blend is affected by the immersion time and nature of water. In general, the strain rate and the impact strength of the polymer blend have increased after immersion whereas the Young's modulus has decreased for the same conditions.

13.3.8 Compatibilization of Rubber Based Blends by Filler Modification

Nanoclays are an attractive substitute to traditional compatibilizers because they can be compounded very easily. They also stabilize various crystalline phases of polymers, and improve thermal and mechanical properties. Blending of rubbers is a very regular practice and inclusion of nanoclays to reinforce the rubber blend is apparently a common approach to get a rubber compound with superior mechanical and physical properties. Polymer nanocomposites are commonly used in various applications. Though it has many superior mechanical properties, low toughness is quiet an important problem which should be solved. In some cases, dispersion and addition of clay into rubber phase may offer synergistic effects leading to very fair balance of mechanical behavior. Compatibilization with clay is efficient in many polymer combinations including toughened nanocomposites like PET, PA and PP matrixes.

Fillers are usually used in rubber compound for a number of purposes such as enhancing mechanical properties, increasing of volume etc. Therefore, in this study NR/SBR blend of equal contents are added with various types of fillers namely, HAF, Hisil (highly fined silicon), CB, clay and titanium dioxide (TiO_2) at fixed amount of 40 phr (part per hundred parts of rubber). In this study, styrene butadiene rubber and natural rubber were blended in different ratios namely, NR/SBR (0/100), NR/SBR (25/75), NR/SBR (50/50), NR/SBR (75/25) and NR/SBR (100/0). All the samples were exposed to gamma irradiation dose to 250 kGy. The improvement in the physico-mechanical properties, mechanical properties and thermal properties was followed as a function of irradiation dose and blend ratio. The SBR/NR (50/50) blend with reasonable properties were filled with 40 phr of Hisil, HAF, carbon black (high abrasion furnace), TiO_2 titanium dioxide and clay; the reinforcing ability was found to follow the order: Hisil > HAF carbon black > Clay > TiO_2. This research also showed the effect of gamma irradiation on improving the above mechanical, physical and rheological properties in the presence of filler and. Moreover, radiation dose at 100 kGy is sufficient enough to obtain the desired properties. This prepared composites can be used in many industrial applications such as radio controlled model race car tires to footwear applications. The SBR component adds the toughness whereas the natural rubber gives superior resilience and energy return when used in footwear applications.

13.3.9 Fly Ash Filler with Elastomers

Ramesan [17] studied the mechanical and electrical properties of SBR blended with fly ash (FT). This work was focused on the use of waste fly ash (FA) as filler in the reinforcement of chlorinated styrene–butadiene rubber (CSBR) and the processing characteristics, thermal, morphological, and mechanical properties, solvent resistance, electrical resistivity measurements were studied. Rheometric properties like optimum cure time and scorch time minimizes with rise in filler content, but the maximum torque rises up to 30 phr for FA particles. A significant increase in glass transition temperature has been detected using differential scanning calorimetric study. The mechanical properties like tensile and tear strength, hardness, modulus, compression set, and heat build-up increases, but resilience and elongation at break decreases with increase of filler. Using scanning electron microscopy the fractography of the composite blend has been studied. Variation in the bound rubber formation in FA/CSBR composite were studied. The bound rubber content decreases with rise in extraction temperature, and the activation energy was also calculated using the Arrhenius plot. The electrical properties like dielectric constant, alternating current conductivity, and dissipation factor are advanced than that of CSBR, and the values raises with increase in the content of FA in the composite blend. The glass transition behaviour, cure characteristics, morphology of fractured surfaces, cross-link density and bound rubber content of FA-filled CSBR have been studied. Cure characteristics of CSBR composites revealed that there is a reduction in optimum cure time with increase in fillers, whereas the rheometric torque and CRI raises with raise in concentration of fillers. SEM study showed that sample with 30 phr of FA gives better dispersion of filler in the polymer system. Increase in the values of dielectric behaviour was due to the increase in interfacial interaction between the polymer and the filler.

13.3.10 Calcium Carbonate/Rubber Composites

Calcium carbonate ($CaCO_3$) is one of the most inexpensive and recyclable natural capitals, with the maximum capitals on earth and is widely used in polymer processing. $CaCO_3$ nanoparticles existed are used for strengthening of PP/SEBS and PP/carboxylated SEBS (C-SEBS) blends [18]. The composite with SEBS showed morphology with the SEBS domains and $CaCO_3$ particles independently dispersed within the PP matrix.

To use $CaCO_3$ as reinforcing agent, surface treatment with coupling agents or organic acids is absolutely necessary. Song et al., studied the factors affecting the mechanical properties of styrene-butadiene-styrene (SBS) block copolymer composites filled with $CaCO_3$, after its surface treatment with liquid polybutadiene (LB). The above authors, concluded that the mechanical properties of SBS can be improved remarkably through reinforcing with $CaCO_3$ subjected to the above surface treatment, proving the filler surface with high content of 1,2-double bonds. When SBS, $CaCO_3$ and LB were openly blended, subordinate aggregation of $CaCO_3$ took place and the mechanical properties of the composite were lower. In integral blend method, LB was function as a plasticizer. Covering of $CaCO_3$ with stearic acid leads to important decrease of its superficial tension, which results in decreased interfaces and, expectantly, subsidizes to limited accumulation. The part separation of ternary phase composites is influenced by the melt rheology of the system, the compounding techniques, as well as by the surface characteristics and mutual wettability of the fillers and polymer components. In PP/EPDM/$CaCO_3$ composites, the surface treatment of filler was found to result in separate dispersion of phases (EPDM and $CaCO_3$), whereas encapsulation occurred when untreated calcium carbonate was used.

In another related work, the properties of PP/EPR/coated nano-$CaCO_3$ composites, with and without compatibilizers, were studied. The results indicated that good dispersion of nanoparticles in PP/EPR depends on their surface coatings of stearic and other fatty acids. In both cases, morphology is the core-shell structure, the use EPR-g-MA copolymer did not improve the interface between PP/EPR and nanoparticles but propylene ethylene copolymer would be preferentially localized at the interface of PP and EPR/nano-$CaCO_3$ phases generating an upgraded observance, which will confirm a improved cohesion of the whole material. It is believed that the synergistic consequence of together, EPR elastomer and $CaCO_3$ nanoparticles, must account for the balanced utilization of the ternary composites.

13.4 Rubbers in Tire Applications

Polymer-layered silicate nanocomposites/clays have enormous industrial scope in the field of tires for some superior properties such as higher flexibility with good tensile strength, improved flame-retardant property, tread for lower rolling resistance, increased traction, good antifatigue properties and higher gas barrier property. The applications of the PNCs are found in tire inner liner, tire inner tube, off-the-road (OTR) tire tread, and conveyer belts. Industry experts evaluated that the tire industry would contain eight various polymer types, but styrene butadiene rubber (SBR), butadiene rubber (BR), natural rubber (NR), halogenated butyl rubber (HBR) and isoprene rubber (IR) are the foremost polymers used for tires manufacturing. In all these polymers, SBR is the general PNC material being used in tires. Reinforcement of fillers in rubbers is very important for tire applications. Nano-carbon black was first used as reinforcing filler for rubber in 1904, and from then carbon black reinforced rubber nanocomposites was widely used in several tire products such as TB Tread, sidewall, PC Tread and inner liner. In addition to the carbon black, nanoclay was also used as nano-reinforcing filler and it was a promising one for industrialization due to its greater cost/performance ratio, easy manufacturing process and outstanding exfoliation.

Elastomeric nanocomposites are having large scope in automotive industry exclusively for tires application because of their less weight, low rolling resistance, and superior performance in terms of fuel economy. The key drivers for these materials are big demand for fuel efficiency, enhanced durability, strict automotive standards for safety and noise reduction. For example, replacement of nanocomposite inner liners for traditional inner liners decreases permeability by 50% and hence results in a decrease of 2 kg per truck approximately and an improvement in fuel efficiency by (2%). China has become the world's leading manufacturer of automobiles and tires as well as consumer of latex rubber and automobile. The first production line of PNC-enabled tires with ability of approximately 10,000 t/year was established in China.

13.5 Challenges and Further Research

Since carbon nanomaterials such as CNTs and graphene exhibit unique properties including large surface area and high functional and mechanical properties, they are capable to reinforce and functionalize elastomers. There are a number of challenges there in this research area which are mentioned below.

CB is generally used to reinforce elastomers, but its production depends mainly on petroleum feedstock and it has negative impacts on the environment. It is a tough challenge to develop alternatives to CB since they are inexpensive and therefore more effective in reinforcing elastomers. Although CNTs and graphene are stronger and far stiffer more electrically and thermally conductive than CB, these two materials are more expensive for industrial applications, and therefore new economical fabrication methods must be developed.

It is difficult to handle graphene because of its very low bulk density, which poses a obstacle on the development of elastomer/graphene nanocomposites. The fluffy structure of graphene creates an inhaling hazard for operators during processing.

Elastomeric materials are limited by exceptionally low thermal and electrical conductivity. The accumulated internal heat is an important ageing mechanism for the elastomers used in dynamic loading environment such as conveyors, tyres, rubber roller and belts. Elastomers in aerospace industry requires electrical conductivity to disperse electrical charges in the case of striking of lightning.

Since CNTs and graphene owns good thermal and electrical conductivity, the elastomer conductivity can be highly improved greatly with addition of these fillers. But, the current improvements are not agreeable. The electrical and thermal conductivity of elastomeric composites is determined by the dispersion, conductivity and aspect ratio of fillers and the bonding between the fillers and the matrices. It is a tough task to control the interface and the filler dispersion to meet a designed network for target conductivity.

Smart polymeric blends, nanofiller application in blends are growing interests recently. Controlling the migration of additives will get some applications in food packaging and biomedical industries. Nanaoclays and nanosilvers are the materials which are gaining interests because of their efficient properties. Nowadays, shape-memory polymers are also in emerging issues for academy and also industry. However, the industrial application of these materials is limited because of the manufacturing and scale-up difficulties. These difficulties tend to be reduced by using smart polymer blends and new compatibilizers. Self-adjusting smart materials are guaranteed for rapid growing in electronic industry. The properties of polymers can be upgraded by using multi-layered smart pellet additives in the blends. But, still there is some need to increase the adhesion between filler and polymer interphase. This improvements can be achieved by surface modification of the fillers using suitable compatibilizers.

References

1. Woods RJ, Sprinks JWT (1994) An introduction to radiation chemistry, 3rd edn. Wiley, pp 126
2. Araby S, Meng Q, Zhang L, Zaman I, Majewski P, Ma J (2015) Elastomeric composites based on carbon nanomaterials. Nanotechnology 26(11):112001
3. Al-Hartomy A, Al-Ghamdi A, Farha A, Dishovsky N, Mihaylov M, Ivanov M (2015) Influence of carbon black/silica ratio on the physical and mechanical properties of composites based on epoxidized natural rubber. J Compos Mater 24(11)
4. Jovanovic S, Jovanovic V, Markovic G, Konstantinovic S, Cincovic M (2011) Nanocomposites based on silica-reinforced ethylene–propylene–diene–monomer/acrylonitrile–butadiene rubber blends. Compos B Eng 42(5):1244–1250

5. Magaraphan R, Thaijaroen W, Ochakun RL (2003) Structure and properties of natural rubber and modified montmorillonite nanocomposites. Rubber Chem Technol 76(2):406–418
6. Varghese S, Gatos KG, Apostolov AA, Kocsis JK (2004) Morphology and mechanical properties of layered silicate reinforced natural and polyurethane rubber blends produced by latex compounding. J Appl Polymer Sci 92(1):543–551
7. Rajasekar R, Nayak GC, Malas A, Das CK (2012) Development of compatibilized SBR and EPR nanocomposites containing dual filler system. Mater Des 35:878–885
8. Gatos KG, Thomann R, Kocsis JK (2004) Characteristics of ethylene propylene diene monomer rubber/organoclay nanocomposites resulting from different processing conditions and formulations. Polymer Int 53(8):1191–1197
9. Pongdhorn S-O, Thepsuwan U, Thaptong P, Sirisinha C (2014) Comparison of reinforcing efficiency of carbon black, conductive carbon black, and carbon nanotube in natural rubber. Adv Polymer Tech 33(4):21422
10. Zhan YH, Liub GQ, Xia HS, Yan N (2011) Natural rubber/carbon black/carbon nanotubes composites prepared through ultrasonic assisted latex mixing process. Plast Rubber Compos Macromol Eng 40(1):32–39
11. Wang Y, Yan H, Huang Z, Zhang T (2012) Mechanical, dynamic mechanical and electrical properties of conductive carbon black/piezoelectric ceramic/chlorobutyl rubber composites. Polym Plast Technol Eng 51(1):105–110
12. Liu YB, Lia L, Wang Q (2010) Reinforcement of natural rubber with carbon black/nanoclay hybrid filler. Plast Rubber Compos Macromol Eng 39(8):370–376
13. Alberdi-Muniain A, Gil-Negretea N, Kari L (2012) Influence of carbon black and plasticisers on dynamic properties of isotropic magneto sensitive natural rubber. Plast Rubber Compos Macromol Eng 41(7):310–317
14. Bai X, Wanb C, Zhanga Y, Zhai Y (2011) Reinforcement of hydrogenated carboxylated nitrile–butadiene rubber with exfoliated graphene oxide. Carbon 49(5):1608-1613
15. da Costa HM, Ramos VD, da Silva WS, Sirqueira AS (2010) Analysis and optimization of polypropylene (PP)/ethylene–propylene–diene monomer (EPDM)/scrap rubber tire (SRT) mixtures using RSM methodology. Polymer Test 29(5):572–578
16. Hameed AM (2012) Effect of water absorption on some mechanical properties of unsaturated polyester resin/natural rubber blends. Jordan J Phys 5(3):119–127
17. Ramesan MT (2012) Fabrication and characterization of conducting nanomaterials composed of copper sulphide and polyindole. J Polym Compos 33:2169–2176
18. Patel MG, Desai KR, Patel HS (2011) Synthesis and characterization of coloured polyureas. Int J Polym Mater 54:767–773
19. Sridhar V, Xiu ZZ, Xu D, Lee SH, Kim JK, Kang DJ et al (2009) Fly ash reinforced thermoplastic vulcanizates obtained from waste tire powder. J Polym Compos 29:1058–1066
20. Sombatsompop N, Thongsang S, Markpin T, Wimolmala E (2004) Fly ash particles and precipitated silica as fillers in rubbers and untreated fillers in natural rubber and styrene–butadiene rubber compounds. J Appl Polym Sci 93:2119–2130
21. Mahmood N, Khan MS, Khan AU, Stockelhuber KW, Heinrich G (2010) Purification, surface modification of coal ash silica and its potential application in rubber composites. J Appl Polym Sci 117:1493–1501

Chapter 14
Interfacial Compatibilization of Multilayered Products

V.K. Abitha, S. Lakshmi Priya, Sainath Jadhav and Sabu Thomas

14.1 Introduction

Each and every entity in the cosmos has surfaces and interfaces. Surface is defined as a part of material in contact with either a gas or a vacuum. Interface is defined as that part of a material in contact with condensed phase, be it liquid or solid. Surfaces of any substance are different from their interior. Exterior of surface or interfacial tension is one uncomplicated demonstration. The area under discussion of polymer surfaces and interfaces is sundry and interdisciplinary [1, 2]. Chemist, physicists, material scientists and engineers are all on the go contributors. Polymer surfaces and interfaces participate an important role in polymer technology, all the way through processes of wetting, adsorption and adhesion furthermore through their effects on the kinetics of phase separation and processing of polymers. Polymer blends and composites typically contain very finely divided phases which are full of interfaces [3]. As interfaces are frequently weak mechanically, they pose special problems in manufacture of strong, tough plastics, adhesives, elastomers, coatings and fibers. All materials have surfaces or interfaces. When we speak of polymer materials there are basic classes of interfaces which are as follows: dilute polymer solution colloidal interface, symmetric interface, asymmetric polymer interface and composite interface. Polymeric materials provide and exceptionally varied class of interfacial system. Polymers can have interface with air or solid substrates, but also can form internal interfaces, for example the interface

S.L. Priya · S. Thomas (✉)
International and Interuniversity Centre for Nanoscience and Nanotechnology, Kottayam, Kerala, India
e-mail: sabuchathukulam@yahoo.co.uk

S. Jadhav
Department of Polymer and Surface Engineering, Institute of Chemical Technology, Matunga, Mumbai, India

V.K. Abitha
School of Chemical Sciences, Mahatma Gandhi University, Kottayam, Kerala, India

J.K. Kim et al. (eds.), *Multicomponent Polymeric Materials*,
Springer Series in Materials Science 223, DOI 10.1007/978-94-017-7324-9_14

between two incompatible homopolymers. Among the internal interfaces, that are of interest to polymer scientist are melt interfaces between incompatible or partially incompatible homopolymers, incompatible homopolymers, polymer solution interfaces and interface between homopolymers, polymers mixed with common solvent. Besides air/melt and melt/solid surfaces that are of great technological importance, Polymer layers are prepared by adsorption onto a substrate from solution. Compatibilization of multiphase polymer blends of more than two compounds has become an area of considerable research interest, which is mainly a result of commercial quest for new materials. The typical multilayer systems contain multi layers each designed to provide specific tailor properties to meet customer demands for appearance and durability [4].

14.2 Thermodynamics of Multilayered Products

Layered structures are formed in a number of applications ranging from magnetic to structural and the stability of individual layers is often critical to the performance of the material. Stable layered microstructures are also very important in many low to moderate applications. Multilayer sensors in magnetic storage, copper lines in integrated circuits and multilayer mirrors and in soft X-ray optics are some current examples. In each of these examples the layers must resist coarsening and breakdown which is driven thermodynamically by free energy of the interfaces. Broadly speaking, the thermodynamic driving forces that can destroy the layering in a multilayer include chemical energy, elastic strain energy, and interfacial free energies [5–8]. The drive to reduce the sum total of these energies leads to instabilities in the layered structure, either through mixing of the layers, atomic rearrangement, or breakdown of the layering through capillary forces. Consider first the chemical energy between two layered elements, "A" and "B" in an "A/B" multilayered system. When the layers contain elements that have positive heats of mixing, and therefore are immiscible (silver/nickel or copper/niobium, for example), no intermixing that might damage the layering will occur. When the two elements have negative heats of mixing (niobium/silicon or copper/zirconium), these elements will mix and form a solid solution or compound, with the potential to destroy the layering. In addition to chemical energy, elastic strain energy can also control the stability of layered structures. Under loading, concentrations in stress and elastic strain energies can arise in the thinner sections of the layers, driving material from the thin sections to the thickest parts of the layers, where stresses are lowest [9]. Thus, diffusion should induce growth of perturbations, damaging the layering.

When polycrystalline layers are chemically stable and contain only small elastic strains the ratio of grain boundary free energies and interfacial free energies control the stability of layered polycrystalline materials, for polycrystalline multilayered materials there exist an equilibrium groove angle at the point where A/A interfaces meets the A/B interfaces. This groove angle is determined by the ratio of grain boundary free energy to interfacial free energy γ_{gb}/γ_{int} through the relationship.

$$2\cos\theta = \gamma_{gb}/\gamma_{int}$$

When γ_{gb}/γ_{int} is greater than grain boundaries meet the interfaces between layers. For large values of γ_{gb}/γ_{int} the grain boundary grooves can extend through layers and the microstructure is unstable [10].

When γ_{gb}/γ_{int} is small, there is little grooving at grain boundary interface junction and the layering stable. In polycrystalline multilayer systems the value of γ_{gb}/γ_{int} will vary both layer to layer and from grain to grain. While evaluating the stability of a multilayer the simplest approach is to neglect such variations by choosing interfacial and grain boundary free energies that are averaged for a given layer to examine general trends in layer stability. The more accurate and considerably more difficult approach is to consider the specific values of interfacial and grain boundary free energies that are averaged for a given layer to examine general trends in layer stability. The more accurate and considerably more difficult approach is to consider the specific values of interfacial and grain boundary free energies that vary from grain to grain due to differences in crystallographic orientation of grains within a given layer [11, 12].

14.2.1 Prediction of Stability in Multilayers

Consider an immiscible elemental system A/B. The angle of groove of one layer into an adjacent layer depends on the ratio of grain boundary free energies to interfacial free energy (γ_{gb}/γ_{int}) in each layer. Grain boundary free energies tend to scale with an elements melting temperature. For this reason the layers of the higher melting temperature material. Because of the higher associated γ_{gb}/γ_{int} ratio, one would expect that at equilibrium, deeper grooving and associated pinch-off will occur at grain boundaries in the high T_m layers. For materials with vastly different melting temperatures, the attainment of an equilibrium groove configuration may be inhibited by the difference in diffusivity, at a given temperature, the layer with a higher melting temperature will have the lower diffusivity, at a given temperature the layer with a higher melting temperature will have the lower diffusivity. This behaviour has been observed in a number of systems [13–15].

14.3 Compatibilization

Compatibilization is the process used to modify the interphase of immiscible blends. It helps to reduce the interfacial energy and increases the stability of blends. It gives polymer better performance capabilities. Compatibility is ability of two different phases to exist in harmonious manner. Compatibility can be understood using example of Velcro. In Velcro, one part is has a rough structure while the other

Table 14.1 Types of blends

Immiscible and compatible	• Different structure • Different polarity • 2 phases • High interfacial energy
Miscible and compatible	• Similar structure • Similar polarity • Single phase • No interfacial energy
Immiscible and compatible	• Similar structure • Similar polarity • 2 phase • Low interfacial energy

part has smooth structure. If both parts were of same type then couldn't have adhere to each other. Thus, the two parts are compatible with as they unite to form a single entity [16].

When we blend two plastics together, the blend can be categorized into three types (Table 14.1).

14.3.1 Interfacial Compatibilization

In a co extrusion process, different polymeric layers are extruded together to form multilayer structure. As we use different polymers to produce these multilayer structure, interface of the two polymeric layers must have a strong adhesion to each other. Thus, they must compatible with each other. During the process, interfacial defects get generated and generally it is difficult to completely eliminate these defects. The interfacial phenomena like interfacial diffusion, interfacial slippage and interfacial reactions play important role in the properties [17] (Fig. 14.1).

Fig. 14.1 Interaction between two layers due to compatibilizer

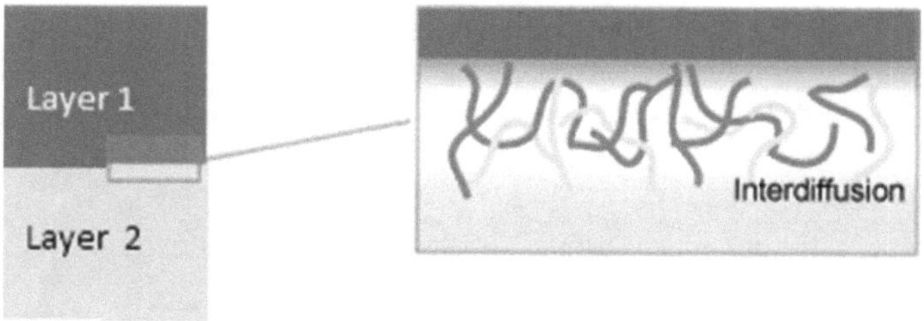

Fig. 14.2 Inter diffusion between two polymeric layers

14.3.1.1 Inter Diffusion

- Polymer interdiffusion have major effect on the properties of polymers across the interface (Fig. 14.2).
- Interdiffusion is function of temperature, composition, compatibility, molecular weight, orientation and molecular structure of polymers.
- Interdiffusion process helps to enhance the adhesion between two polymer layers and makes the interface stable.
- During diffusion, interfacial chemical gradient leads to flow of polymers of two adjacent layers. The gradient is the result of difference in polymer molecular weight and chemical structure.
- In a multilayered structure, the two adjacent layers should not similar concentration.

14.3.1.2 Interfacial Slip

- It is important phenomena the shear resistance of the multilayered structure (Fig. 14.3).
- This phenomenon happens when certain shear stress is applied on the structure, adhesion between the layers breaks and structure is disturbed.
- After extrusion, if the multilayered is quenched, development of irregular interfacial entanglements could be seen which leads to decrease in fracture toughness.
- Annealing of the structure rather helps to increase the fracture toughness.
- To avoid the slip, melt polymers could be added to the structure before extrusion which will help to reduce the instabilities of the interfacial flow.

Fig. 14.3 Interfacial slip between two polymeric layers

14.3.1.3 Interfacial Reaction

- In case of multilayered structure, delamination of layer needs to be avoided.
- Generally, chemical reactions are carried out at interface to increase polymer-polymer adhesion.
- This leads physiochemical affinity in the multilayered structure.

14.3.2 Method of Compatibilization

- Addition of linear or star shaped AB block copolymer for blends of A and B.
- Addition of linear random copolymer.
- Co-reaction in blend, e.g. transesterification. It has the disadvantage that it is difficult to control and can lead to completely different (co)polymers.
- Crosslinking of blend ingredients. This is similar to the IPN technology since it also involves the use of crosslinked polymers to maintain an initially formed morphology.
- Interaction between the constituent polymers after modification by ionic interactions or hydrogen bonds.
- Addition of a co-solvent. Two immiscible polymers may dissolve in a common solvent. After evaporation of the solvent, the interfacial area is so large that even weak polymer-polymer interactions will stabilize the system.
- High stress shearing may result in mechanical interlocking of the two phases.
- Reactive Compatibilization involves the generation of the desired of block copolymers or graft copolymers to interact with different parts of blends. The compatibilization can take place in molten or liquid state.

In reactive compatibilization, functional groups on the immiscible blends interact with each other in solution or melt state. The presence of functional groups like alcohol, carboxylic, amines, etc. lead to formation of hydrogen bonding, ionic or covalent bond. This will reduce phase separation between two immiscible blends and compatibilize. In case, no reactive functional groups are present on the immiscible blends these functional groups are added to the blends. As compared to non reactive compatibilization, the reactive compatibilization has more efficiency. The functional groups are either present or can be easily grafted on to the blends. Also, compatibilizers in non reactive compatibilization are specific for each polymeric blends [18].

14.4 Compatibilizers

Compatibilizers are polymers or copolymers that are used to bind the two phases together. They are polymers which when added to blend modifies interfacial properties as well as stabilizes the morphology. Consider two phases A and B, a

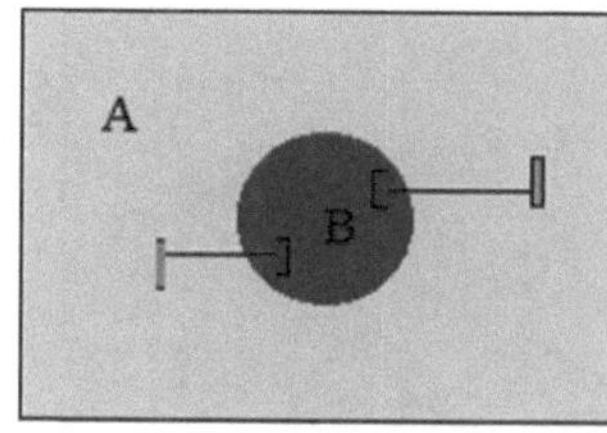

Fig. 14.4 Interfacial slip between two polymeric layers

compatibilizer has to bring this two phases together. Compatibilizer has two parts: one part will have affinity towards Phase A and other part will have affinity towards Phase B. Compatibilizer can be classified as reactive, non reactive, block, random and graft copolymers (Fig. 14.4).

The commonly used compatibilizers include modified polyolefins and styrenic block copolymers. Ethylene/propylene copolymers are common polyoelfins used for compatibilization [19].

14.5 Morphology—Interfacial Adhesion of Multilayered Polymer Products

The final performance of a product depends very much on morphology and so in polymer blends the morphology must be optimized for the desired performance. The morphology of the blend can be classified into two-dispersed or co-continuous. Within each of these two classifications, several categories like droplet, fiber or lamellar dispersion can be identified. Scanning electron microscopy is a powerful tool to investigate the morphology of multilayer films. Ma et al. [20] prepared Ethylene vinyl acetate copolymer/thermoplastic polyurethane (EVA/TPU) blending foams using TPU-grafted-EVA (EVA-g-TPU), as a compatibilizer and studied the morphology using SEM. Phase separation between the blends are seen as large particles in the image. Finer morphology and narrow size distribution of dispersed phase could be observed in compatible blends (Fig. 14.5).

If there is a strong contrast in refractive indices of the constituent polymers in the multilayered polymer particles [21, 22], strong light scattering can be expected from the resulting films. The morphology of the particles could be controlled by varying the reaction temperature, cross-linking density, monomer feeding rate, and glass transition temperatures of core- and shell-forming polymers [23]. Alteheld et al. [24] synthesised polymer particles comprising layers with alternating high [polystyrene (PS)] and low [poly(heptafluorobutyl methacrylate) (PHFBMA)] refractive index. They produced spherical particles with a well-defined morphology by introducing a copolymerization approach by adding a phase transfer catalyst (cyclodextrin) and employing a mixed initiator approach. They produced four-layer 1400 ± 120 nm size particles by this method, Fig. 14.6. This method can be used to synthesize polymeric spherical dielectric resonators.

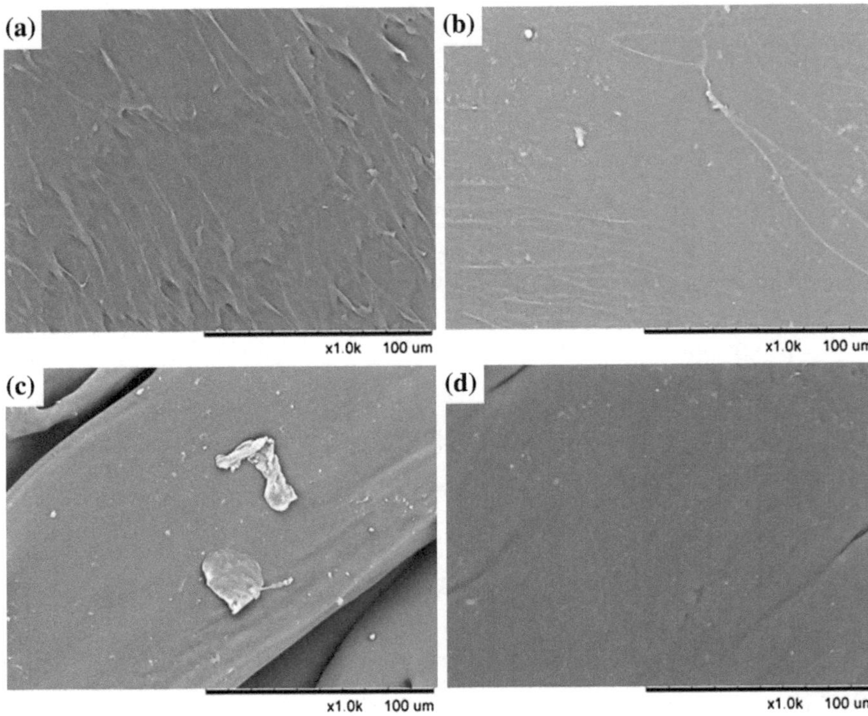

Fig. 14.5 SEM images of the resin samples: **a** EVA; **b** TPU; **c** EVA/TPU (S-1); **d** EVA/EVA-g-TPU/TPU (S-5) [20]

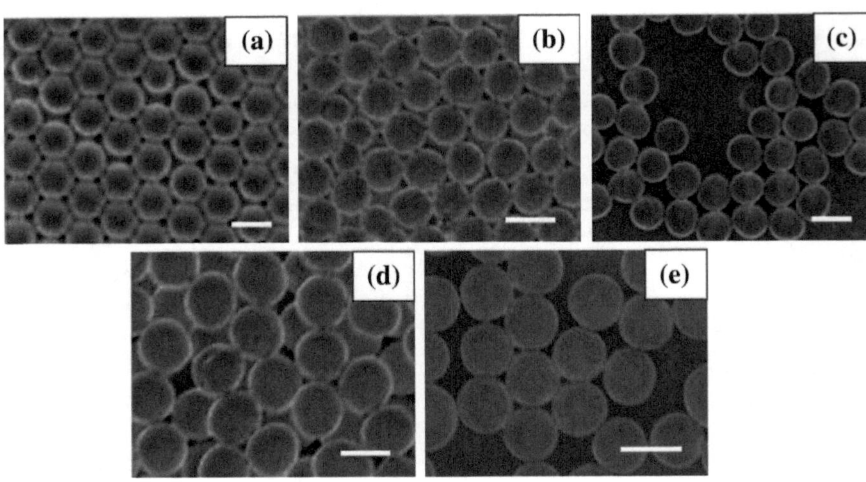

Fig. 14.6 Typical SEM images of microspheres synthesized in different polymerization stages: **a** PS-rich cores (sample 1a), scale bar is 200 nm; **b** particle from (**a**) engulfed with PHFBMA-rich layer (sample 2a), scale bar is 400 nm; **c** particles from (**b**) engulfed with PS-rich layer (sample 3a), scale bar is 1.0 μm; **d** particle from (**c**) engulfed with PHFBMA-rich layer (sample 4a), scale bar is 1.0 m; **e** particle from (**d**) engulfed with PS-rich fourth layer (sample 5a), scale bar is 1.5 μm [24]

Yousefi and Ait-Kadi [25] prepared stretched–extruded PP/EVOH blend films and studied the influence of differently functionalized polypropylene (PP) and styrene-ethylenebutene-styrene copolymer (SEBS) on barrier properties. The effect of post extrusion unidirectional melt stretching was studied. The use of effective compatibiliser resulted in lamellar morphology of blends which showed better interfacial adhesion and higher barrier properties. Kim and Chun [26] found that the barrier properties of linear low density polyethylene (LLDPE)—ethylene vinyl alcohol copolymer (EVOH) blends can be improved by generating a laminar morphology of the dispersed phase in the matrix phase as found by the SEM analysis.

The effect of the reactive compatibilizer on the interfacial morphology can be examined by scanning electron micrography (SEM) [27]. Lee and Kim [28] found that a well-developed laminar structure was formed for low density polyethylene (LDPE)/ethylene-vinyl alcohol copolymer (EVOH)/liner low density polyethylene with 0.1 wt% grafted maleic anhydride (LLD-g-MAH) blends resulting in low oxygen permeability. Large predeformed domains in the extruder outlet with short residence time under high screw rpm and low viscosity ratio in the die favored the formation of these structures.

Xiang et al. [29] prepared crack-free hydrogen-bonded poly(acrylic acid) (PAA)/poly-(ethylene oxide) (PEO) multilayer thin films by LBL technique. The softest assembly was obtained at pH 3 and the lowest oxygen transmission rate was obtained for 367 nm thick 20 BL PAA3/PEO3 nanocoating.

At 0% strain, the 20 bilayer PAA_3/PEO_3 nanocoating is very smooth and stretching up to 25% has little influence on the morphology. A few shallow lines were seen perpendicular to the stretch direction which became more pronounced at 100% strain level which turned out to be creases rather than cracks, Fig. 14.7. This is due to the plastic deformation of the films induced by stretching. And they also noted that the plastic deformed films have similar gas barrier properties independent on the extent of deformation.

Bondon et al. [30] prepared multilayer films of ethylene-vinyl alcohol copolymer (EVOH) and polyamide 6 (PA6), sandwiched between random polypropylene (PP) layers by reactive co extrusion. Polypropylene grafted maleic anhydride (PP-g-MA) was used as tie-layer to enhance the optical and mechanical properties of the films. They studied the interfacial morphology for the two reactive polymer pairs after 5 min of annealing similar to the stress relaxation experiments. Interface in the PA6/PP-g-MA bilayer was flat while an irregular and rough interface appeared between EVOH and PP-g-MA. The relaxation time was found to be extended for EVOH from stress relaxation studies. This could therefore be related to this interfacial roughness and to the copolymer triggered between the neighboring layers. This difference in morphologies is an indication of grainy effect in coextruded films and also the importance of interfacial reaction and copolymer structure, as in Fig. 14.8.

Qin et al. [31] prepared multilayer LBL films of poly (sodium 4-styrenesulfonate) (PSS) grafted single walled carbon nanotubes (SWNT) and a diphenylamine-4-diazoresin (DR) by cross linking reaction between the two.

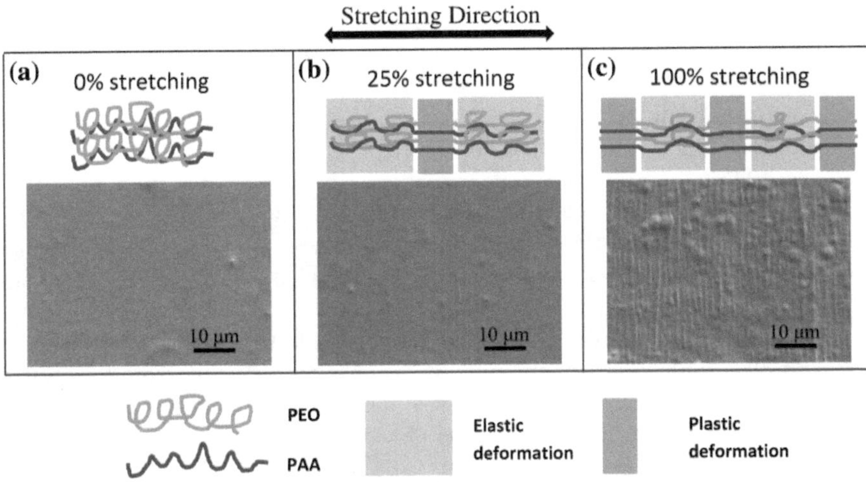

Fig. 14.7 Schematic showing the influence of different strain levels 0 (**a**), 25 (**b**), and 100% (**c**) on structure and morphology of PAA_3/PEO_3 assembly (FESEM surface images are *below* each schematic) [29]

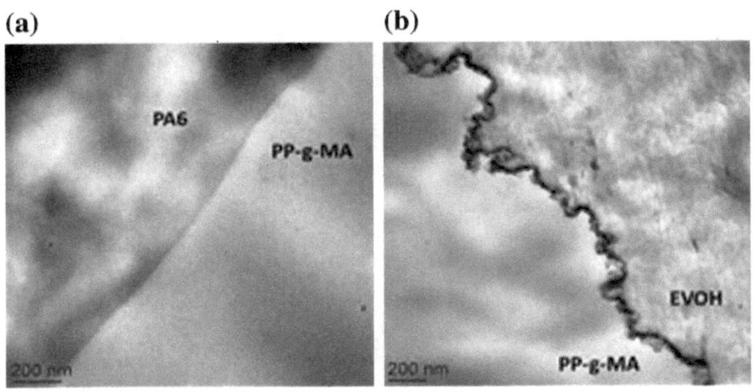

Fig. 14.8 TEM micrographs of reactive interfaces in bilayer samples of PP-g-MA with **a** PA6 and **b** EVOH [30]

Initially an ionic bond was formed between the diazonium ions and sulfonate ions which were converted to covalent bonds upon UV irradiation which was supported both by Raman spectroscopy and AFM images, Fig. 14.9. Before irradiation the force of attraction between the films was electrostatic attraction. By irradiation, N2 was released and the ionic bonds were converted to covalent bonds. The bilayers contracted and stronger signals were seen in the Raman spectra due to the greater exposure of SWNTs to the film surface. This was also confirmed by the presence of individual or smaller bundles of SWNTs in the AFM image.

Liu et al. [32] fabricated multilayer films containing Pd nanoparticles by layer-by-layer assembly of $PdCl_4^{2-}$ and polycation, QPVP-Os(a quaternized poly

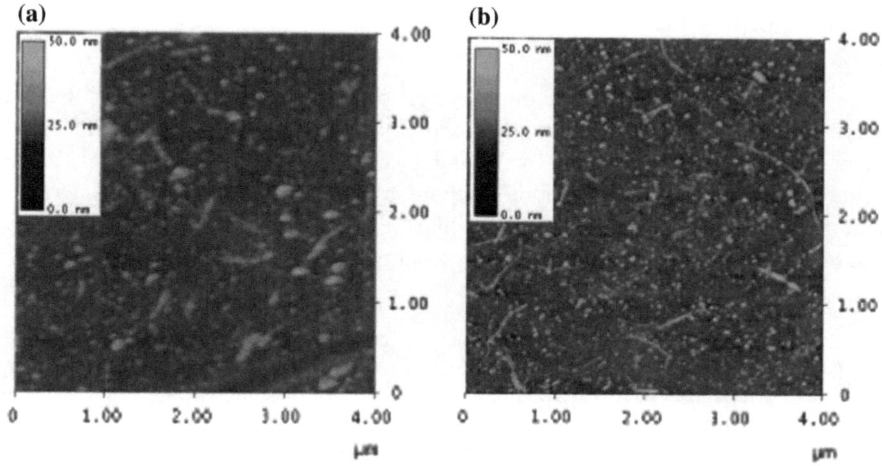

Fig. 14.9 AFM height images of a $(DR/SWNT\text{-}PSS)_{14}$ film **a** before and **b** after UV irradiation [31]

(4-vinylpyridine) complexed with $[Os(bpy)_2Cl]^{2+/+}$), on 4-aminobenzoic acid-modified glassy carbon electrodes. AFM images showed a fine dispersion of nanoparticles and a decrease in the surface roughness in the blend morphology. These films do not have a sharp interface between the nanoparticle and the polymer and are widely applicable as catalysts and in device fabrications. Séon et al. [33] studied the topography of films formed by noncovalent host-guest interactions between poly(N-hydroxypropylmethacrylamide) PHPMA bearing β-cyclodextrins (β-CD) hosts and hydrophobic guests adamantine (Ad), Ferrocene (Fc) and pyrene (Py). As the strength of the interaction between polymers increases, the morphology changes from a droplet like structure to a continuous rough structure. Intermediate strength interactions (involving Fc groups) lead to a smooth continuous film whereas too weak interactions (involving Py derivatives) favour the formation of isolated aggregates.

Yang et al. [34] prepared thin film assemblies of branched polyethylenimine (PEI) and poly(acrylic acid) (PAA), deposited using the layer-by-layer technique. Altering the pH of PEI and PAA results in large thickness variations while cross-linking these films with glutaraldehyde (GA) inhibited polymer inter diffusion. AFM images showed that the surface morphology of PEI/PAA assemblies can be controlled by the pH i.e., 8 bilayers of PEI at pH 10 and PAA at pH 4 produce a 305 nm thick film with an oxygen transmission rate below 0.005 $cm^3/(m^2$ day). Connal et al. [35] also observed a pH dependence of the morphology of poly (acrylic acid) (PAA) based core cross-linked star (CCS) polymer assembled into multilayered films.

14.6 Conclusion

Interfacial compatibilization of multilayered products and their morphology and thermo dynamical approach has discussed. Interfacial compatibilization of multilayered products can be done by adding the suitable compatibilizers. There are different types of compatibilization methods by considering the type of interfaces and multilayer. Thermodynamics of multilayered products include chemical energy, elastic strain energy, and interfacial free energies.

References

1. Sperling LH (1997) Polymeric multicomponent materials: an introduction. Wiley-Interscience, New York
2. de Gennes P-G (1979) Scaling concepts in polymer physics. Cornell University, Ithaca
3. Wool RP (1995) Polymer interfaces: structure and strength. Hanser, Munich
4. Helfand E (1982) In: Sole K (ed) Polymer compatibility and incompatibility. Midland Macromolecular Institute, New York
5. Van Heerden D, Gavens A, Foecke T, Weihs TP (1999) Mater Sci Eng A261:212
6. Van Heerden D et al (2001) Met Trans 32A:2363
7. Gavens A, Van Heerden D, Foecke T, Weihs TP (1999) Met Trans 30A:2959
8. Bewlay BP, Jackson MR, Lipsitt HA (1996) Met Trans 7A:3801
9. Sridhar N, Rickman JM, Srolovitz DJ (1997) Acta Mater 45:2715
10. Josell D, Carter WC (1996) In: Merchant HD (ed) Creep and stress relaxation in miniature structures and components. Minerals, Metals and Materials Society, p 271
11. Udin H, Shaler AJ, Wulff J (1949) Trans AIME 185
12. Murr LE (1975) Interfacial phenomena in materials and alloys. Addison-Wellesley, Reading
13. Josell D, Carter WC, Bonevich JE (1999) Nanostruct Mater 12:387
14. Lee HJ, Kwon KW, Ryu C, Sinclair R (1999) Acta Mater 47:3965
15. Snoeck E et al (1995) J Magn Magn Mater 151
16. Frensch H, Harnischfeger P, Jungnickel B-J (1989) In: Utracki LA, Weiss RA (eds) Multiphase polymers: blends and ionomers. ACS symposium series, 395. American Chemical Society, Washington, DC
17. Utracki LA (ed) (1994) Encyclopaedic dictionary of commercial polymer blends. ChemTec Publications, Toronto
18. Utraki LA (2002) Polymer blends handbook. Kluwer Academic Publishers, Dordrecht
19. Teh JW, Rudin A, Keung JC (1994) Adv Polym Technol 13:1
20. Ma J, Shao L, Xue C, Deng F, Duan Z (2014) Compatibilization and properties of ethylene vinyl acetate copolymer (EVA) and thermoplasticpolyurethane (TPU) blend based foam. Polym Bull 71:2219–2234
21. Pham H, Gourevich I, Oh JK, Jonkman JEN, Kumacheva E (2004) Adv Mater 16:516–520
22. Siwick B, Kalinina O, Kuma-cheva E, Miller DRJ (2001) J Appl Phys 90:5328–5338
23. Durant YG, Sundberg EJ, Sundberg DC (1997) Macro-molecules 30:1028–1032
24. Alteheld A, Gourevich I, Field LM, Paquet C, Kumacheva E (2005) Multilayer polymer particles with periodic modulation in refractive index. Macromolecules 38:3301–3306
25. Yousefi AA, Ait-Kadi A (2005) Effects of compatibilizer and post-extrusion stretching on properties of plastic barrier films. Iranian Polym J 14(5)
26. Kim SW, Chun YH (1999) Barrier property by controlled laminar morphology of LLDPE/EVOH blends. Korean J Chem Eng 16(4):511–517

27. Chapleau N, Huneault MA, Li H (2007) Biaxial orientation of polylactide/thermoplastic starch blends. Int Polym Process XXII:5
28. Lee SY, Kim SC (1997) Laminar morphology development and oxygen permeability of LDPE/EVOH blends. Polym Eng Sci 37(2)
29. Xiang F, Ward SM, Givens TM, Grunlan JC (2014) Super stretchy polymer multilayer thin film with high gas barrier. ACS Macro Lett 3:1055–1058
30. Bondon A, Lamnawar K, Maazouz A (2015) Influence of copolymer architecture on generation of defects in reactive multilayer coextrusion. Key Eng Mater 651–653:836–841
31. Qin S, Qin D, Ford WT, Zhang Y, Kotov NA (2005) Covalent cross-linked polymer/single-wall carbon nanotube multilayer films. Chem Mater 17:2131–2135
32. Liu J, Cheng L, Song Y, Liu B, Dong S (2001) Simple preparation method of multilayer polymer films containing Pd nanoparticles. Langmuir 17(22):6749
33. Séon L, Parat A, Gaudière F, Voegel J-C, Auzély-Velty R, Lorchat P, Coche-Guérente L, Senger B, Schaaf P, Jierry L, Boulmedais F (2014) Influence of the interaction strength between supramolecular complexes on the topography of neutral polymer multilayer films. Langmuir 30:6479–6488
34. Yang Y-H, Haile M, Park YT, Malek FA, Grunlan JC (2011) Super gas barrier of all-polymer multilayer thin films. Macromolecules 44:1450–1459
35. Connal LA, Li Q, Quinn JF, Tjipto E, Caruso F, Qiao GG (2008) pH-responsive poly(acrylic acid) core cross-linked star polymers: morphology transitions in solution and multilayer thin films. Macromolecules 41:2620–2626

Chapter 15
Multilayer Nanowires and Miscellaneous Multilayer Products

A.R. Ajitha, P.S. Sari, Hanna J. Maria and Sabu Thomas

15.1 Significance of Polymer Multilayer Products

Multilayer Polymer Films have significantly enhanced performance properties, allowing films to be made thinner, stronger, and with better sealing properties and are used in a variety of industries. Recent progresses have opened up a range of new possibilities for Engineers and polymer scientists related to the field of polymer processing, polymer films, optics, polymer film capacitors and polymer packaging. In food packaging, for example, polymer laminates are used not only to protect food, but also to retain aroma and flavors, and to extend shelf life. Many diverse applications of multilayer films have been recommended and a range of applications have, indeed, been realized. Electrodes, nanowires and food coatings, coatings for cell and tissue culture, and artificial red blood cells, pharmaceutical applications etc. are examples of potential and actual applications of multilayer films. Notes in the present work will be focused mainly on nanowires, pharmaceutical applications. A brief report on the miscellaneous applications is also included.

A.R. Ajitha · P.S. Sari · S. Thomas (✉)
International and Inter University Centre for Nanoscience and Nanotechnology, Mahatma Gandhi University, Kottayam, Kerala, India
e-mail: sabuchathukulam@yahoo.co.uk

H.J. Maria · S. Thomas
School of Chemical Sciences, Mahatma Gandhi University, Kottayam, Kerala, India

P.S. Sari
LIMATB Laboratory, Université de Bretagne Sud, Rue St Maudé, 56100 Lorient, France

J.K. Kim et al. (eds.), *Multicomponent Polymeric Materials*,
Springer Series in Materials Science 223, DOI 10.1007/978-94-017-7324-9_15

15.2 Polymers as Multilayer Nanowires

In the late 1990s, the field of nanowires underwent a significant expansion and became one of the most active research areas within the nanoscience community [1]. A nanowire is a nanostructure, it's diameter is in the order of a nanometer and length to width ratio being greater than 1000. Alternatively, nanowires can be defined as structures that have a thickness or diameter constrained to tens of nanometers. Nanowires, nanotubes, nanorods etc. have great importance in many applications including optical and electronic nanodevices and chemical and biological sensors. Application of these nanostructured materials into functional nanodevices requires controlled patterning at micro- and nanometer scale.

Polymer nanowires are sensitive chemical and biological sensors, field-effect transistors, interconnects in electronic circuitry, and tools for studying one-dimensional charge transport in materials [2]. Polymers are more mechanically flexible and less expensive to produce and process, that is why it have applications in many nanodevices. The common polymers used to prepare nanowires are Polypropylene (PP), Polypyrrole (PPy), (poly(methyl methacrylate (PMMA))), polystyrene (PS), polydimethylsiloxane (PDMS), (poly(3,4ethylenedioxythiophene) poly(styrenesulfonate) (PEDOT/PSS), polyethylene naphthalate (PEN), polyethylene terephthalate (PET) etc.

Now a days conducting polymers are emerging as a promising material for synthesis of nanostructured materials and devices. Because they exhibit electrical, electronic, magnetic, and optical properties similar to metals or semiconductors while retaining their flexibility, ease of processing, and modifiable electrical conductivity. Conducting polymers have many advantages over semiconductor and metallic nanowires and CNTs. A variety of conducting polymers have uses in sensor materials, including biosensors, because their properties can be tailored to detect a wide range of chemical compounds [3]. Among conducting polymers Polyaniline is one of widely used because of its simple and reversible doping/dedoping chemistry, stable electrical conduction mechanism, and high environmental stability. Another extensively used conducting polymer used in nanowires is polypyrrole, which is more conductive, stable and biocompatible than other polymer materials [4].

15.2.1 General Preparation Methods

There are number of techniques used for the fabrication of nanowires. Few among them are track edge method, printing techniques, photochemical lithography, edge lithography, nanolithography, slip casting, micromolding, electrospinning, stamping and electrodeposition etc. However the widely used methods are electrospinning and electrodeposition because of its simplicity, cost effectiveness and industrial viability.

15.2.1.1 Electrospinning

Electrospinning technique is recognized as a powerful method for fabrication of polymer wires. Here, electric force is used to overcome the surface tension of a surface charged droplet of polymer solution resulting in the ejection of thin polymer from the droplet to a counter electrode. During this process, the solvent evaporates and solid polymer nanowires precipitate on the counter electrode. Because the jet ejection can be continuous, the produced polymer wires can have arbitrary length. Now a days, electrospinning method in combination with sol-gel chemistry, has been developed, to fabricate continuous inorganic/polymer hybrid nanowires. By the removal of the organic polymer, inorganic nanowires, which are typically polycrystalline, can be produced [5] (Fig. 15.1).

15.2.1.2 Stamping or Microtip Writing

Another novel technique for fabrication of polymer wires is Stamping or microtip writing where a patterned and aligned polymer nanowire arrays on a material (a wafer-level substrate) is obtained. Designed pattern on a spin-coated polymer film on plasma etching results in the formation of aligned polymer nanowire arrays. This is a new technique for large scale fabrication of organic nanowires (NWs). Fang et al. expanded the technique for fabricating aligned NWs of any polymer with a control over the growth density and length to meet the needs of future device fabrications Instead of using surface roughness first created on the polymer film surface, they deliberately sputtered a thin layer of metal nanoparticles on the substrate, which serve as the “nanomasks” for ion irradiation, resulting in the initial roughness for further etching [7].

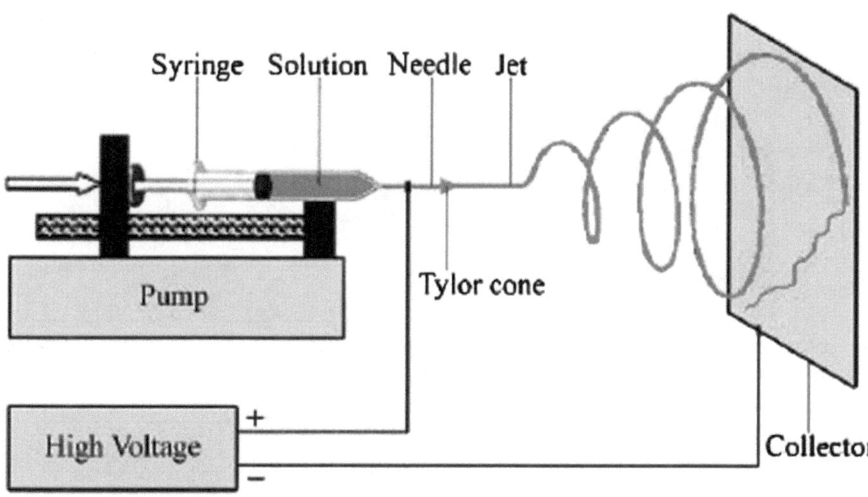

Fig. 15.1 Sketch of a typical laboratory electrospinning setup [6]

15.2.1.3 Electrodeposition

This is the widely used method for the fabrication of nanowires. Ramanathan et al. recently reported a facile technique for synthesis of micron/submicron width metallic and conducting polymer wires by electrodeposition within channels between two electrodes on the surface of silicon wafers. Here, the electrode chip/wafer was mounted on a probe station and the contact was established using metallic pins. A three-electrode setup was used for the deposition. The deposition and growth of the nanowire chains are based on well-known electrochemical oxidative polymerization, starting with monomers and dopants. The procedure was a single-step deposition process for each nanowire, and multiple-nanowire arrays of different materials can be deposited on the same wafer sequentially. They could synthesize nanowires of conducting polymers like polyaniline and polypyrrole having high aspect ratio and dendrite free nanowires. The technique is capable of producing arrays of individually addressable nanowire sensors, with site-specific positioning, alignment, and chemical compositions [3]. Penner and co-workers have also used the electrodeposition method for the preparation of nanowires using electro-beam lithography [8].

15.2.1.4 Nanoskiving

Nanoskiving involves the sectioning of spin-coated multilayer films with an ultra-microtome. Conjugated polymers are widely used in this method because these materials can form thin films (by spin-coating) and are tough enough to be sectioned by a diamond knife without fracture. Also they can adhere to the embedding matrix (usually epoxy). The combination of forming thin films and nanoskiving can create structures with high aspect ratios and cross-sectional dimensions of 100 nm or less.

Lipomi et al. have developed a technique for the fabrication of conjugated polymer nanowires that requires only a spin-coater and an ultramicrotome they fabricated electrically conductive nanowires of two conjugated polymers poly(2-methoxy-5-(2′-ethylhexyloxy)-1,4 phenylenevinylene) (MEHPPV) and poly(benz-imidazobenzophenanthroline ladder) (BBL). This is a form of edge lithography. In this method they followed six major steps to fabricate the nanowires (Fig. 15.2).

15.2.1.5 Micromolding

Micromolding is one method which can be effectively used to fabricate nanowires nanowires with high aspect ratio although this method is costly and time consuming. Here, a polydimethylsiloxane (PDMS) mold is produced with the desired length, radius and pitch (distance between nanowires) using a photolithography process. Then an antisticking layer (PDMS or paraffin) is poured into the mold and cured for a few hours and is peeled off. Figure 15.3 shows the process involved in micromolding.

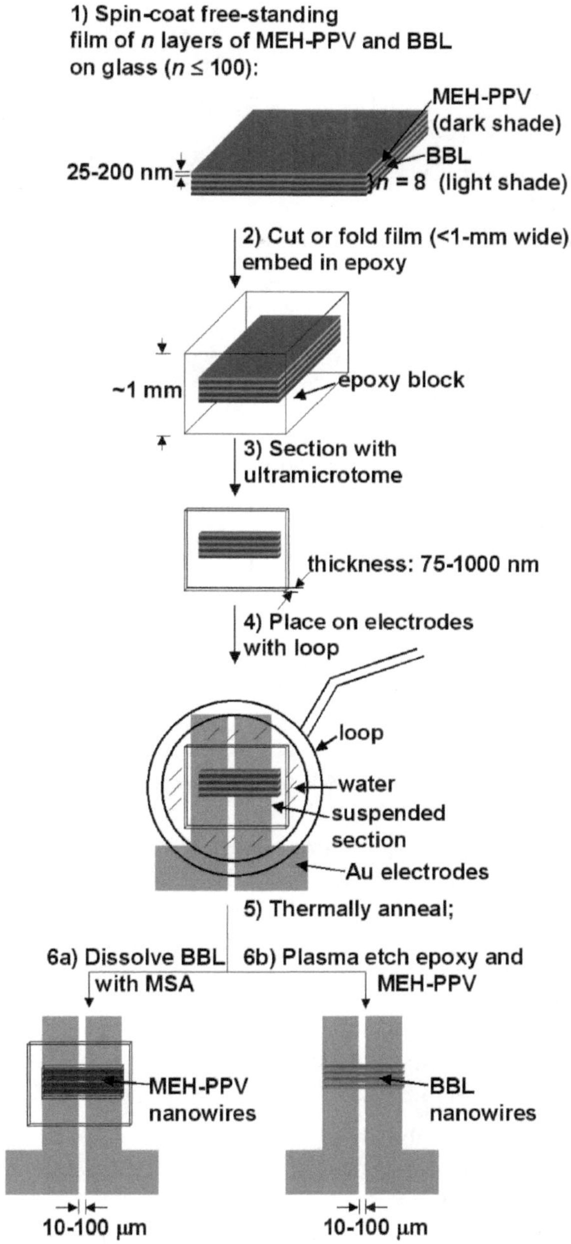

Fig. 15.2 Summary of the procedure used for fabrication of multiple nanowires of MEH-PPV or BBL [2]

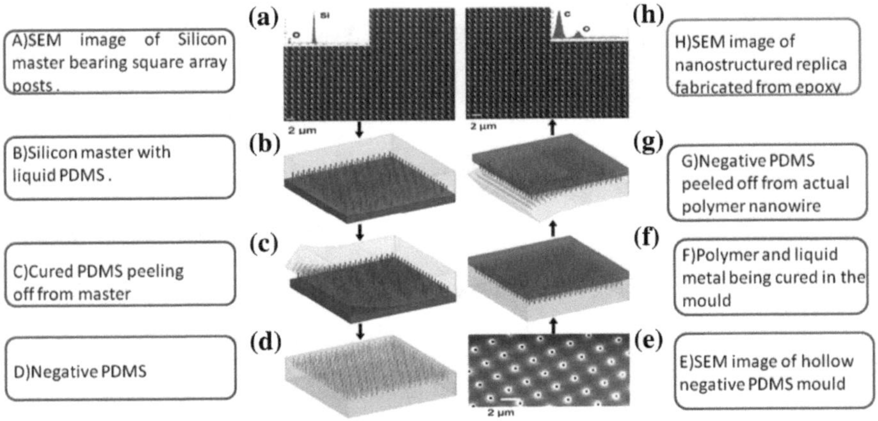

Fig. 15.3 Different steps involved in micromolding process for fabricating polymer nanowires [9]

15.2.2 Alignment of Nanowires

After processing of nanowires it is necessary to do the alignment of nanowires for further application have great importance in the field of application. For example if the surface of the layer is not uniform all over it will making the protein binding impossible and this will act as a barrier to improving the sensitivity of the nanowire device. Several methods are used for aligning nanowires among which the 'mechanical break junction method 'is the widely used one for aligning polymer nanowires. Kumar et al. were reported that, it works in a way that allows the material to be synthesized and aligned at the same time. The drawback of this method is that only one nanowire can be grown at a time. The dielectrophoresis is another method used for aligning the nanowires and in which the alignment is easier and greater amount of nanowires can be aligned and tested at the same time [9].

15.2.3 Characterization

In the fabrication of Polymer nanowires, characterization part has great importance. To find the application of a particular nanowire in nanodevices, characterization is an indispensable part.

Morphological Characterization, Fourier Transform infrared spectroscopy (FTIR), Raman spectroscopy, Energy dispersive spectrum etc. are the major techniques used for the characterization of nanowires some of which are explained below.

15.2.3.1 Morphological Characterization

Morphological Characterization is a direct method to analyze the quality of the material that is fabricated. Generally Scanning Electron Micrograph (SEM), Transmission electron micrograph (TEM), Atomic Force Micrograph (AFM) etc. are used for the morphological characterization.

The SEM images of nanowires are used to compare the morphological features while using different polymers, fabrication techniques, composition etc. For example, in the study of Fang et al., the SEM images (Fig. 15.4) shows the morphological differences of the nano wires fabricated using different polymers, including PMMA, PS, PDMS, PEDOT/PSS, PEN, PET, Durafilm and Kapton film.

In their study about Individually Addressable Conducting Polymer Nanowires array using Polyaniline (PANI) Ramanathan et al. [3] could observe from the SEM images (Fig. 15.5a) a continuous, well-confined, and nondendrite nanowire. SEM images of polypyrrole nanowires grown using a three electrode system is given by kumar et al. in Fig. 15.5b [9].

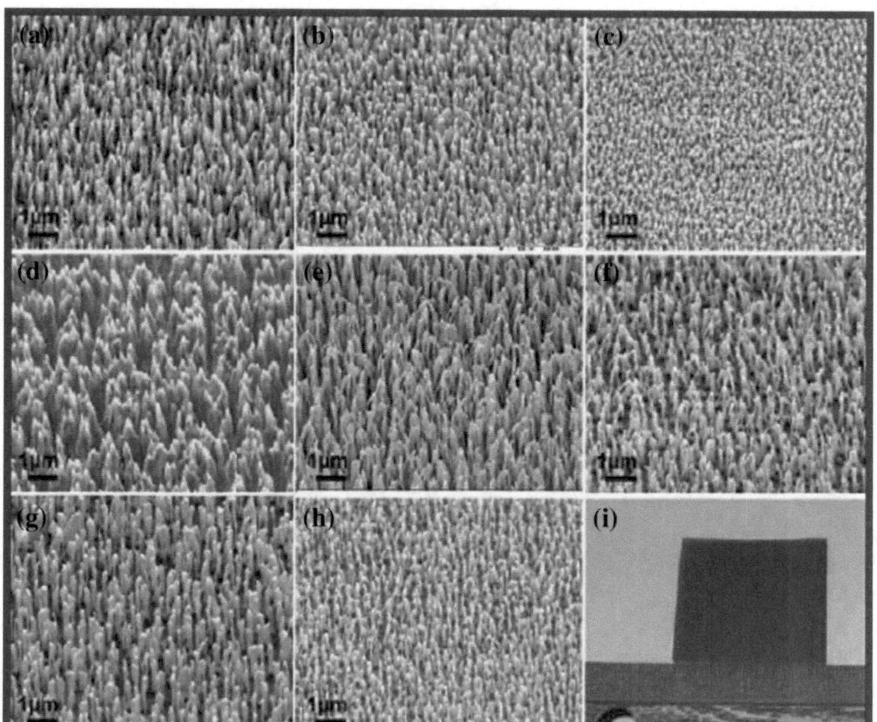

Fig. 15.4 SEM images of fabricated polymer nanowire arrays: **a** PMMA, **b** PS, **c** PDMS, **d** PEDOT/PSS, **e** PEN, **f** PET, **g** Durafilm, and **h** Kapton film. **i** Optical image of 4 cm × 4 cm Kapton film covered with uniform PNW array [7]

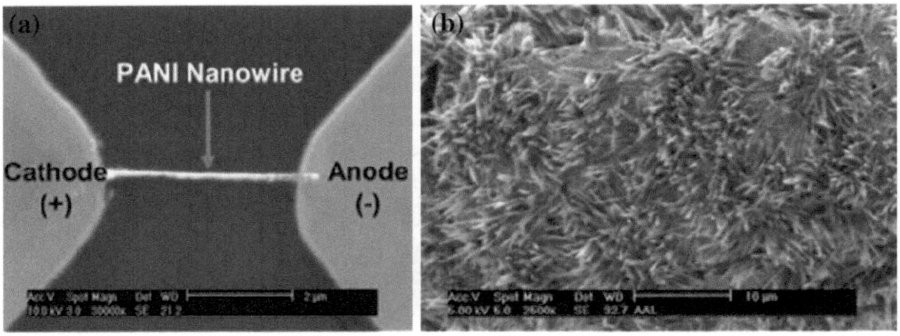

Fig. 15.5 SEM image of **a** PANI nanowire **b** Polypyroll nanowire

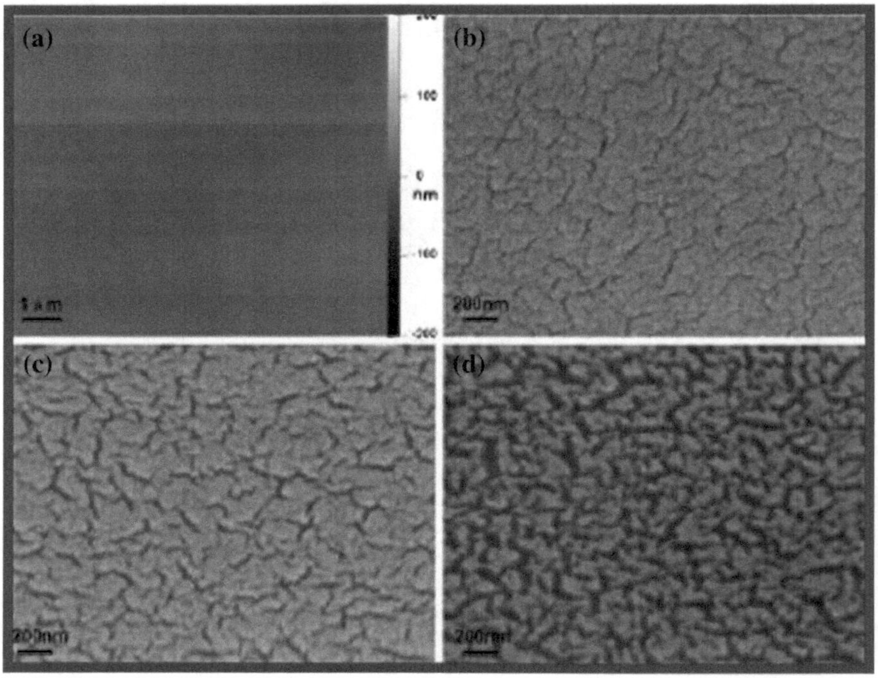

Fig. 15.6 Showing undetectable surface roughness change. Note that the lines in the horizontal direction are due to the tip scanning. **b–d** SEM images of Kapton films after ICP etching for **b** 5, **c** 10, and **d** 15 s [7]

Fang et al. used the AFM micrograph to understand the roughness of the nanowire surface. Figure 15.6a shows the atomic force microscopy (AFM) image of the Kapton film coated with 5 nm of Au before Inductive Coupled Plasma etching. They reported that the morphology of the film did not change after Au coating, and the roughness was less than 5%. Figure 15.6b–d shows the SEM images of the Kapton films at three different etching times.

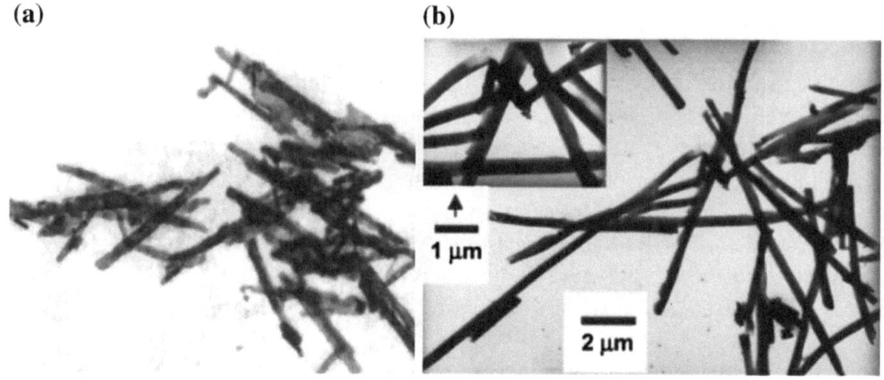

Fig. 15.7 TEM images of **a** polypyrrole/polyaniline copolymer nanowires and **b** polypyrrole-carbon nanotube composite nanowires [10]

Size of the nanowires can study clearly using the Transmission electron micrograph. Choi et al. reported the TEM studies of Li et al. and Wang et al. Li et al. worked with polypyrrole/polyaniline copolymer nanowires and the TEM micrograph is shown on Fig. 15.7a. While Wang et al. were studied about polypyrrole-carbon nanotube composite nanowires and the TEM images are shown in Fig. 15.7b.

15.2.4 Applications

One-dimensional nanostructures have unique electrical, mechanical and optical characteristics, so these nanostructures have great attention in many applications. Nanowires have unique characteristics similar to carbon nanotubes; nanorods etc. and are different from the bulk materials. We can apply these nanostructures as field effective transistors or highly sensitive chemical and biological sensor [11]. The one dimensional structures have high surface-to-volume ratio, ability to be integrated into electronic devices, novel sub-wavelength optical phenomena, they have large tolerance for mechanical deformations, also they can interface with other microscopic and nanoscopic systems in nature, the decoupling of length scales associated different physical phenomena in the radial and axial directions, and these properties have led to many applications of these small dimensional structures [5].

Polymer nanowires are widely used in sensing application. Polymer nanowires mainly the conducting polymer nanowires are widely in sensor materials, including biosensors (example; protein bio sensing), because their properties can be tailored to detect a wide range of chemical compounds. Due to the efficient properties of conducting polymers it have many advantages over semiconductor and metallic Nanowires and CNTs [3]. Also we know that nanowire is a small-dimensional geometry, nanowires composed of conjugated polymers have a high aspect ratio of area to volume and is therefore permits the adsorption/desorption process from its surface. These characteristics allow electrical response and recovery rates that are superior to those of devices based on thin films or fibrous networks. Incorporation of molecular recognition elements into conjugated polymer nanowires is relatively straight forward by synthesis. Therefore the conducting polymer nanowires have great importance in sensor application and are well suited for chemical sensing [2].

Nanowires have large surface/volume ratio and a Debye length comparable to the nanostructure radius; the electronic properties of these nanostructures are strongly influenced by surface process and therefore great attention in bio-application as bio-sensors. Conducting polymers have high stability at room temperature and have good conductivity due to the π conjugated electrons. Therefore the nanowire of conducting polymers can be used as a transducer in biological sensors. Another advantage of conducting polymer wire is that the biomolecules can be immobilized in a single step. The electrochemically prepared nanowires with controlled thickness using lower potential have an excellent enzyme-entrapping capability. Kumar et al. reported that, nanosized polypyrrole, which allows even a very low current to pass through, making the device even compatible with the human body. According to them, electrochemically prepared pyrroles are more conductive and stable when compared to the chemically prepared ones [9].

Polymers are highly selective, they can be used for the immobilization of gases and biological molecules and they also exhibit highly reversible redox behavior with a distinguishable chemical memory. Conducting polymer nanowires (CPNWs) have many advantages are now are used as molecular electronic architectures and devices [4]. Identification and quantification of chemical and biological species are of great importance in many fields such as medical, pharmaceutical, environmental etc. And the devices made by one dimensional nanomaterial (nanotubes, nanowires, or nanorods etc.) are highly selective for sensing a particular material, very simple to handle and shows fast reply in the case of sensing. Since nanowires have small cross sectional area, it can accelerate the current response axially Also nanowires have large surface area hence they are highly sensitive. Nanodevices made of nanowires can translate the chemical change into an electrical signal easily and directly. There by it can simplify the device configuration [12].

Now a days the nanowires founds uses in the field of electronic application also, since they can accumulate charge in the massive of the material rather than the two dimensional materials. Nanowires of conjugated polymers are widely used in electronic devices. Since these polymers have electrical properties similar to the traditional semiconducting materials [2]. Which ensures good electrical properties during detection. These 1-D nanostructures such as nanowires, nanorods, nanotubes are chosen by industries since they high response to external stimulus, thereby it can also used for real time monitoring applications [9]. Polymer nanowires can now used for studying one-dimensional charge transport or as field-effect transistors, actuators, or interconnects [2].

15.3 Multilayer Films for Pharmaceutical Applications

The pharmaceutical application of multilayer polymer mainly includes encapsulation of drugs, enzymes, proteins and therapeutics. Polymer multilayer capsules (PMLC) are fabricated by means of Layer-by-Layer (LbL) assembly which has unwrapped exciting opportunity for the design of multifunctional drug carriers with acute control over their properties such as size, composition, porosity, stability, surface functionality, colloidal stability, etc. [13, 14]. Different strategies are followed for the formation of these capsules. First, alternating polymeric layers are deposited onto core microparticles. Secondly, the core microparticles are decomposed into low molecular weight degradation products which can freely diffuse through the LbL membrane [15]. In LbL assembly the films can be formed through the alternate adsorption of materials through complementary interactions electrostatic interactions [16, 17] and hydrogen bonding [18, 19]. The responsiveness of multilayer polymer to specific stimuli which can be physicals, chemical or biological nature, make a way to encapsulation, release, and targeting.

Encapsulation is achieved by direct or indirect techniques. In the first case a template is used for encapsulation and second method by co precipitation which makes use of external stimuli such as temperature, light, and magnetic fields for disturbing the permeability change.

There are different methods of encapsulation. During co-precipitation, molecules are added and, therefore, they are entrapped into the template interior, which are created by a direct precipitation method. Subsequently, the template is covered with a polyelectrolyte shell, and then removed, producing capsules [20, 21]. It is attractive due to ease of fabrication. In Adsorption method, molecules are added to already pre-formed templates, and adsorption takes place due to electrostatic interactions, hydrogen bonding, etc. One of the biggest disadvantages of this approach is limited loading capacity. Another approach for incorporation of molecules into a template is the so-called solvent-exchange; it is based on different solubilities of molecules in various solvents [22].

pH and light are other tools for encapsulation reversible shrinking/expansion method [23]. The pH induced method works by means of electrostatic interaction. Microcapsules swell by increasing the pH of the solution [24]; their pores increase and molecules are penetrating inside the capsules. Upon decreasing pH the process is reversed, the pores close and encapsulated molecules are entrapped into the capsule interior. For light induced encapsulation special molecules are used. For example, azobenzene molecules, which undergo transitions from cis- to trans-configuration upon exposure to light, close the pores in a cooperative way thus entrapping molecules inside capsules [25]. Non-uniformity of loading is a potential down side of this, otherwise elegant approach. UV-light was also demonstrated as a source of encapsulation [26]. Photoacid generation was shown as another method of light induced encapsulation and release [27]. It is essential to note that both pH and light induced encapsulations have a significant disadvantage: microcapsules obtained using such a method are mechanically weak.

Temperature [28] and its thermo-chemistry [29] have an important influence on the morphology of polyelectrolytes, while salt induces a similar effect [30]. Upon overcoming a critical threshold (the glass transition temperature of the polyelectrolyte multilayer complex in the case of temperature, and critical salt concentration in the case of salt) [31]. The relationship between hydrophobic and electrostatic interactions induces irreversible shrinking (or expansion). These methods are extremely attractive because they allow production of mechanically strong capsules.

Targeting represents an important functionality because it influenced successful release of drug encapsulated. Typically, targeting is accomplished by attaching antibodies to the surface of capsules [32, 33]. Another important aspect is the delivery. Mechanically strong capsules can successfully deliver molecules, while weak microcapsules fail delivery. Recent studies showed that microcapsules appear to be deformed after uptake into the interior of cells. Therefore, encapsulation methods which allow strengthening of capsules (for example, thermal/salt shrinking based encapsulation) are valuable tools for successful delivery. Measurement of enhancement is made by the AFM colloidal probe technique [34], which allows measurement of mechanical properties [35]. Peculiarly, nanoparticles were shown to enhance mechanical properties of microcapsules [36], presumably through formation of additional bonds in the polyelectrolyte network.

Once encapsulation and delivery are accomplished the contents of capsules need to be released. Release methods can be logically divided into three main categories: physical, chemical and biological. Methods of release are extremely diverse [37–39]; and it can be initiated even from films. Within each of the three main categories there are several methods to induce release including light, magnetic field, Electric fields, nanoparticle adsorption, Biological stimuli, chemical stimuli such as pH and salt.

Hsu et al. [40] designed and characterized films composed of naturally derived materials for controlled release of proteins. Traditional drug delivery strategies rely on synthetic or semisynthetic materials or utilize potentially denaturing assembly conditions that are not optimal for sensitive biologics (Fig. 15.8).

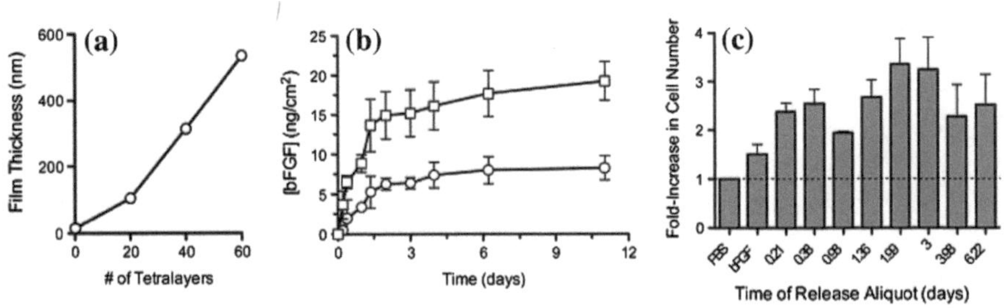

Fig. 15.8 Film characteristics of (chitosan/PMLA/bFGF/PMLA)n films including its growth curve (**a**), release profiles in PBS, pH 7.4 at 37 °C for 20 (*circles*) and 60 tetralayers (*squares*) (**b**), and proliferative activity of PBS, 10 ng/mL of as-received bFGF and film-released bFGF on NIH3T3 cells (**c**) [40]

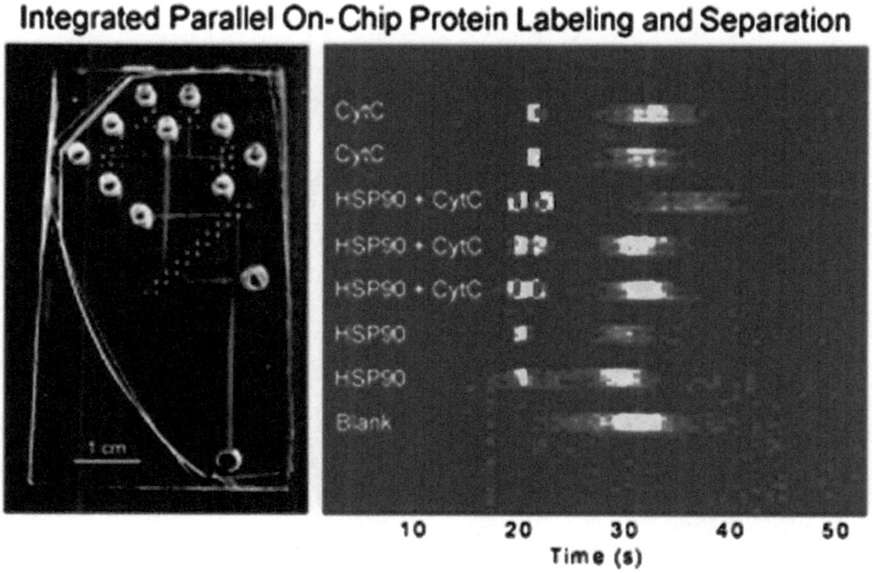

Fig. 15.9 Integrated parallel on chip protein and separation [46]

They investigated the ability of PMLA to assemble into multilayers with PAH, chitosan, PLL, and LPEI and found the characteristic layer dependent growth and sustained release behavior. Bilayer and tetralayer films using a model protein, lysozyme, showed tunable release from 30 min to more than 5 days. They found that the differences in their release kinetics and overall film assembly could be explained by the relative stabilities of electrostatic complexation. By using a PMLA/PAA blend for the polyanionic component, we were able to tune release out to more than 3 weeks. To explore biomedical applicability of this film architecture, we demonstrated the sustained release of a growth factor, bFGF, for nearly 2 weeks and found it retained its biological activity.

The film-assembly process can be described as follows. A polycation is adsorbed onto a polyanionic surface, generally until binding sites are saturated. For films made of polyelectrolytes, electrostatic interactions both drive and limit polymer adsorption. After the substrate is rinsed to remove loosely bound material, a polyanion is adsorbed. In the process, soluble complexes are formed between oppositely charged polyelectrolytes on the film surface and the sign of charge of the film surface is reversed. Vodouhê et al., for example, have studied multilayer films made of poly(lysine), a polycation, and hyaluronic acid, a polyanion [41]. The investigators sought to characterize the accessibility of the antiproliferative agent paclitaxel (Taxol®) on the surface of the film in the absence of chemical modification of the polymers or the drug. The dose available to cells seeded onto films in vitro was 'tuned' by controlling the number of 'inert' polyelectrolyte layers between the drug (inside the film) and the cells (outside the film). In such cases, the film acts as a diffusive barrier. The film may also protect encapsulated material from the surrounding environment or elements of the surrounding environment from encapsulated material.

A current active area of research within the drug-delivery segment of the multilayer film community is vaccine development. Typical differences between investigators are the polymers selected for film fabrication, the means selected for film functionalization and the target indications. The utilization of coated micro particles or capsules for immunization holds the promise of the same particles serving both as immunogens and adjuvants. This topic and other relevant ones are covered in the recent review article by De Koker et al. [42].

15.4 Multilayer Polymer Films for Medical Applications

The control of the physical, chemical and biochemical properties of an implant surface is one of the most important issues in the design of biomedical devices since the first interaction between a foreign body (implant) and the biological environment occurs at the interface.

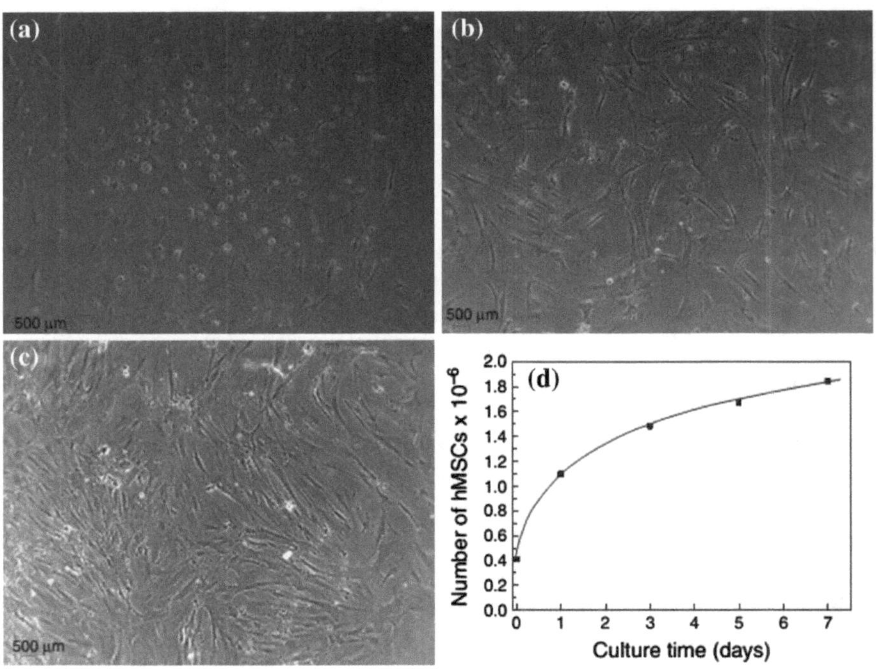

Fig. 15.10 Human bone marrow stromal cells (hMSCs) successfully adhere to and grow on photo-cross-linked layer-by-layer (LbL) films containing diphenylamine-4-diazoresin and pectin over time. HMSCs were cultured on LbL films for 1 day (**a**), 3 days (**b**), and 6 days (**c**). Quantification of cell density versus culture time is shown in (**d**) [48]

The multilayer polymer coatings are finding applications in the field of medical mainly as protective covering for implanted medical devices. Surface of the material can be modified by various functionalization such as physical, chemical or biological to get enhanced tissue assimilation. Thus, the surface may develop into biologically active via additional insertion of biomolecules. The fabrication of multilayer film by layer by layer (LbL) method has had a significant impact on advancing the fields of medical. LbL assembly is significantly based on the interactions of multivalent species with complementary functionalities of each layer. This can be used to impart cell adhesive properties and biocompatibility to an existing substrate that may, itself, lack these properties but have other desirable mechanical and physical characteristics. Conformal coating of these substrates with LbL films allows one to maintain the desired characteristics of the substrate while enabling their use in cellular and tissue engineering applications, ranging from bone tissue engineering to cancer therapy. LbL films containing natural and synthetic polymers, as well as therapeutic molecules, have been widely explored. The mechanical and chemical properties of these films can be tuned by changing film architecture and assembly conditions, which directly relates to film growth.

Polyelectrolyte multilayer films developed by the alternated adsorption of oppositely charged polyelectrolyte layers make up a novel and promising technique to modify surfaces in a controlled way [43]. These polyelectrolyte films offer large potentialities in the design of bioactive biomaterials aimed at inducing different biological responses at the cellular level. One also has to consider that such materials are developed with the perspective to be introduced on the surface of human repair devices or at least to be put in close contact with various biofluids, which also implies interactions and adsorption of proteins on the terminating polyelectrolyte layers. However, only few quantitative data are available for protein adsorption onto such films.

Biological applications of Multilayer Polyaniline/Gold Nanoparticle Films were studied in detail by Tian et al. [44] they found that the PANI/MSAGNP multilayer films obtained via the LBL technique which can be used for DNA hybrization detection, are very stable. The films can electrocatalyze the oxidation of NADH and can be utilized to detect DNA hybridization either by an electrochemical method or by surface plasmon enhanced fluorescence spectroscopy (SPFS).

Hogg et al. [45] used multilayer polymer film for protective packaging for long-term implantable medical devices like Cardiac pacemakers. These films protect the human body from potentially nonbiocompatible materials and also protect the device from body fluids. Previously materials like glass titanium etc. were used as protective packaging for long term biomedical implants. These layers were deposited in a single chamber process, combining CVD and PECVD technologies. Due to the high stability of process parameters by means of the automated deposition system, a conformal and uniform coating at ambient temperature for smart medical implants could be achieved.

Yu and coworkers created a multilayer polymer microfluidic device with integrated on-chip labeling and parallel electrophoretic separation [46]. Microchannels were distributed in two different layers and connected through interlayer through-holes in the middle layer. A single set of electrophoresis reservoirs and one fluorescent label reservoir address parallel analysis units for up to eight samples. Individual proteins and a mixture of cancer biomarkers have been successfully labeled on-chip and separated in parallel with this system. This integrated on-chip labeling microdevices show great potential for low-cost, simplified, rapid, and high-throughput analysis. Multilayer poly (methyl methacrylate) (PMMA) microchip CAE systems with integrated on-chip fluorescence derivatization function showed in Fig. 15.9. On these multilayer devices there are eight parallel lanes, allowing up to eight different samples to be labeled on-chip and separated simultaneously. As a demonstration of the capabilities of these systems, different mixtures of model biomarkers have been simultaneously labeled on-chip and separated in parallel.

The LbL assembled multilayer polymer films are also used in cell engineering and tissue culture [47]. The mechanical and chemical properties of LbL films, which can be appropriately tuned to control cellular behaviors such as adhesion, proliferation, and differentiation, will also be examined. Multilayer film architectures used for cellular engineering applications may be composed of natural and synthetic polymers. Many natural polymers such as collagen44 and fibronectin45 are components of the extracellular matrix (ECM) and serve as adhesion proteins for cells in vivo; when incorporated into an LbL film architecture these proteins may improve film biocompatibility. Several synthetic polyelectrolytes, including PSS, PAA, PAH, poly(l-lysine) (PLL), and poly(glycolic acid) (PGA), have also been shown to be biocompatible and promote cellular adhesion by electrostatic interactions. The properties of these polyelectrolytes can often be adjusted in order to assemble films that are better suited for their intended purpose. These properties, such as degree of ionization of weak polyelectrolytes, may be altered by changing chemical properties such as pH and ionic strength of the solvent in which the polymer is dissolved.

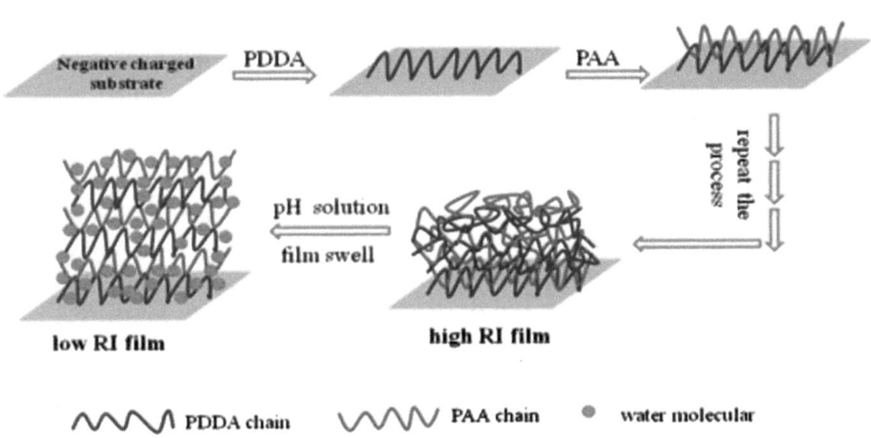

Fig. 15.11 Schematic principle of TFBG based pH sensor [49]

LbL film-coated surfaces can be used successfully as cell culture substrates for simple cultures as well as more complex cellular co-cultures. A recent study examined human bone marrow stromal cell (hMSC) culture on photo-cross-linked LbL films assembled from the synthetic polycation, diphenylamine-4-diazoresin, and the natural anionic polysaccharide, pectin [48]. HMSCs exhibited healthy morphologies and steady proliferation on these LbL film-coated surfaces, resulting in a 4.5-fold increase in cell density over 7 days, as seen in Fig. 15.10.

15.5 Multilayer Polymer Films for Sensor Applications

The properties of polymers can exhibit reversible and irreversible changes to an external stimuli for example pH, temperature, ionic strength, light irradiation, mechanical forces, electric and magnetic fields, external additives (ions, bioactive molecules, etc.), or a combination of them. The capability of facile manipulation of the solubility, hydrodynamic volume, and chain configuration and conformation of responsive polymers by external stimuli has indeed enabled the development of responsive polymeric systems with novel functions. Intuitionally, responsive polymers should play an important role in detection and sensing applications. This behavior makes them used for the application of sensors. Multilayered polymer

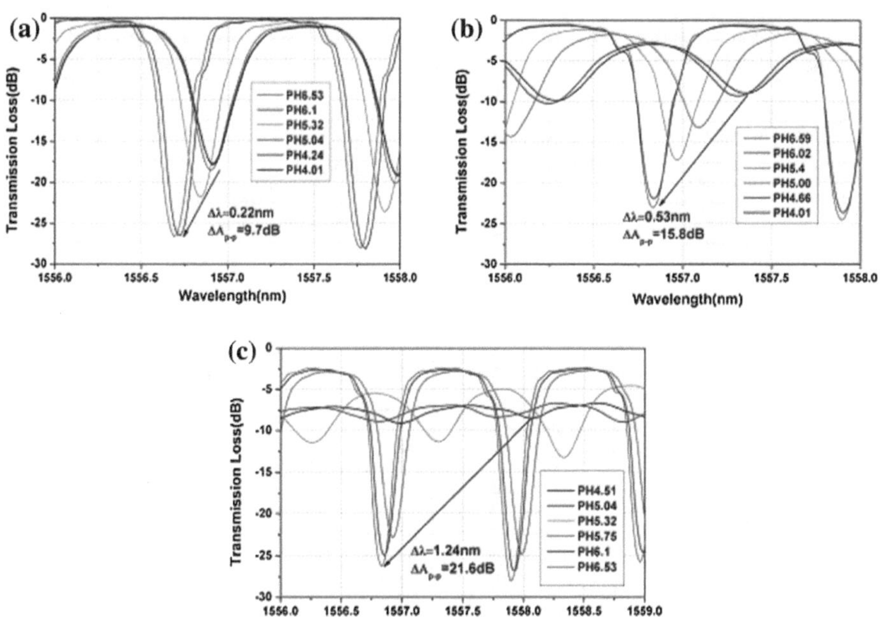

Fig. 15.12 Evolution of optical spectra of TFBG pH sensors with different coatings [**a** (PDDA/PAA)4, **b** (PDDA/PAA)6, **c** (PDDA/PAA)10] [49]

thin-films have proven to be one of the most promising approaches for optical sensor solutions. Their properties and sensitivity can be tailored especially nanostructures are in use. Numerous optical instruments can be manufactured using these materials such as, lenses, prisms, optical fibers, etc.

Shao et al. [49] developed a new fiber-optic pH sensor based on a TFBG with PDDA/PAA multilayer sensing films. The PDDA/PAA coating has a high refractive index (relative to the material of the fiber, i.e., silica), so not only the resonance wavelength but also the coupling strength change with the pH of the solution. The principle of TFBG based pH sensor is given in Fig. 15.11. As the swelling state of the PDDA/PAA multilayer sensing film changes with the pH of the solution, the RI of the sensing film will also change accordingly. As a result, the transmission spectra of the fiber-optic sensors will also be impacted.

Figure 15.12 gives the spectral evolution of fiber-optic sensors with pH for different thicknesses of sensing film. For each film, the pH of the aqueous solutions was adjusted by HCl or NaOH solution (in the range from 3.4 to 7.24). It can be seen that the pH increases, the coupling strength of the cladding mode resonances also increases. The reason is that with increasing pH, more carboxyl groups will be ionized and the sensing film becomes more hydrophilic. This in turn leads to swelling of the sensing film to allow more water inside it, and reduces the RI. Due to the effective RI decrease of the sensing film the cladding modes become more strongly guided and less lossy, thereby recuperating some cumulative coupling strength along the grating.

Figure 15.13 shows the wavelength shift and amplitude changes of the resonance around 1557 nm. For the sensors with 4, 6 and 10 bilayers of (PDDA/PAA) film, the wavelength shifts are 0.22, 0.53 and 1.24 nm, while the relative amplitude change is 9.7, 15.8 and 21.6 dB when the pH increases from 4.01 to 6.59. It is further observed that for certain ranges of the thickness of sensing film, the sensitivity (the slope of the curves in the Figures) also becomes larger.

A polyelectrolyte sensing film has been successfully deposited on TFBG surfaces by the LbL electrostatic self-assembly technique to fabricate a novel

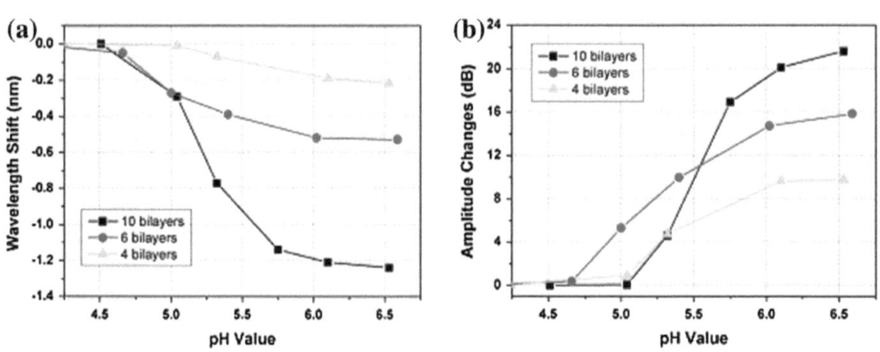

Fig. 15.13 **a** Wavelength shift and **b** amplitude change of one resonance around 1557 nm for three sensors with coatings having different numbers of bilayers [49]

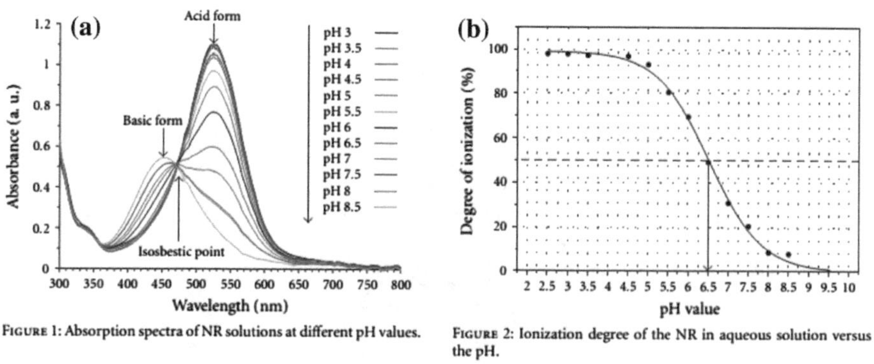

Fig. 15.14 **a** Absorption spectra of NR solutions at different pH values, **b** Ionization degree of the NR aqueous solution versus pH [50]

fiber-optic pH sensor. The main benefit of this kind of coating is its high RMS porosity, which makes the dynamic response very fast, of the order of 10 s. A disadvantage of this sensor is the limited linear operating range of pH (from 4.66 to 6.02) due to the high RI of the coating.

In another work layer-by-layer electrostatic self-assembled multilayer (LbL) structures were fabricated using poly(allylamine hydrochloride) (PAH) and Neutral Red (NR) as cations, and poly(acrylic acid) (PAA) as polyanion [50]. These LbL films, due to the colorimetric properties of the NR, are suitable for sensor applications such as pH sensing in the physiological range. In the (PAH + NR/PAA) LbL structure, it has been observed a very important influence of the pH of the solutions in the properties of the resultant films (Figs. 15.14, 15.15 and 15.16).

PAH + NR/PAA LbL coatings was studied and optimized for pH sensing. As far as all the materials are weak electrolytes, the pH of the aqueous solutions used for building up the films plays a key role in the final properties of the assemblies.

It was found that the concentration of NR molecules adsorbed into the LBL films and the bilayer thickness as well strongly depends on the pH of the polyelectrolyte solutions used for the fabrication of the coatings. An optimization of the absorption linear coefficient was performed achieving a maximum when the pH of the polyelectrolyte solutions is fixed to 4.0. Therefore, in order to obtain sensitive PAH + NR/PAA LbL coatings with a maximum adsorption of NR, the nanostructured films have to be fabricated at pH4.

Yang et al. [51] successfully utilized an alternating electrostatic adsorption approach to incorporate molecular recognition elements into polymer films for chemical sensing applications. The process is highly efficient and reproducible for fabricating multilayer molecular films with high uniformity and stability. SAW devices coated with multilayer films displayed high sensitivity and selectivity to organic vapors studied.

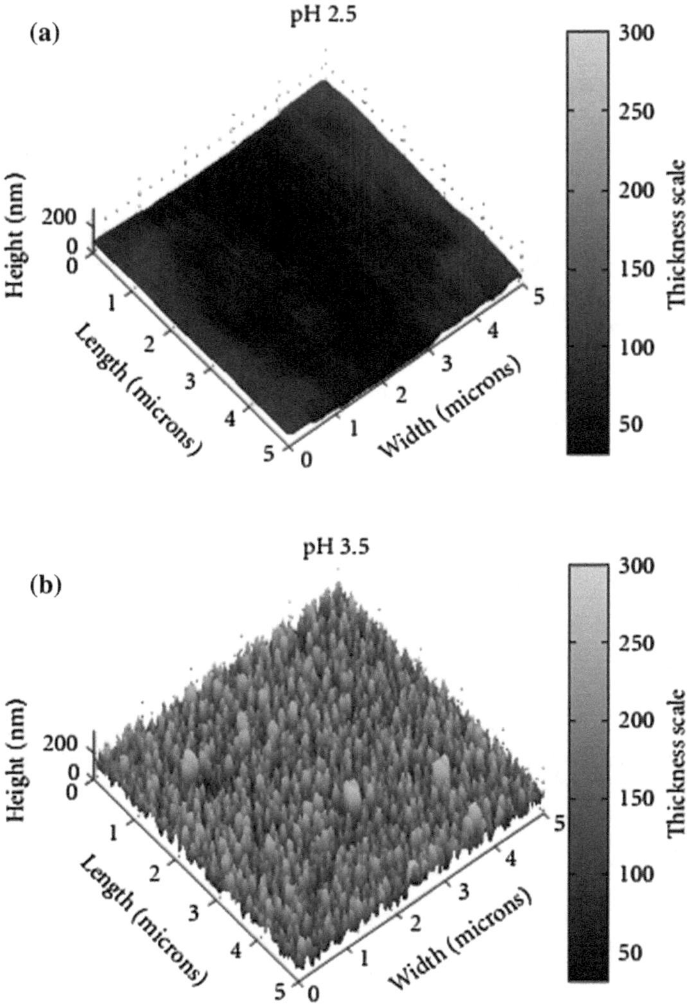

Fig. 15.15 AFM pictures of two (PAH + NR/PAA) 25 structues built up at different pH values **a** pH 2.5 **b** pH 3.5 [50]

Conductivity sensors using the conducting polymer poly(pyrrole) and its derivatives have been prepared with a novel electrode design which allows the probing of resistance changes within zones of a single sensor. It was found that the application of principal component analysis to the sensor responses allowed methanol, ethanol and propanol to be distinguished. The use of layered conducting polymers improved the discrimination of the sensor. A sensor consisting of four electrode pairs and two polymer layers is capable of separating the response of certain alcoholic beverages. It is possible to use a sandwich of conducting polymers as part of a single sensor to separate and classify various vapour types. In order to

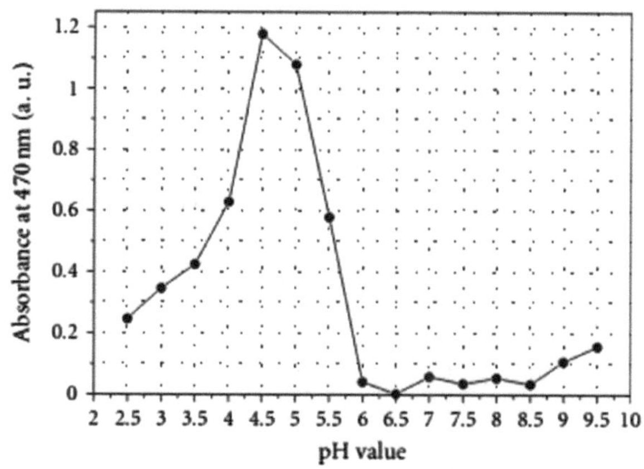

Fig. 15.16 Comparison between the average bilayers thickness and the absorbance at 470 nm. This comparison shows evidence that the composition of the (PAH + NR/PAA) changes as pH is increased [50]

maximize the discrimination of the sensor it is necessary to use a time-resolved response. The problems associated with drift of individual array components appear to be minimized with this type of device as measurements are essentially always based on the difference in behaviour of a single polymer or polymer sandwich. By refining the particular properties of the polymer layers it should be possible to develop sensors with enhanced discrimination of particular target analytes.

15.6 Multilayer Polymer Films in Agricultural Field

The use of polymers was made of in the plastic greenhouses, polytunnels, soil fumigation film, irrigation drip tape/tubing, nursery pots and silage bags, plastic mulch film etc. A variety of polymers were used in this area, including, polyethylene (PE), Polypropylene (PP), Ethylene-Vinyl Accetate Copolymer (EVA), Poly-vinyl chloride (PVC) and Polycarbonate (PC) and poly-methyl-methacrylate (PMMA) in often. The use of multilayer polymer film is advantageous than monolayer polymer film in terms of better performance and cost effectiveness. The requirements for the selection of multi-layers are unlimited. The key idea is that the price/performance ratio is better achieved with the synergy of the layers than with a single material. The key properties for these applications are durability, optical (ultraviolet, visible, near infrared, and middle infrared) properties, and the antidrip or antifog effect. In the last few years the amount of plastic used in the agricultural application are continuously increasing. The main characteristics of green house covering films are optical and mechanical properties, resistance to photo oxidation and recyclability. Adam et al. [52] studied about the formulation of polyethylene multilayer film used as greenhouse cover. This film is a sandwich structure made of a polyethylene layer and expanded EVA layer and a second polyethylene layer. The multilayer film shows better resistance to a simulated erosion process and its thermal, surface and mechanical are not altered. Dintcheva et al. [53] investigated the possibility of using post consumed polymer for making multilayer film used as green house cover. The multilayer film was formulated as three layers, in which recycled PE was sandwiched between two layers of virgin PE. Aimed at meeting the needs of specific applications, multilayer starch-polyesters films have also been developed. Multilayer films based on plasticized wheat starch (PWS) and various biodegradable aliphatic polyesters have been prepared through flat film coextrusion and compression molding. Poly(lactic acid) (PLA), polyesteramide (PEA), poly (epsilon-caprolactone) (PCL), poly(butylene succinate adipate) (PBSA), and poly (hydroxybutyrate-co-valerate) (PHBV) were chosen as the outer layers of the stratified "polyester/PWS/polyester" film structure.

15.7 Multilayer Polymer Films for Miscellanious Applications

Polymer multilayers are finding application as roofing material after the development of lamination process which combines various properties of individual layers. The use of multi layer polymer provides high quality roofing membranes with low maintenance costs, and obtaining the roofs with prolonged service life. For thermal resistant roofs fiberglass/polystyrene, poly (vinyl chloride) were used and for improved resistances to UV light ethylene-propylene diene rubber (EPDM) roof. In order to get polymer roof with relatively low density and low thermal conductivity cellular or foamed products are used. A fully bonded ply roof from composites of wood particles and ebonite vulcanized natural rubber (NR) with EPDM layer by melt-lamination technique has recently been developed [54]. The interfacial

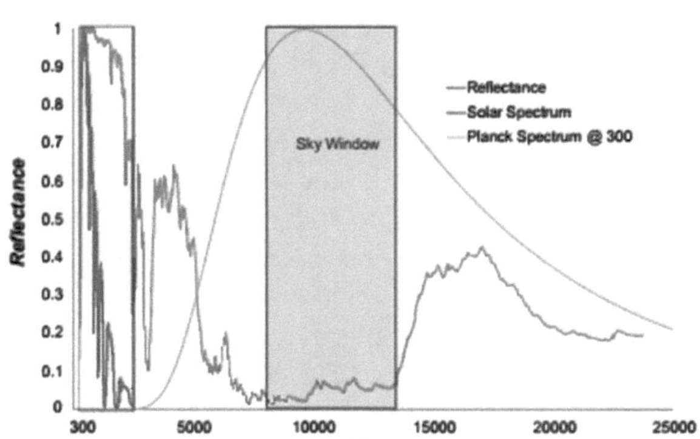

Fig. 15.17 Spectral reflectance of the super-cool roof material across the solar and thermal infrared showing key cutoff wavelengths including the boundaries of the sky window (blue shading) and the solar spectrum (yellow shading). Spectral response in the unshaded zones is where down-welling sky radiation occurs. The solar spectrum (blue) and Planck spectrum at 300 K (green) are plotted for reference [55]

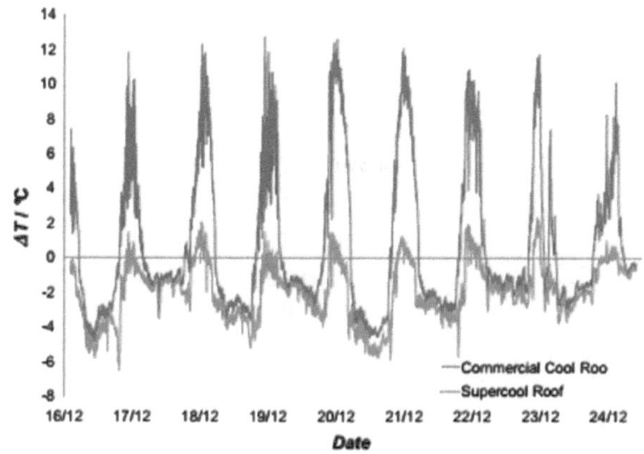

Fig. 15.18 Performance of the super-cool material and the side-by-side existing commercial cool roof during extended exposure near a main transit city road over a 10 days summer period. Temperature of each is plotted relative to ambient [55]

adhesion between the wood/NR composite and solid EPDM layers was improved through N-(b-Bis-(3-triethoxylpropyl) tetrasulfan (Si69) coupling agent.

A novel multilayer polymer composite material system and process for encapsulation of thin film solar cells into profiled roofing elements were developed in view of building integrated photovoltaic (PV) applications. The multilayer material included texturized hybrid nanocomposites for enhanced light trapping and a glass fiber reinforced composite for structural integration. At production volumes beyond 100,000 parts per year, the financial payback time of the modules was found to be less than 4 years.

One of the interesting application of multilayer polymer film is super-cool roofing with thermal performance well above that of existing cool roofs [55]. Stringent spectral requirements at solar and infrared (IR) wavelengths are needed, leading to quite limited choices for suitable coating materials and layer structure. Metal alone, except silver, cannot provide the required level of solar reflectance of above 96% and the thermal emittance of common metals is far too low to cool. Placing silver under a glass or polymer material with very low solar absorptance may cool well as high emittance Er results. However, options such as PMMA and most glasses absorb too much solar radiation. Low iron glass and various polyesters including PET absorb very little solar energy if thin, but their IR spectral response is not ideal for this task. Cooling is enhanced if IR spectral response in the thermal emission band involves a mix of moderate reflectance at those wavelengths where the atmosphere irradiates the earth under clear sky conditions and very high absorptance hence emission in the range from 7.9 μm < λ < 14 μm, called the "sky window," which is largely free of incoming radiation as it views the cold of space.

Super-cool surfaces require solar reflectance and sky window absorbance to be close to 100%. The admix of IR reflectance and absorbance in the down-welling atmospheric radiation band is less stringent but together with the sky window setting determines the value of overall thermal remittance which dictates the heat output rate at the roof temperature Tr. This radiative rate is an important practical consideration. Its final choice depends on total rate of heat input including the contribution of absorbed down-welling atmospheric radiation. We concluded that the desired combination of solar and IR criteria could be met using two or more specially chosen polyesters on a silver layer. At the set thickness, their IR absorbance had to be very strong in slightly different sections of the sky window with moderate transmittance at incoming wavelengths. A suitable surface was produced and set up outdoors. It remained subambient throughout a hot summer day including under the peak intense solar intensity of 1060 Wm^{-2}, with ambient at 27 °C, and high IR intensity from the atmosphere of 400 Wm^{-2} and cities, and helps eliminate peak power demand problems from operation of many air conditioners. Added feedback benefits from cool roofs are not yet widely appreciated, but recent reports have shown they are substantial. Examples include ventilation with cooler air and higher performance of adjacent chillers when in cooler air. Adjacent cooler air also lowers temperature gains from convective exchange for a subambient roof. Air above established roofs is much warmer in the daytime.

A thin polyethylene cover to suppress convective exchange but transmit IR was used separated from a stack of two thin oxides deposited over 200 nm of silver on silicon. The demonstration here is the first using polymers and an open surface, and is suited to basic roofing. Spectrally suitable polymers happened to be available commercially as coextruded combinations of many bilayers. This has an additional advantage as it acts as an all-dielectric mirror and reflects better than metals at blue wavelengths. The best stacks let through some NIR solar energy and a considerable amount of atmospheric radiation so modification was needed.

The complete solar-IR spectral reflectance of the resulting surface is in Fig. 15.17. The $R(\lambda)$ value is very low as desired across the sky window zone. The silver reduces heat gained from the atmosphere and raises the Air Mass 1.5 albedo to 0.97. The solar reflectance was also measured from normal incidence out to 85° incidence (Fig. 15.18).

Another interesting application is multilayer polymer mirrors. Weber et al. successfully developed multilayer polymer mirror. Using highly birefringent polymers, we have found that multilayer mirrors can be constructed that maintain or increase their reflectivity with increasing incidence angle. The reflective characteristics of these mirrors require a generalization of Brewster's law. This generalization has enabled the development of a new class of multilayer interference optics with design freedoms that can result in unprecedented means for transporting, filtering, and reflecting light. Multilayer polymeric interference mirrors were pioneered in the late 1960s, and even though the large birefringence of oriented polyethylene terephthalate (PET) was known at the time, the use of materials with large optical birefringence in a multilayer mirror (polymeric or otherwise) has not been reported.

Such mirrors use multiples of birefringent polymer pairs, one with high index and one with low index, for example polyethylene terephthalate (PET)/naphthalene icarboxylate and polyethylene napthalate (PEN)/THV (a 3 M fluoro-thermoplastic containing tetrafluoroethylene, hexafluoropropylene, and vinylidene fluoride). Further spectral details and the layer structure of polymer mirror stacks, in which pair thicknesses are graded to set the breadth of spectral and angular response, are in. The product used Vikuiti Enhanced Specular Reflector (ESR) is all polyester and believed to consist of PET/ECDL pairs with ECDEL a Kodak copolyester using 1,4-cyclohexane dicarboxylic acid, 1,4-cyclohexane dimethanol, and polytetramethylene ether glycol. The overall thickness of this polymer mirror is 67 ± 4 μm and consists of 300 layers of each polymer plus PET outer layer [17], so average thickness per layer is near 110 nm. The average final density of the stack is 1.29 g cc^{-1}. Assuming the bulk density ratios of PET (1.38 g cc^{-1}) to ECDEL (1.13 g cc^{-1}) persists in the final stack, the layer thickness for PET averages 141 nm and ECDEL 79 nm. These thicknesses are, however, graded with individual layer thicknesses on one side of the stack about 1/4 that of those on the opposite side. This feature extends the high reflectance band to 1 μm. PET is uniaxial with vertical component of refractive index 1.51 and in-plane index 1.66, while ECDEL is isotropic with index 1.52. Layer thickness profile and in-plane index differences provide most of the very high solar reflectance from 0.4 to 1 μm. These two polymers in combination plus each one's total thickness of 42.3 and 23.7 μm for PET and ECDEL, respectively, also yield the required strong sky window absorptance, and moderate transmittance at downwelling IR bands. The bare stack thus needed to have its solar reflectance raised in the NIR, while its transmitted IR had to be reflected. Simply adding a bottom silver coating achieved both solar and IR goals.

A novel thermally cross-linkable hole transporting polymer has been successfully developed and applied for OLEDs. This copolymer could be completely cross-linked to form smooth insoluble films through reaction of the benzocyclobutene (BCB) units at 200 °C. Fluorescent devices with copolymers as HTL exhibit similar behavior before and after cross-linking. A comparison of the polymer to vacuum deposited small molecule TPD shows only a slight change in device performance, suggesting that replacement of the normally vacuum-deposited layer by our solution processible polymeric system is feasible. Using this novel thermally cross-linkable polymer, a highly efficient solution processed multilayer green emitting phosphorescent devices with EQE) 10.4% at 350 cd/m^2 was demonstrated.

Graphene/PANI multilayer films have been successfully fabricated by the combination of LBL assembly and chemical reduction [56]. The graphene/PANI multilayer films grow linearly in their thicknesses with the number of bilayers, yielding smooth and compact composite films with roughness less than 6 nm. In addition, the thickness, conductivity and transmittance of the multilayer film can be easily tuned by changing the alternate deposition steps. The composite films were also tested to be promising electrode materials for electrochromic devices even without using conductive transparent supporting electrodes such as indium tin oxide (ITO). The electrochomic device based on (GO/PANI)15 film showed improved electrochemical stability comparing with the similar device with a conventional ITO electrode. The method developed here can be extended to fabricate other graphene/conducting polymer multilayer films for electronic, energy, and sensor applications.

15.8 Conclusion

We know that in our daily life polymers have great importance in every field. In which multilayered polymers have wide application. This chapter discusses mainly about polymer nanowires and the application of multilayered polymers in various products in different fields. First part of this chapter discusses the preparation methods and the characterization techniques of polymer nanowires. The sensing application of nanowires is also discussed nicely. The second part of this chapter discusses the role of multilayer polymers in the field of medical, cosmetics, electronics, pharmaceutics and sensors. Multilayer polymers have great contribution in every field of our day to day life.

References

1. Yang P, Yan R, Fardy M (2010) Semiconductor nanowire: what's next? Nano Lett 10 (5):1529–1536
2. Lipomi DJ et al (2008) Fabrication of conjugated polymer nanowires by edge lithography. Nano Lett 8(7):2100–2105
3. Ramanathan K et al (2004) Individually addressable conducting polymer nanowires array. Nano Lett 4(7):1237–1239
4. Liu H et al (2004) Polymeric nanowire chemical sensor. Nano Lett 4(4):671–675
5. Dasgupta NP et al (2014) 25th anniversary article: semiconductor nanowires–synthesis, characterization, and applications. Adv Mater 26(14):2137–2184
6. Zhao L-M et al (2011) Preparation and application of chitosan nanoparticles and nanofibers. Brazilian J Chem Eng 28(3):353–362
7. Fang H et al (2009) Controlled growth of aligned polymer nanowires. J Phys Chem C 113 (38):16571–16574
8. Yun M et al (2004) Electrochemically grown wires for individually addressable sensor arrays. Nano Lett 4(3):419–422
9. Kumar MA, Jung S, Ji T (2011) Protein biosensors based on polymer nanowires, carbon nanotubes and zinc oxide nanorods. Sensors 11(5):5087–5111
10. Choi WS, Lim G, An T (2011) Fabrication of conducting polymer nanowires. INTECH Open Access Publisher
11. Kameoka J et al (2004) Polymeric nanowire architecture. J Mater Chem 14(10):1503–1505
12. Song E, Choi J-W (2013) Conducting polyaniline nanowire and its applications in chemiresistive sensing. Nanomaterials 3(3):498–523
13. Johnston AP, Cortez C, Angelatos AS, Caruso F (2006) Layer-by-layer engineered capsules and their applications. Curr Opin Colloid Interface Sci 11(4):203–209
14. Decher G (1997) Fuzzy nanoassemblies: toward layered polymeric multicomposites. Science 277:123
15. De Cock LJ, De Koker S, De Geest BG, Grooten J, Vervaet C, Remon JP, Sukhorukov GB, Antipina MN (2010) Polymeric multilayer capsules in drug delivery. Angew Chem Int Ed 49 (39):6954–6973
16. Decher G, Hong JD, Schmitt J (1991) Makromol. Chem. Macromol Symp 46:321–327
17. Decher G (1997) Science 277:1232–1237
18. Decher G, Schlenoff J (eds) (2003) Multilayer thin films. Wiley-VCH, Weinheim
19. Kharlampieva E, Kozlovskaya V, Sukhishvili SA (2009) Adv Mater (Weinheim, Ger) 21:3053–3065

20. Volodkin DV, Petrov AI, Prevot M, Sukhorukov GB (2004) Langmuir 20:3398–3406
21. Petrov AI, Volodkin DV, Sukhorukov GB (2005) Biotechnol Prog 21:918–925
22. Bai JH, Sebastian B, Yein TS, Dieter T (2011) ACS Appl Mater Interfaces 3:1665–1674
23. Zhao QH, Li BY (2008) Nanomed Nanotechnol Biol Med 4:302–310
24. Mauser T, Dejugnat C, Sukhorukov GB (2004) Macromol Rapid Commun 25:1781–1785
25. Bedard M, Skirtach AG, Sukhorukov GB (2007) Macromol Rapid Commun 28:1517–1521
26. Katagiri K, Koumoto K, Iseya S, Sakai M, Matsuda A, Caruso F (2009) Chem Mater 21:195–197
27. Koo HY, Lee HJ, Kim JK, Choi WS (2010) J Mater Chem 20:3932–3937
28. Leporatti S, Gao C, Voigt A, Donath E, Mohwald H (2001) Eur Phys J E Soft Matter Biol Phys 5:13–20
29. Shao L, Lutkenhaus JL (2010) Soft Matter 6:3363–3369
30. Gong X, Gao CY (2009) Phys Chem Chem Phys 11:11577–11586
31. Köhler K, Shchukin DG, Möhwald H, Sukhorukov GB (2005) J Phys Chem B 109:18250–18259
32. Cortez C, Tomaskovic-Crook E, Johnston APR, Scott AM, Nice EC, Heath JK, Caruso F (2007) ACS Nano 1:93–102
33. Gao X, Zhang F, Chen ZC, Lin XF (2011) Chem. J. Chin. Univ. 32:957–963
34. Fery A, Weinkamer R (2007) Polymer 48:7221–7235
35. Lulevich VV, Nordschild S, Vinogradova OI (2004) Macromolecules 37:7736–7741
36. Bedard MF, Munoz-Javier A, Mueller R, del Pino P, Fery A, Parak WJ, Skirtach AG, Sukhorukov GB (2009) Soft Matter 5:148–155
37. Ji Q, Miyahara M, Hill JP, Acharya S, Vinu A, Yoon SB, Yu JS, Sakamoto K, Ariga K (2008) J Am Chem Soc 130:2376–2377
38. Rosenbauer EM, Wagner M, Musyanovych A, Landfester K (2010) Macromolecules 43:5083–5093
39. Sultan Y, DeRosa MC (2011) Small 7:1219–1226
40. Hsu BB et al (2014) Multilayer films assembled from naturally-derived materials for controlled protein release. Biomacromolecules 15(6):2049–2057
41. Vodouhê C et al (2006) Control of drug accessibility on functional polyelectrolyte multilayer films. Biomaterials 27(22):4149–4156
42. De Koker S, Hoogenboom R, De Geest BG (2012) Polymeric multilayer capsules for drug delivery. Chem Soc Rev 41(7):2867–2884
43. Ladam G, Schaaf P, Decher G, Voegel JC, Cuisinier FJG (2002) Biomol Eng 19:273
44. Tian S, Liu J, Zhu T, Knoll W (2004) Polyaniline/gold nanoparticle multilayer films: assembly, properties, and biological applications. Chem Mater 16(21):4103–4108
45. Hogg A, Uhl S, Feuvrier F, Girardet Y, Graf B, Aellen T, Keppner H, Tardy Y, Burger J (2014) Protective multilayer packaging for long-term implantable medical devices. Surf Coat Technol 255:124–129
46. Yu M, Wang Q, Patterson JE, Woolley AT (2011) Multilayer polymer microchip capillary array electrophoresis devices with integrated on-chip labeling for high-throughput protein analysis. Anal Chem 83(9):3541–3547
47. Shukla A, Almeida B (2014) Advances in cellular and tissue engineering using layer-by-layer assembly. Wiley Interdisciplinary Rev Nanomed Nanobiotechnol 6(5):411–421
48. Plewa A, Niemiec W, Filipowska J, Osyczka A, Lach R, Szczubialka K, Nowakowska M (2011) Photocrosslinkable diazoresin/pectin films—synthesis and application as cell culture supports. Eur Polym J 47:1503–1513
49. Shao L-Y, Yin M-J, Tam H-Y, Albert J (2013) Fiber optic pH sensor with self-assembled polymer multilayer nanocoatings. Sensors 13:1425–1434
50. Goicoechea J et al (2007) Study and optimization of self-assembled polymeric multilayer structures with neutral red for pH sensing applications. J Sens 2008
51. Yang X, Johnson S, Shi J, Holesinger T, Swanson B (1997) Polyelectrolyte and molecular host ion self-assembly to multilayer thin films: an approach to thin film chemical sensors. Sens Actuators B 45:87–92

52. Adam A et al (2005) Studies of polyethylene multi layer films used as greenhouse covers under Saharan climatic conditions. Polym Testing 24(7):834–838
53. Dintcheva NT, La Mantia FP (2009) Formulation of agricultural films by using recycled polymers: resistance to the photoxidation. Spec Top Mater Sci Technol Ital Panorama 173
54. Yamsaengsung W, Sombatsompop N (2009) Interfacial adhesion and molecular diffusion in melt lamination of wood sawdust/ebonite NR and EPDM. Polym Compos 30(3):248–256
55. Gentle AR et al (2015) A Subambient Open Roof Surface under the Mid-Summer Sun. Advanced Science 2(9).
56. Sheng K et al (2011) Layer-by-layer assembly of graphene/polyaniline multilayer films and their application for electrochromic devices. Polymer 52(24):5567–5572